BUSINESS

- Demand functions, 33, 47, 212
- Supply functions, 34, 50, 212
- Profit as a function of production rate, 52
- Marginal analysis, 69
- Quality control, 85
- Two-step production, 83
- Maximum yield per input, 103
- Minimum cost, 103, 111
- Maximum profit, 104, 111
- Inventory control model, 105, 114
- Discount rates, 114
- Marginal cost, 71, 138
- Marginal revenue, 71, 138
- Input versus output, 122
- Rate of net investment, 156, 160, 164
- Cost of nth item of production, 156
- Efficiency rate model, 160
- Optimization of land productivity, 206
- Defective item decisions, 206
- Inventory demand, 210
- Marginal physical productivity, 211
- Rent, 211
- Consumer's surplus, 212
- Producer's surplus, 212
- Equilibrium price, 213
- Lagrange multipliers, 286
- Maximum profit subject to production constraints, 289
- Maximum profit subject to capital constraints, 290
- Minimum cost subject to production constraints, 290
- Minimum cost, 292
- Shadow prices, 294
- Optimum production, 294
- Cost function determination, 308
- Fixed cost, 309

PHYSICS AND ENGINEERING

- Sound, 41
- Celestial mechanics, 100, 102, 114
- Snell's empirical law of refraction, 101
- Hydrodynamics, 110, 123, 169
- Disturbance in a medium, 121
- Residue of a solution, 122
- Tracking a storm, 123
- Half-life of radioactive material, 137
- Thermodynamics and work, 138
- Laminar flow in fluids, 157
- Work, 198
- Solids of revolution, 198
- Hooke's law, 199
- Newton's Second Law of Motion, 200
- Escape velocity of a rocket, 200
- Newton's Universal Gravitation Law, 200
- Kinetic energy, 201
- Van't Hoff's equation, 239
- Phenomena dependent upon change in temperature, 239
- Tension, 250
- Sinusoidal curve, 354
- Simple harmonic motion, 355
- Sound, 356

PSYCHOLOGY

- Learning curve, 109, 114, 155, 204, 206
- Prediction of group behavior, 158
- Deviation of group or individual achievement from a goal, 160
- Probability, 203
- Group measurement, 205
- Psychological testing, 210
- Excitation level experiments, 238
- Hullian model of learning, 251

APPLIED CALCULUS

Raymond F. Coughlin

Temple University

Allyn and Bacon, Inc.   Boston, London, Sydney

Copyright © 1976 by Allyn and Bacon, Inc.
470 Atlantic Avenue, Boston, Massachusetts 02210

Portions of this book first appeared in ELEMENTARY APPLIED CALCULUS: A SHORT COURSE, *by Raymond F. Coughlin, Copyright* © 1974 *by Allyn and Bacon, Inc.*

All rights reserved. Printed in the United States of America. No part of the material protected by this copyright notice may be reproduced or utilized in any form or by any means, electronic or mechanical, including photocopying, recording, or by any information storage and retrieval system, without written permission from the copyright owner.

Library of Congress Cataloging in Publication Data

Coughlin, Raymond F.
    Applied calculus.

    An expanded version of the author's Elementary applied calculus.
    Bibliography: p.
    Includes index.
    1. Calculus. I. Title.
QA303.C835   515   75-19072
ISBN 0-205-04890-0

TO MOM AND DAD

Contents

Preface vii

CHAPTER 1 Preliminaries   1

1-1 Elementary Set Theory   1
1-2 Functions   12
1-3 Linear Functions and Straight Lines   19
1-4 Quadratic Functions   24
1-5 Some Additional Types of Functions   30
1-6 An Introduction to Exponential and Logarithmic Functions   37
1-7 Inverse Functions and Composition of Functions   44

CHAPTER 2 Limits   49

2-1 The Limit of a Function and Continuity   49

CHAPTER 3 The Derivative   63

3-1 Definition of the Derivative   63
3-2 Finding Derivatives   72
3-3 Product and Quotient Rules   77
3-4 Composition and the Chain Rule   82
3-5 Higher Derivatives   88

CHAPTER 4 Applications of the Derivative   91

4-1 Maxima and Minima   91
4-2 The Second Derivative Test   105
4-3 Implicit Differentiation   115
4-4 Related Rates   121
4-5 The Mean Value Theorem   126
4-6 L'Hôpital's Rule   129
4-7 Antiderivatives   133

CHAPTER 5 The Integral   140

5-1 Area and the Definite Integral   140
5-2 The Fundamental Theorem of Calculus   149
5-3 Techniques of Integration I: Substitution   161
5-4 Techniques of Integration II: Integration by Parts   164
5-5 Techniques of Integration III: Partial Fractions   168
5-6 Tables of Integrals   175

CHAPTER 6 Applications of the Integral   180

6-1 Numerical Integration: Trapezoidal Rule   180
6-2 Numerical Integration: Simpson's Rule   186
6-3 Infinite Limits and Improper Integrals   189
6-4 Volume and Work   194
6-5 Probability   203
6-6 Applications from Economics   210

Contents

CHAPTER 7 Exponential and Logarithmic Functions — 217

- 7-1 The Definition of the Natural Logarithm Function   217
- 7-2 Properties of $y = \ln x$ 221
- 7-3 Definition of the Number e 226
- 7-4 The Definition of the Exponential Function $y = \exp x$ 229
- 7-5 Properties of $y = e^x$ 231
- 7-6 Other Bases   234

CHAPTER 8 Elementary Differential Equations — 237

- 8-1 Differential Equations Revisited   237
- 8-2 Separation of Variables   241
- 8-3 Integrating Factors   247

CHAPTER 9 Functions of Several Variables — 253

- 9-1 Examples and Graphs   253
- 9-2 Definition of a Function of Two Variables   259
- 9-3 Limits and Continuity   267
- 9-4 Partial Derivatives   271
- 9-5 Maxima and Minima   277
- 9-6 Lagrange Multipliers   286
- 9-7 Double Integrals   296
- 9-8 Method of Least Squares   304

CHAPTER 10 Trigonometric Functions — 311

- 10-1 Angles   311
- 10-2 Definition of the Trigonometric Functions   315
- 10-3 Graphs of the Trigonometric Functions   323
- 10-4 Identities   326
- 10-5 Derivatives of the Trigonometric Functions   332
- 10-6 Integration of the Trigonometric Functions   339
- 10-7 Techniques of Integration IV: Trigonometric Substitution   343
- 10-8 Inverse Trigonometric Functions   346
- 10-9 Applications   353

APPENDIX

- A **Basic Integration Formulas**   362
- B **Tables of Natural Logarithms and the Exponential Function**   365
- C **Mathematical Induction**   373
- D **A Brief Review of Algebra**   375

CHAPTER TESTS   389

SOLUTIONS TO SELECTED EXERCISES   395

SELECTED BIBLIOGRAPHY   421

INDEX   423

Preface

APPLIED CALCULUS is an expansion of my text, ELEMENTARY APPLIED CALCULUS: A SHORT COURSE. Whereas "A SHORT COURSE" is probably best suited for a one-semester course, this text can be used either as a one-semester or year-long sequence in which more thorough treatments of exponential, logarithmic, and trigonometric functions are desired. I am indeed grateful to the many users and reviewers of "A SHORT COURSE" who suggested and encouraged the writing of this expanded version.

The primary purpose of both books is to provide a straightforward, intuitive presentation of calculus, interwoven with numerous realistic mathematical models. Whereas other similar texts may be at about the same level of rigor and include as many or even more so-called "applications," this text focuses only on applications that are truly significant and meaningful, derived from actual data covered in basic courses and literature in such fields as management, economics, and the life sciences. The applications are meant to augment the mathematics and to help explain the significance of the calculus. For this reason I have avoided grouping the applications in their own isolated sections, choosing rather to include a small number in almost every section. Thus the reader is constantly reminded that calculus is a living and vibrant discipline.

Once the mathematical concepts have been introduced and explained, an application or two is presented. These applications are then explained via calculus without obscuring the basic purpose of learning the mathematics. For example, on page 69, rather than simply mentioning the concept of marginal cost, the significance of this one application of the derivative is explained in terms of the concept of mass production. The result is that the student can see and appreciate a significant "real world" situation in which the cost function is increasing yet concave down—that is, as production increases, so does total cost; but the cost per unit, or marginal cost, can actually decrease. A similar example is on page 358 where the predator-prey relationship is described. This predator-prey model is also developed in other places in the text (for example, pages 102 and 124).

It is evident from reviewer responses that there are vast differences of opinion as to how much review material should be included in a text at this level. There are those who prefer to move as quickly as possible into the calculus because of time requirements and because most students are required to have had a previous course in either algebra or finite mathematics, or both. On the other hand, it cannot be denied that many students for whatever reasons are coming to this course with a less than satisfactory knowledge of basic algebra. In this text, the common stumbling blocks encountered in high school mathematics are reviewed in Chapter 1 with additional review material in the appendix. It is hoped that this arrangement will provide as much flexibility as possible, depending upon particular instruction preferences and student needs.

Another side of this debate centers on what emphasis should be placed on topics considered by some to be too difficult or of little interest to the student. For example, I have chosen to include a brief discussion of the Riemann integral. Some may prefer to skip this material.

Others, myself included, feel that the significance of the Fundamental Theorem of Calculus is lost if no mention of the definition of the integral is given. The problems (page 149) associated with this topic are meant to be straightforward enough to diminish the difficulty factor. If the student tries his hand at just a few of these problems his appreciation of the Fundamental Theorem is greatly enhanced.

Another topic which might be considered optional by some and required by others is the definition of the exponential and logarithmic functions in Chapter 7. These functions are intuitively introduced in Chapter 1 and used throughout the text. With these functions available the array of illustrative examples and exercises increases manyfold. This is especially true when one demonstrates the product, quotient, and chain rules. However, it is possible to skip the more rigorous definitions in Chapter 7 and go right to the remaining material. Inclusion of this material provides a powerful mathematical application of the calculus.

My style entails introducing concepts with an intuitive discussion and giving the student a geometric grasp of the concept. Accompanying each discussion of a new topic are numerous detailed examples of various applications. The exercises contain many routine problems as well as a vast amount of additional mathematical models.

I have tried to avoid presenting dogmatic truisms to be digested without explanation. Since the inquisitive reader often requires an argument in order to be convinced of the validity of some theorems, proofs are, therefore, included where the conclusion of a theorem is not intuitively clear — provided that the proof is accessible to the intended audience. Thus there are two criteria I've used in judging whether to include a proof of a theorem. The proof is included (1) if the result is not so intuitively obvious that its insertion would unnecessarily interrupt the discourse and (2) if the details of the proof are within the reach of the student. For example, the theorem governing the derivative of a sum is not included, whereas the product formula is proved.

I have provided elementary flowcharts for the more important and complicated computations, such as the calculation of the limit of a rational function and the first and second derivative tests.

One further suggestion, one which we have used quite effectively at Temple University, is to encourage students to expand on some of the mathematical models in the text and present a research paper to the class. Many of the models in the text have sufficient substance to merit such elaboration. For example, the equations of Lotka and Volterra, mentioned in Section 8-2 but developed throughout the text, are a pervading force in the study of ecology. One can find a vast amount of material well within the reach of the calculus student.

I wish to express my appreciation to the members of the Allyn and Bacon, Inc. staff, especially Carl Harris, Garen Wickham, Gene Thornton, and John Coleman. I would also like to extend special thanks to my wife, Judy, whose hard work and valuable suggestions are a major part of this text.

Raymond Coughlin

1

Preliminaries

1-1 ELEMENTARY SET THEORY

We shall use the word *set* quite frequently, and it will imply what it does everyday: a collection, a grouping, or a list of objects. The objects in sets can be anything—numbers, cars, thoughts, even sets—and they are called *elements*, or members of the set. A set's name is usually a capital letter:

$$A, B, C, D, E, \ldots, S, T, \ldots$$

The elements of a set will usually be denoted by lower-case letters,

$$a, b, c, d, \ldots, s, t, \ldots$$

Also, the notation

$$x \in S$$

means that "x is an element of S" or that "x is a member of S". Similarly, the notation

$$x \notin S$$

means that "x is not an element of S". We now list some examples of sets.

EXAMPLE 1-1
> The numbers 1, 2, 3.

EXAMPLE 1-2
> The words in your vocabulary.

EXAMPLE 1-3
The letters in the word "mug".

EXAMPLE 1-4
The letters in the word "gum".

EXAMPLE 1-5
The people taking calculus at your school.

EXAMPLE 1-6
The rivers of North America.

EXAMPLE 1-7
The planets of our solar system whose names begin with "m".

EXAMPLE 1-8
All sets containing two elements.

Notice that the elements of a set can be physical objects (e.g., people, planets) or nonphysical objects (e.g., numbers, words). A set is not a physical object even though its members may be physical objects. It is good to regard a set as a concept, something which comes into being only as a thought. We speak of a set of objects when we want to think of these objects as having some property or properties in common.

A set must also be well-defined, that is, we must always be able to tell whether or not an element is or is not in a set. Thus "all intelligent people" or "all large buildings" are not sets.

There are two basic ways of defining sets: We can list the elements as in the first example above, or we can give a defining property of the elements in the set as is done in the remaining examples.

If we list the elements we will use brackets to enclose the elements. For example, if we let A be the set in Ex. 1-1, then

$$A = \{1, 2, 3\}$$

Similarly the sets in Examples 1-3 and 1-4 are

$$B = \{m, u, g\} \quad \text{and} \quad C = \{g, u, m\}$$

We say that $B = C$ because these sets have exactly the same elements. The other sets must be defined by giving the defining properties of the elements. We will use the notation

$$S = \{x \,|\, P(x)\}$$

which is read "S equals the set of all x such that $P(x)$ is true". Thus the sets in the remaining examples may be written as follows:

$D = \{x \mid x \text{ is either the planet Mercury or Mars}\}$
$E = \{x \mid x \text{ is a word in your vocabulary}\}$
$F = \{x \mid x \text{ is a person taking calculus and } x \text{ is enrolled at your school}\}$
$G = \{x \mid x \text{ is a river in North America}\}$
$H = \{x \mid x \text{ is a set and } x \text{ contains two elements}\}$

The listing method may also be used in conjunction with an ellipsis, ..., which implies that the elements not listed are in some well-defined sequence.

EXAMPLE 1-9

The set $N = \{1, 2, 3, ...\}$ is the set of *natural* or *counting* numbers.

EXAMPLE 1-10

The set $J = \{..., -3, -2, -1, 0, 1, 2, 3, ...\}$ is the set of *integers*.

Note that all the elements of the set N in Ex. 1-9 are elements of the set J in Ex. 1-10. We say that N is a *subset* of J.

Definition If S and T are sets, then T is a *subset* of S (written $T \subset S$) if and only if for every $x \in T$ we have $x \in S$.

EXAMPLE 1-11

Let $S = \{1, 2, 3\}$, $T_1 = \{1\}$, $T_2 = \{2\}$, $T_3 = \{3\}$, $T_4 = \{1, 2\}$, $T_5 = \{2, 3\}$, $T_6 = \{1, 3\}$, $T_7 = \{1, 2, 3\}$. Then $T_i \subset S$ for each $i = 1, 2, 3, 4, 5, 6, 7$. Note that $S = T_7$. Here are seven subsets of S. Actually S has one more subset, the set which contains no elements, $\emptyset$, called the *empty set* or the *null set*. Thus $\emptyset \subset S$.

EXAMPLE 1-12

The set $Q = \{x \mid x = a/b, a \in J, b \in J, b \neq 0\}$ is the set of *rational* numbers. Note that $N \subset Q$ and $J \subset Q$.

There are many numbers which are not rational. These are called *irrational* numbers. For instance, $\sqrt{2}$ is an irrational number. The proof of this is outlined in the Exercise 1-1, Problem 22. The set of real numbers, R, is the set of all rational numbers together with the set of all irrational numbers.

We will use two basic ways of describing real numbers: One is by decimal expansions; the other is by a "real line".

Every rational number has a decimal expansion that either terminates, e.g., $\frac{1}{4} = 0.25$, or repeats, e.g., $\frac{1}{3} = 0.333...$ and $\frac{2}{11} = 0.181818...$. Every irrational number has a nonrepeating, nonterminating decimal expansion, e.g., $\pi = 3.14159...$, and $\sqrt{2} = 1.41214...$.

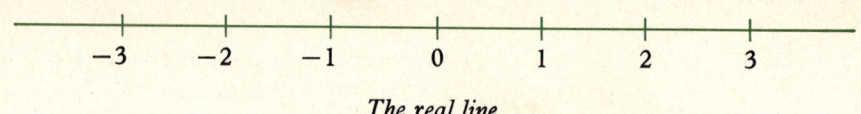

The real line

FIGURE 1-1

The second description, the "real line", is geometrical. Draw a straight line and mark two points on it, one for the *origin* (label it 0, zero); and the other, 1. The distance from zero to one is our unit length. The point which is one unit length to the right of one will be 2 and the point one unit length to the left of zero will be -1. In this way we have a correspondence between the points equally spaced on the line and the integers. Now consider the real number x as a distance and plot it on the line x units to the right or left of zero depending upon whether x is positive or negative.

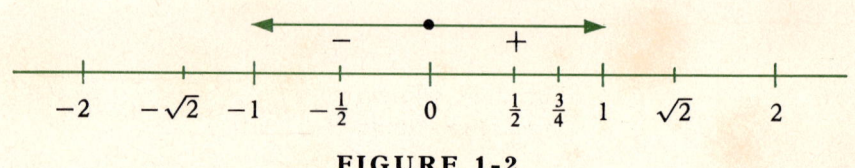

FIGURE 1-2

Hence every real number corresponds to one and only one point on the line. If the number b is to the right of the number a on the real line, then b is greater than a, and we write $b > a$ or $a < b$, which is read "a is less than b". Let us now state the familiar rules governing the symbols ">" and "<".

Rule 1 If $a > 0$ and $b > 0$, then $ab > 0$ and $a+b > 0$.

Rule 2 If a is any real number, then exactly one of the following is true: either $a > 0$, $a < 0$, or $a = 0$.

Rule 3 If $a > 0$ and $b < 0$, then $ab < 0$.

Rule 4 If $a < 0$ and $b < 0$, then $ab > 0$ (*the product of two negative numbers is positive*).

Rule 5 If $a > b$, then $a+c > b+c$.

Rule 6 Suppose $a > b$. Then if $c > 0$, then $ac > bc$; and if $c < 0$, then $ac < bc$. (Recall that if an inequality is multiplied by a negative number, the "sense" of the inequality is reversed.)

Rule 7 If $0 < a < b$ or if $a < b < 0$, then $1/b < 1/a$.

CHAP. 1   Preliminaries

The two sets just mentioned, $\{x \mid x < \frac{7}{3}\}$ and $\{x \mid x < \frac{1}{3}\}$, are examples of particular types of sets called intervals. An interval can be either the set of all real numbers between two given real numbers, or the set of all real numbers greater than a given number or less than a given number. We give these types of sets special attention.

Definition 1. The *open interval* from a to b is defined by
$$(a, b) = \{x \mid a < x < b\}$$
2. The *closed interval* from a to b is defined by
$$[a, b] = \{x \mid a \leqslant x \leqslant b\}$$
3. The *half-open* intervals from a to b are defined by
$$[a, b) = \{x \mid a \leqslant x < b\} \qquad (a, b] = \{x \mid a < x \leqslant b\}$$
4. The *half lines* are defined by
$$(a, \infty) = \{x \mid a < x\} \qquad [a, \infty) = \{x \mid a \leqslant x\}$$
$$(-\infty, a) = \{x \mid x < a\} \qquad (-\infty, a] = \{x \mid x \leqslant a\}$$

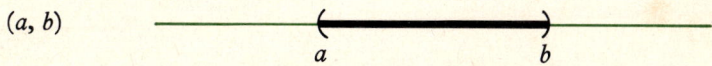

FIGURE 1-5

EXAMPLE 1-13

If $5 < 10$ then, according to Rule 5, $5+6 < 10+6$; to Rule 6, $5 \cdot 2 < 10 \cdot 2$, as well as $5(-2) > 10(-2)$; and to Rule 7, $\frac{1}{10} < \frac{1}{5}$.

A look at the real line should suffice to convince yourself of the validity of these rules. We will also have occasion to use the symbols $\leq$ and $\geq$, which are read: "less than or equals" and "greater than or equals", respectively. The rules for these relations are much the same as for $<$ and $>$.

EXAMPLE 1-14

Consider the inequality $3x-5 < 2$. Clearly zero is a solution, since $3 \cdot 0 - 5 < 2$, while 3 is not, since $3 \cdot 3 - 5$ is not less than 2. We utilize the above rules to solve the inequality for the solution set. We have

$$3x - 5 < 2$$
$$3x < 7$$
$$x < \tfrac{7}{3}$$

Hence the solution set is $\{x \mid x < \tfrac{7}{3}\}$. On the real line we can give a geometrical description of this set.

Solution to $3x - 5 < 2$, $\{x \mid x < \tfrac{7}{3}\}$

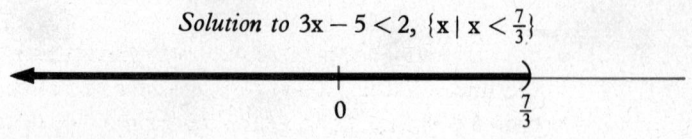

FIGURE 1-3

EXAMPLE 1-15

Consider $5 - 3x > 4$. To solve for the solution set we have

$$5 - 3x > 4$$
$$-3x > -1$$
$$x < \tfrac{1}{3}$$

Hence the solution set is $\{x \mid x < \tfrac{1}{3}\}$. Geometrically, this set is described in Fig. 1-4.

Solution to $5 - 3x > 4$, $\{x \mid x < \tfrac{1}{3}\}$

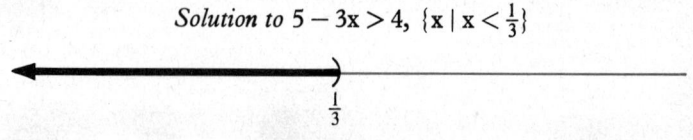

FIGURE 1-4

The symbol ∞ is read "infinity" and the last four sets are also referred to as: the open interval from a to ∞, the closed interval from a to ∞, the open interval from $-∞$ to a, and the closed interval from $-∞$ to a. Geometrically, we can specify these sets on the real line as in Fig. 1-5.

Hence the solution sets of Examples 1-14 and 1-15 can be given also by $(-∞, \frac{7}{3})$ and $(-∞, \frac{1}{3})$.

Two fundamental set operations which will be used throughout the text are *union*, ∪, and *intersection*, ∩.

Definition If S and T are sets, then $S \cup T$ is the set of all elements in S or in T, and $S \cap T$ is the set of all elements in S and in T.

EXAMPLE 1-16

Let $A = \{1, 2, 3\}$, $B = \{2, 3, 4\}$. Then
$$A \cup B = \{1, 2, 3, 4\} \quad \text{and} \quad A \cap B = \{2, 3\}$$
If $C = \{4, 5, 6\}$, then
$$A \cup C = \{1, 2, 3, 4, 5, 6\} \quad \text{and} \quad A \cap C = \emptyset$$

EXAMPLE 1-17

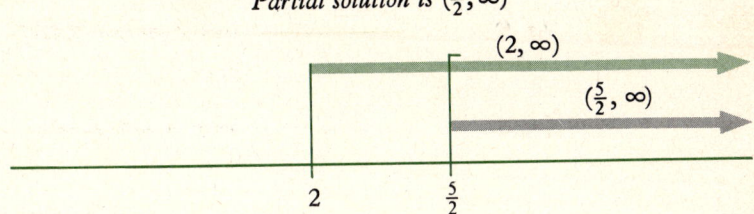

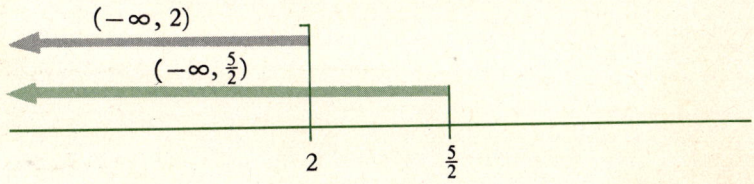

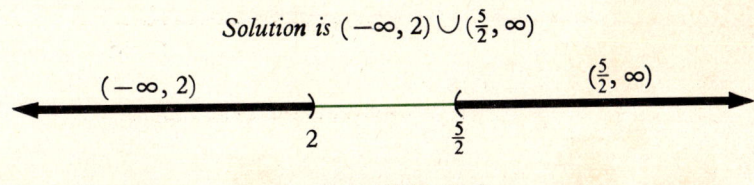

FIGURE 1-6

Consider the inequality $(x-2)(2x-5) > 0$. From Rules 1 and 4 there are two cases to the solution.

(1) $\qquad x - 2 > 0 \qquad$ and $\qquad 2x - 5 > 0$

so that $\qquad x > 2 \qquad$ and $\qquad x > \tfrac{5}{2}$

thus $\qquad x \in (2, \infty) \cap (\tfrac{5}{2}, \infty) = (\tfrac{5}{2}, \infty)$

(2) $\qquad x - 2 < 0 \qquad$ and $\qquad 2x - 5 < 0$

so that $\qquad x < 2 \qquad$ and $\qquad x < \tfrac{5}{2}$

thus $\qquad x \in (-\infty, 2) \cap (-\infty, \tfrac{5}{2}) = (-\infty, 2)$

The solution set is the *union* of the two partial solution sets, $(\tfrac{5}{2}, \infty)$ and $(-\infty, 2)$, because all the elements in each set are solutions. The solution set therefore contains all of the elements in each set, that is, $(-\infty, 2) \cup (\tfrac{5}{2}, \infty)$. Geometrically, we can depict these sets on the real line as in Fig. 1-6.

★ EXAMPLE 1-18

Consider the inequality $(3x-4)(x-5) \leq 0$. Again we have two cases.

(1) $\qquad 3x - 4 \geq 0 \qquad$ and $\qquad x - 5 \leq 0$

$\qquad x \geq \tfrac{4}{3} \qquad$ and $\qquad x \leq 5$

$\qquad x \in [\tfrac{4}{3}, \infty) \cap (-\infty, 5] = [\tfrac{4}{3}, 5]$

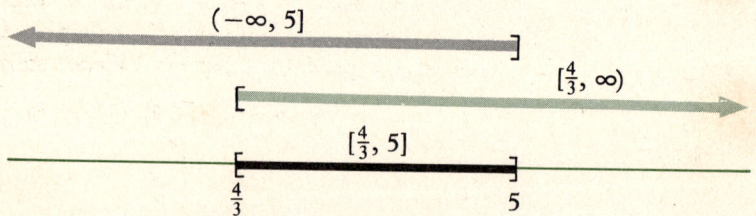

FIGURE 1-7

(2) $\qquad 3x - 4 \leq 0 \qquad$ and $\qquad x - 5 \geq 0$

$\qquad x \leq \tfrac{4}{3} \qquad$ and $\qquad x \geq 5$

$\qquad x \in (-\infty, \tfrac{4}{3}] \cap [5, \infty) = \varnothing$

Hence the solution set is $[\tfrac{4}{3}, 5]$. (See Figs. 1-7 and 1-8.)

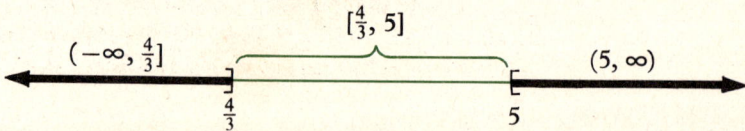

FIGURE 1-8

Definition If a is any real number we define the *absolute value* of a, $|a|$, to be

$$|a| = \begin{cases} a & \text{if } a > 0 \\ 0 & \text{if } a = 0 \\ -a & \text{if } a < 0 \end{cases}$$

EXAMPLE 1-19
Observe that $|2| = 2$, $|0| = 0$, $|-2| = -(-2) = 2$. Consequently $|-8| = -(-8) = 8$, which also is $|8|$. Therefore $|-8| = |8|$ or $|-a| = |a|$. In other words, if $x < 0$, then $|x| = -x > 0$.

EXAMPLE 1-20
We solve the equality $|2x - 3| = 4$ by first noting that if $2x - 3 \geqslant 0$, i.e., if $x \geqslant \frac{3}{2}$, then

$$|2x - 3| = 2x - 3 = 4$$

which implies that $x = \frac{7}{2}$. Next if $2x - 3 < 0$, i.e., if $x < \frac{3}{2}$, then

$$|2x - 3| = -(2x - 3) = 4$$

which implies that $x = -\frac{1}{2}$. The solution set is $\{\frac{7}{2}, -\frac{1}{2}\}$.

EXAMPLE 1-21
To solve the inequality $|x| < 4$, we use two cases.
1. If $x \geqslant 0$, then $|x| = x < 4$, so a partial solution is $0 \leqslant x < 4$, i.e., $[0, 4)$.
2. If $x < 0$, then $|x| = -x < 4$, which implies that $x > -4$; so we get $(-4, 0)$ as a partial solution set. Hence the solution set is

$$(-4, 0) \cup [0, 4) = (-4, 4)$$

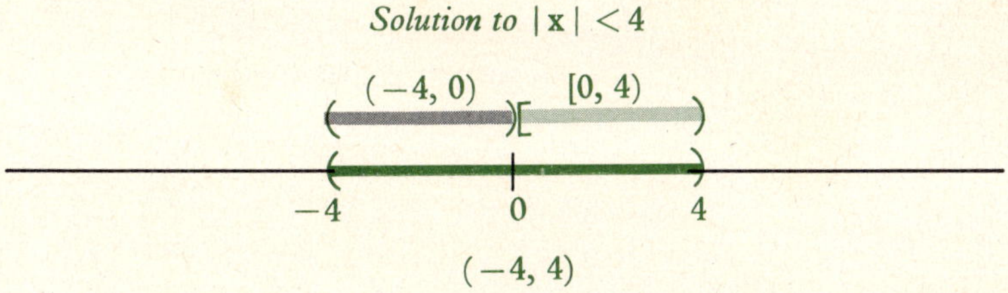

Solution to $|x| < 4$

FIGURE 1-9

Note that we can interpret the solution to the inequality $|x| < 4$ to be all x within 4 units of zero, which is all $x \in (-4, 4)$, which implies that $-4 < x < 4$. One can readily generalize to prove Theorem 1-1.

Theorem 1-1 *If a is any real number greater than zero, then $|x| \leqslant a$ is true if and only if $-a \leqslant x \leqslant a$. Similarly $|x| \geqslant a$ if and only if $x \geqslant a$ or $x \leqslant -a$.*

EXAMPLE 1-22

Solve the inequality $|3x-4| \leq 6$. From Theorem 1-1 we have the inequality being equivalent to $-6 \leq 3x-4 \leq 6$, which implies that $-2 \leq 3x \leq 10$, so $-\frac{2}{3} \leq x \leq \frac{10}{3}$ and our solution set is $[-\frac{2}{3}, \frac{10}{3}]$.

Solution to $|3x - 4| \leq 6$

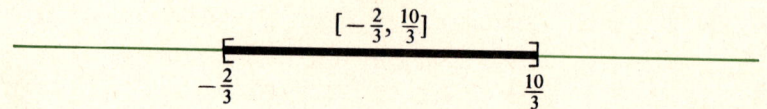

FIGURE 1-10

EXERCISE 1-1

1. Determine which of the following are sets.
   (a) $\{x \mid x = 2n - 1, n \text{ is a natural number}\}$
   (b) The set of all intelligent people.
   (c) The set of all healthy animals.
   (d) The set of all students on the Dean's list with a D average.
   (e) $\{1, 2, a, b, a, b\}$

2. Determine the elements of the given sets.
   (a) $\{1, 2, 3, 4, 5\}$
   (b) $\{x \mid x^2 - 1 = 0, x \text{ is a natural number}\}$
   (c) $\{x \mid x = 1/n, n \text{ is a natural number less than 6}\}$
   (d) $\{1, 3, 5, 3, 4, 5, 1\}$

3. Use the listing method to specify the given sets.
   (a) $\{x \mid x^2 = 9, x \text{ is an integer}\}$
   (b) $\{x \mid x^4 = 1, x \text{ is an integer}\}$
   (c) $\{x \mid (x-1)(x-2)(x-3)(x-4) = 0, x \text{ is a natural number}\}$
   (d) $\{x \mid x \text{ is a positive prime number less than 20}\}$

4. In the following use the defining property method to specify the following sets.
   (a) $\{2, 4, 6, 8, 10\}$
   (b) $\{\frac{1}{2}, \frac{3}{2}, \frac{5}{2}, \frac{7}{2}\}$
   (c) $\{2, 3, 4, 5, \ldots\}$
   (d) $\{-3, 0, 3, 6, \ldots, 15\}$
   (e) $\{4, 16, 64, 256\}$
   (f) $\{x \mid x \text{ is your brother}\}$

5. Which of the following sets are equal?
   (a) $A = \{x \mid x \text{ is a letter in the word "follow"}\}$
   (b) $B = \{x \mid x \text{ is a letter in the word "flow"}\}$
   (c) $C = \{x \mid x \text{ is a letter in the word "wolf"}\}$
   (d) $D = \{l, o, l, f, w\}$

6. Which of the sets in Problems 3, 4, and 5 are subsets of others?
7. Let $A = \{1, 2, 3, 4\}$, $B = \{3, 4, a, b\}$, $C = \{a, b, c, d\}$. Find $A \cup B$, $B \cup C$, $A \cup C$, $A \cap B$, $B \cap C$, $C \cap A$.
8. The notation $A \not\subset B$ means that A is not a subset of B. If $A \not\subset B$, then there exists an element $x \in A$ such that $x \notin B$. State why $B \not\subset A$, $C \not\subset B$, $C \not\subset A$ in Problem 17.
9. In Problem 17 is $A \subset B$? $B \subset A$? $C \subset D$? $D \subset C$?
10. Use the listing method to specify the sets.
    (a) $A = \{x \mid x \in J \text{ and } |x| \leqslant 5\}$
    (b) $B = \{x \mid x \in J \text{ and } x \in (-3, 4)\}$
11. Find another name for the sets.
    (a) $A = \{x \mid |x| < 0\}$
    (b) $B = \{x \mid |x| \geqslant 0 \text{ and } x \in J\}$
12. Which of the following are true?
    (a) $\frac{1}{3} < \frac{1}{3}$
    (b) $-8 < -10$
    (c) $-3 < -2 < -1$
    (d) $-x \leqslant x$ for any real number x
    (e) $a < 2 < a - 1$ for some real number a
    (f) $x \leqslant |x|$ for any real number x
    (g) $0.333 < \frac{1}{3}$
    (h) $\dfrac{2}{5+8} < \dfrac{1}{7}$
    (i) $(-2)(3) < (2)(-2)$
13. The statement $a \div b = x$ is equivalent to the statement $b \cdot x = a$. Thus $\frac{1}{0} = x$ is equivalent to $0 \cdot x = 1$. But no such x exists and therefore $\frac{1}{0}$ is meaningless. Is $\frac{2}{0}$ meaningless? Why is $\frac{0}{0}$ meaningless? What is $(x-1)/(x-1)$ when $x = 2$? when $x = 1$?
14. One can give an alternate definition of the equality of sets by saying that $A = B$ if and only if $A \subset B$ and $B \subset A$. Suppose $A \not\subset B$, show that $A \neq B$.
15. Let $A = \{a, b, c, d\}$. Which of the following statements are correct?
    (a) $a \in A$ (c) $\{a\} \in A$
    (b) $a \subset A$ (d) $\{a\} \subset A$
16. Which of the sets are empty?
    (a) $\{x \mid x + 1 = 0 \text{ and } x - 1 = 0\}$
    (b) $\{x \mid x \text{ is a rock and } x \text{ is living}\}$
    (c) $\{x \mid x \in x\}$
    (d) $\{x \mid x^2 + 1 = 1\}$
    (e) $\{0\}$
    (f) $\{\emptyset\}$
17. We say that a set is "finite" if the number of elements in the set is a non-negative integer. Otherwise the set is said to be "infinite". Determine

whether the following sets are finite or infinite.
  (a) $A = \{1, 2, 3, \ldots, 100\}$
  (b) $B = \{1, 2, 3, \ldots, 100{,}000\}$
  (c) $C = \{1, 2, 3, \ldots\}$
  (d) $D = \{x \mid 1 \leqslant x \leqslant 10 \text{ and } x \text{ is a natural number}\}$
  (e) $D_1 = \{x \mid 1 \leqslant x \leqslant 10 \text{ and } x \text{ is a rational number}\}$
  (f) $E = \{x \mid \tfrac{1}{2} \leqslant x \leqslant \tfrac{2}{3} \text{ and } x \text{ is a real number}\}$

18. Let $A = [0, 1]$, $B = (1, 3)$, $C = (2, 4]$, $D = (0, 4)$. Find the indicated sets and specify the sets on the real line.
  (a) $A \cup B$
  (b) $A \cap B$
  (c) $B \cup C$
  (d) $B \cap C$
  (e) $A \cap D$
  (f) $A \cup B \cup C \cup D$
  (g) $B \cap D$

19. Which of the sets A, B, C, D in Problem 18 are subsets of the others?

20. In Problem 17 we defined finite and infinite sets. Which of the following sets are finite?
  (a) $(0, 1] \cap [1, 2)$
  (b) $(0, 1)$
  (c) $[0, \tfrac{1}{10}]$
  (d) $\{x \mid x \in J \text{ and } |x| < 5\}$
  (e) $\{x \mid |x| < 5\}$ where x is any real number

21. (a) What does it mean to say that the fraction p/q is in "lowest terms"? Is $\tfrac{4}{6}$ in lowest terms?
  (b) Show that the square of an even number is even by noticing that $(2n)^2 = 4n^2$. Is the square of an odd number even or odd?

22. Suppose that $\sqrt{2} = p/q$ where p and q are integers, $q \neq 0$ and p/q is in lowest terms. Then $2q^2 = p^2$ (why?), so p^2 is even, and hence p is even (why?). Thus $p = 2n$ for some n, and $p^2 = 4n^2 = 2q^2$. Hence $2n^2 = q^2$ and q^2 is even and so is q. Why does this say that p/q is not in lowest terms? How does this prove that $\sqrt{2}$ is not a rational number?

23. Find the solution set for the given inequalities and specify the appropriate intervals on the real line.
  (a) $x(x-5) > 0$
  (b) $x(8-x) > 0$
  (c) $2x(x-2) < 0$
  (d) $(x-1)(x-3) \leqslant 0$
  (e) $(3x+5)(6-x) \geqslant 0$
  (f) $x^2 - 5x + 6 > 0$
  (g) $|x-1| < 3$
  (h) $|3x-5| < 7$
  (i) $|4-3x| < 1$
  (j) $|x-2| > 10$

1-2 FUNCTIONS

The most common underlying notion in mathematics is that of a function. This concept is not new to you, you've dealt with functions for many years. Whenever you think of or derive a correspondence between elements of one set and the objects of another, you create a function or relation between the two sets. Let us look at some everyday examples.

EXAMPLE 1-23
> An ordinary roadmap provides a correspondence between the points on the roadmap and the points on the earth, for there is a definite rule which, determined by the scale on the map, associates points on the map with points on the earth.

EXAMPLE 1-24
> The U.S. Post Office utilizes a very definite rule of correspondence between the set of weights of letters and packages and the cost of mailing these objects.

EXAMPLE 1-25
> At the end of each working day the New York Stock Exchange provides a correspondence between a set of listed firms and their closing prices.

There are three basic elements which comprise a function: a first set, a second set, and a rule that defines a correspondence between the elements of these sets. The mathematical concept of a function must be quite precise for our purposes, so that not every such relationship will be called a function. We must stipulate that no one element in the first set can correspond to two different elements in the second.

Definition   A function $f: A \to B$ is a correspondence between the elements of the set A, called the domain of f, and the elements of the set B, called the range of f, such that every element $x \in A$ corresponds to, or maps to, one and only one element $f(x) \in B$.

The notation $f: A \to B$ is read: "f is a function from A to B", where A is the domain of f and B is the range. Let us emphasize a few basic points about the above definition. Every element of A corresponds to, or maps to, some element of B, whereas it may happen that not every element of B has an element of A mapping onto it. No one element of A can map to two different elements of B, but it may well be the case that two different elements of A map to the same element of B. The term $f(x)$ is read "f of x". We will need to adopt some additional notation.

Definition   If $f: A \to B$ and $a \in A$, then $f(a) \in B$ is called the image of a and we say that a maps onto $f(a)$. If $y = f(x)$, then we say that "f is defined at x" or that "$f(x)$ is defined".

There are many ways of describing functions. A table may provide a listing of the elements in the domain and their images next to them or below them. Daily in the newspaper the stock prices, the contents, and the temperatures of cities are given in tables. Another convenient method is a graph, where the elements of the domain are usually found on a horizontal line and the elements of the range are on a vertical line. Many functions have domains and ranges that are subsets of R, the real number system. The rule of correspondence is usually given by a formula.

EXAMPLE 1-26

Consider the function $f: R \to R$ whose rule is given by $f(x) = 2x+1$. The domain and range is the set of all real numbers. The image of zero is one since $f(0) = 2 \cdot 0 + 1 = 1$. Similarly $f(1) = 3$, $f(2) = 5$, $f(\frac{1}{2}) = 2$, $f(\sqrt{2}) = 2\sqrt{2}+1$.

Since most of our functions will be from the real numbers to the real numbers, i.e., $f: R \to R$, or subsets of the real numbers, we will specify functions by simply giving a formula. Thus we will refer to the function in Ex. 1-26 as the function $f(x) = 2x+1$.

One must always keep in mind that the domain of such a function defined simply by a formula might not be all of the real numbers. We will assume, unless it is otherwise stated, that the domain of a function given by a formula is the largest possible subset of the real numbers. In other words, the domain will be considered to be the set of all real numbers except those for which the formula is not defined.

EXAMPLE 1-27

Consider the function $f(x) = \sqrt{x-1}$. Note that the images of 1, 2, and 3 are defined,

$$f(1) = \sqrt{1-1} = 0 \qquad f(2) = \sqrt{2-1} = \sqrt{1} = 1 \qquad f(3) = \sqrt{2}$$

but if we substitute zero into the formula we get $\sqrt{0-1} = \sqrt{-1}$ which is not a real number. Hence we say that $f(0)$ is not defined. Similarly $f(x)$ is not defined for any real number x such that $x < 1$. Therefore the domain of f is $[1, \infty)$.

Division by zero often causes difficulty for most students. In Exercise 1-1, Problem 13 the reason why $2/0$ is meaningless is discussed. The student should convince himself that $a/0$ is meaningless for every real number a.

EXAMPLE 1-28

If $f(x) = x/(x-2)$ then $f(2)$, equaling $\frac{2}{0}$, is not defined; so 2 is not an element of the domain of f. Therefore the domain of f is $(-\infty, 2) \cup (2, \infty)$, i.e., all real numbers except 2.

Let us now develop the machinery we shall need to discuss graphs of functions and other relationships between sets. We can construct a coordinate system by using our concept of the real line. First draw a horizontal real line, usually called the x-axis, with an origin and a unit

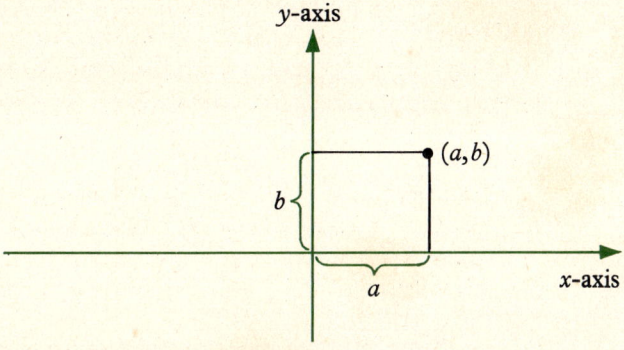

FIGURE 1-11

length. Then draw another real line perpendicular to the first so that the origins of the real lines intersect at the point which we call the origin of our coordinate system. This vertical line is usually called the y-axis. Just as any real number corresponds to a unique point on the line each ordered pair of real numbers (a,b) corresponds to a unique point in the plane.

This correspondence is given as follows: If for a given ordered pair (a,b) we mark off the distance a on the x-axis and the distance b on the y-axis and draw two lines perpendicular to the axes at these points, then the point which is the intersection of these perpendiculars is the point which corresponds to (a,b). We shall not distinguish between the point and the ordered pair.

EXAMPLE 1-29

The points $(1,0)$ $(0,3)$ $(-1,3)$ $(5,-\sqrt{2})$ are plotted in Fig. 1-12.

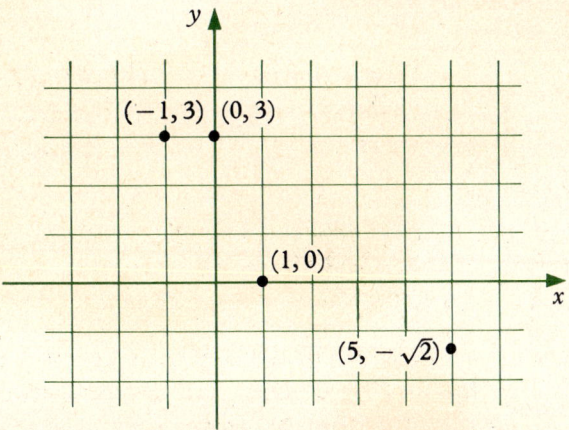

FIGURE 1-12

This type of coordinate system is called a rectangular Cartesian coordinate system after the famous mathematician philosopher René Descartes (1596–1650). Descartes' seemingly simple yet quite profound invention enabled him to draw a "picture" of a function $y = f(x)$ by plotting in a coordinate system the set of points $f = \{(x,y) \mid y = f(x)\}$. We call this set the *graph* of $y = f(x)$.

One method of finding the graph of a function is to crank out many pairs of points which satisfy the equation, plot these points, and then draw a curve through the points. There are a number of glaring imperfections to this method, however.

EXAMPLE 1-30

Consider the function $y = 3x - 4$. When $x = 0$, $y = -4$, so $(0, -4)$ is part of the graph of this function. Similarly $(\frac{4}{3}, 0)$, $(1, -1)$, $(2, 2)$, $(-1, -7)$ and $(-2, -10)$ are part of the graph. If we plot these points and draw a curve through them we get a straight line as on the left in Fig. 1-13. But are we sure that the actual curve is not something like the one on the right?

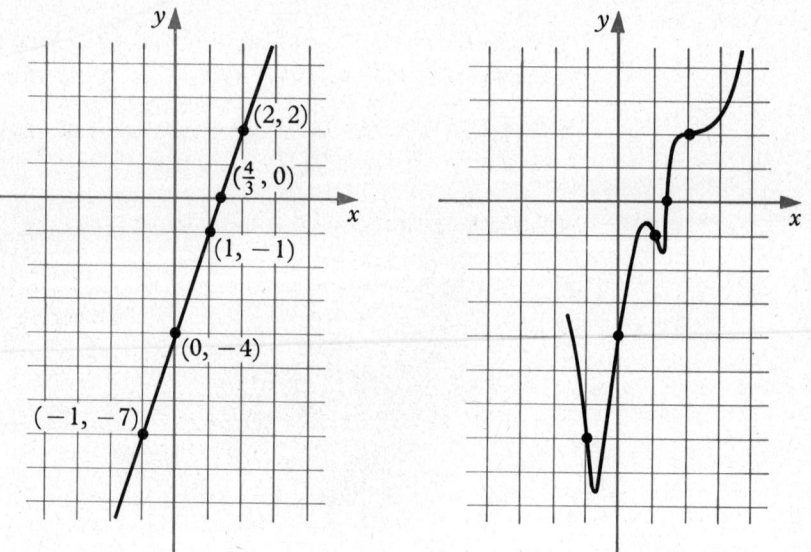

FIGURE 1-13

SEC. 1-2   Functions

The methods of calculus will help us determine the graphs of functions quite exactly.

EXERCISE 1-2

1. Consider the functions defined by the given formulas. Find the implicit domain of each function.

   (a) $f(x) = 2x - 1$
   (b) $f(x) = \sqrt{x+1}$
   (c) $f(x) = \sqrt{3-x}$
   (d) $f(x) = \dfrac{1}{x}$
   (e) $f(x) = \dfrac{1}{x-2}$
   (f) $f(x) = \sqrt{(x-1)(x-2)}$
   (g) $f(x) = \dfrac{1}{(2x-1)(x-3)}$

2. Suppose $f(x) = 2x - 3$ and $g(x) = x^2 - 1$. Find $f(1)$, $f(2)$, $f(2-1)$, $f(\sqrt{2})$, $g(4)$, $g(3)$, $g(4-3)$, $g(4) - g(3)$, $f(a)$, $f(a^2)$, $f(a^2 - 1)$, $g(x+h)$.

3. Plot the points in a coordinate system: $(0,0)$, $(1,5)$, $(-6,7)$, $(2, -\tfrac{3}{2})$, $(-6, -\sqrt{2})$.

4. Draw a graph of the functions by plotting some points and drawing a curve through the points.

   (a) $f(x) = 3x - 5$ (d) $f(x) = x^2$
   (b) $f(x) = -x - 6$ (e) $f(x) = 1 - x^2$
   (c) $f(x) = 4$ (f) $f(x) = |x|$

5. Use a geometrical test to determine whether the graphs are functions.

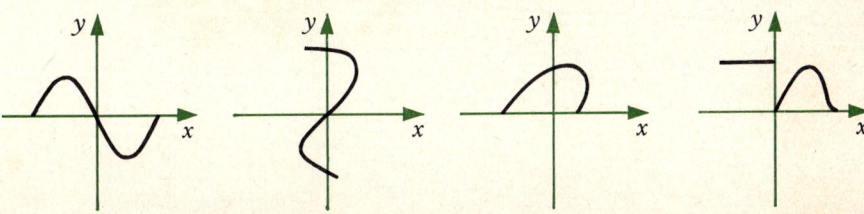

6. The unit lengths chosen on the *x*-axis and *y*-axis do not necessarily have to be equal. Suppose a sales manager accumulates data and he wants to use a graph to indicate that sales have risen sharply in the previous four months. Which of the graphs would he submit to his superiors? How do the two graphs differ?

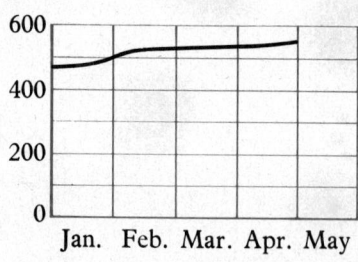 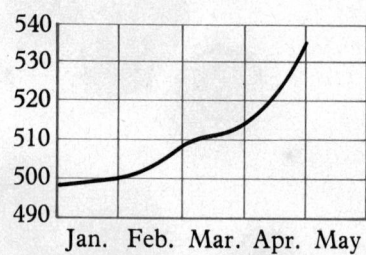

7. Let A and B be two sets. The Cartesian product $A \times B$ is defined by
$A \times B = \{(a, b) \mid a \in A, b \in B\}$.
   (a) Let $A = \{1, 2\}$, $B = \{3, 4\}$. Find $A \times B$.
   (b) Explain why a Cartesian coordinate system can be regarded as $R \times R$ where R is the set of all real numbers.

8. The coordinate axes divide the plane into four sections called *quadrants*. Quadrant I is the set of points above the *x*-axis and to the right of the *y*-axis, i.e.,

$$\{(x, y) \mid x > 0, y > 0\}.$$

Describe the other three quadrants by inequalities. In which quadrants are the points in Problem 3?

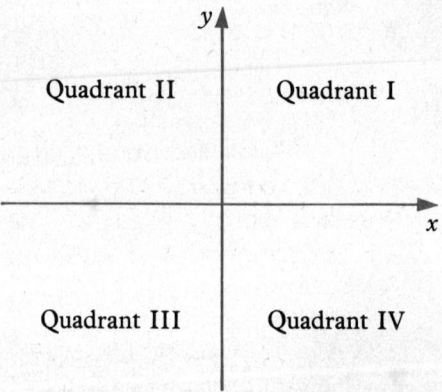

9. If $P_1(x_1,y_1)$ and $P_2(x_2,y_2)$ are two points in the plane, the distance from P_1 to P_2, written as $|P_1P_2|$, is defined to be

$$\sqrt{(x_1-x_2)^2 + (y_1-y_2)^2}$$

Consider the figure below and explain how this definition follows directly from the Pythagorean theorem.

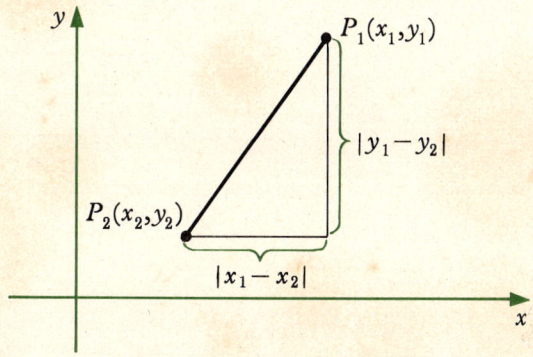

10. Find the distance between the points in Problem 3.

1-3 LINEAR FUNCTIONS AND STRAIGHT LINES

The simplest yet most useful types of functions are linear functions, those whose graphs are straight lines. Their simplicity stems from a straight line being completely determined by just two of its points and the line through them, and from the equation of a linear function having a compact form.

Definition   A linear function is a function given by the equation $f(x) = mx + b$, where m and b are real number constants.

We assume that the graph of $f(x) = mx + b$ is a straight line, and that any nonvertical straight line has an equation of the form $y = mx + b$. A vertical line has an equation of the form $x = a$ for some real number a. The interested reader is invited to prove these remarks.

EXAMPLE 1-31

To graph $y = 5x - 2$ we first find two points on the line, say $(0, -2)$ and $(1, 3)$. Then we plot them and draw a line through them. (See Fig. 1-14.)

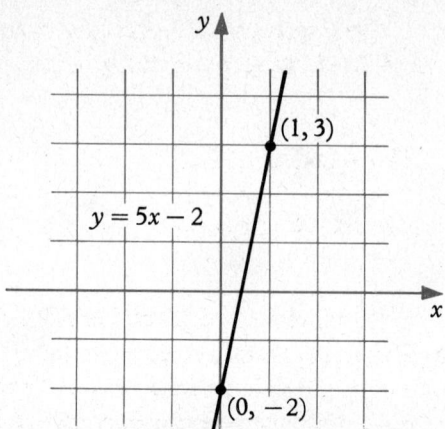

FIGURE 1-14

If $f(x) = mx + b$, then the number m is called the *slope* of f, or of the line, and it measures the steepness of the line. Note that if (x_1, y_1) and (x_2, y_2) are any two points on the line, then

$$\frac{\Delta y}{\Delta x} = \frac{y_2 - y_1}{x_2 - x_1} = \frac{(mx_2 + b) - (mx_1 + b)}{x_2 - x_1} = \frac{m(x_2 - x_1)}{x_2 - x_1} = m$$

where $\Delta y = y_2 - y_1$, the change in y, and $\Delta x = x_2 - x_1$, the change in x.

If $m > 0$, the line rises to the right, and if $m < 0$, it falls to the right. Note also that $f(0) = b$, so the line intersects the y-axis at the point $(0, b)$; hence we call b the *y-intercept*.

If $m = 0$, then the equation becomes $y = f(x) = b$, which is called a *constant function*. The graph of a constant function is a horizontal straight line. The equation of the x-axis is therefore $y = 0$. We also say that a vertical line has no slope. Hence the y-axis, given by $x = 0$, has no slope.

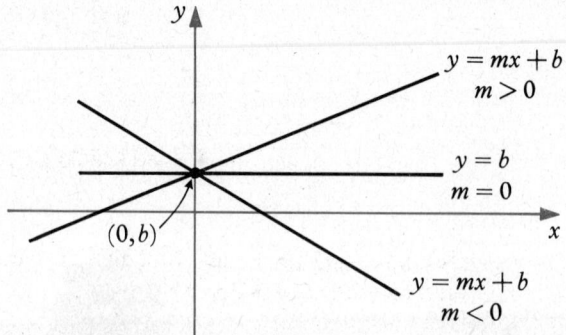

FIGURE 1-15

SEC. 1-3 Linear Functions and Straight Lines

The equations of linear functions, called *linear equations*, have other forms besides the slope-intercept form, $y = mx + b$.

EXAMPLE 1-32

The equation of the line going through the point $(1, -5)$ with slope $\frac{1}{2}$ is given by $(y + 5) = \frac{1}{2}(x - 1)$ which is called the point-slope form. This equation can be put into slope-intercept form,

$$y = \tfrac{1}{2}x - 5\tfrac{1}{2}$$

In general, the equation in point-slope form of the line going through the point (x_1, y_1) with slope m is given by $(y - y_1) = m(x - x_1)$. The general equation of a line is $ay + bx + c = 0$. The slope of this line is $-b/a$, which can be seen by using algebraic manipulations to put the equation in slope-intercept form,

$$y = -\frac{b}{a}x - \frac{c}{a}$$

EXAMPLE 1-33

To find the slope and y-intercept of the line $3x - 4y - 6 = 0$ we put this equation in slope-intercept form,

$$y = \tfrac{3}{4}x - \tfrac{3}{2}$$

and see that the slope is $\tfrac{3}{4}$ and the y-intercept is $-\tfrac{3}{2}$.

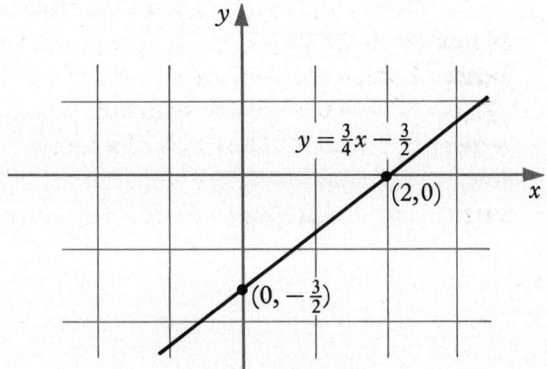

FIGURE 1-16

EXAMPLE 1-34

We now find the equation of the line going through the two points $(1, 2)$ and $(6, -3)$. The slope is

$$m = \frac{-3 - 2}{6 - 1} = \frac{-5}{5} = -1$$

Hence, using the point-slope form, the equation is $y-2=(-1)(x-1)$.

In general, the equation of the line going through the points (a_1,b_1) and (a_2,b_2) is

$$(y-b_1) = \frac{b_1 - b_2}{a_1 - a_2}(x - a_1)$$

It is often necessary to find the point of intersection of two lines. To accomplish this we must find a point whose coordinates satisfy each linear equation; so we must solve the two equations simultaneously.

EXAMPLE 1-35

To find the point of intersection of the two lines $2x-3y-1=0$ and $6x-y+5=0$, we solve the two equations simultaneously:

$$2x - 3y - 1 = 0 \qquad (1)$$
$$6x - y + 5 = 0 \qquad (2)$$
$$-6x + 9y + 3 = 0 \qquad (1')\ \textit{Multiply (1) by } -3$$
$$8y + 8 = 0 \qquad (2)+(1')$$
$$y = -1$$
$$x = -1 \qquad \textit{Substitute y into (1)}$$

Hence the point of intersection is $(-1,-1)$, as seen in Fig. 1-17.

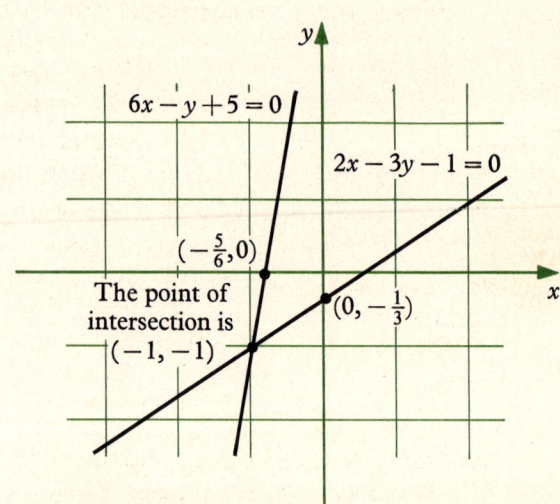

FIGURE 1-17

EXERCISE 1-3

1. Draw the graphs of the linear functions.
   (a) $f(x) = 2x - 1$
   (b) $f(x) = -x + 2$
   (c) $f(x) = \frac{1}{2}x - 5$
   (d) $f(x) = 5$
   (e) $f(x) = x - \sqrt{2}$
   (f) $f(x) = 4x$

2. Find the slope and y-intercept of the following linear functions and draw their graphs.
   (a) $2x - 3y - 1 = 0$
   (b) $x - y + 5 = 0$
   (c) $3y - 5 = 0$
   (d) $2x - 2y + 10 = 0$

3. Find the equation of each line l described by the following:
   (a) The slope of l is 0 and l passes through $(1,1)$.
   (b) l passes through $(-1,1)$ and $(2,6)$.
   (c) l passes through $(1,8)$ and $(1,10)$.
   (d) l passes through $(1,5)$, and the y-intercept is 2.
   (e) l passes through $(6,-3)$ and has no slope.
   (f) l passes through $(1,2)$ and is parallel to $y = 2x - 1$.

4. If a line passes through the point $(a,0)$, then a is called the x-intercept. If $a \neq 0$ and $b \neq 0$, then the intercept form of a line is given by
$$\frac{x}{a} + \frac{y}{b} = 1$$
   Find the x-intercept, y-intercept, and slope of the line in this form.

5. The correspondence between measurements of temperature in degrees Fahrenheit and degrees centigrade is given by the formula $5F - 9C = 160$, where F is on the Fahrenheit scale and C is on the centigrade. Note that C is a linear function of F. Graph this function. What is the slope? Find the corresponding measurements for the given measurements.
   (a) 32°F
   (b) 0°F
   (c) 100°F
   (d) 212°F
   (e) −10°F
   (f) 110°F

6. Let $f(x) = x^2$. Find the slope of the line passing through the points $(2, f(2))$ and $(2.1, f(2.1))$, and through the points $(2, f(2))$ and $(2.01, f(2.01))$.

7. Find an expression for the slope of the line passing through $(2, f(2))$ and $(2.1, f(2.1))$ for the function $y = f(x)$.

8. Find the points of intersection of the following pairs of lines.
   (a) $2x - 3y - 1 = 0$, $3x + y - 1 = 0$
   (b) $y = 2x - 1$, $y = 6x + 2$
   (c) $y - 3 = 2x - 1$, $\frac{x}{2} + \frac{y}{3} = 1$

9. Verify by similar triangles that the slope m of a linear function $f(x) = mx + b$ is constant. (*Hint*: Suppose $P_1(x_1,y_1)$, $P_2(x_2,y_2)$, and $P_3(x_3,y_3)$ are on the line with x_1, x_2, x_3. Construct the triangles $P_1, P_2, P_4(x_2,y_1)$ and $P_2, P_3, P_5(x_3,y_2)$. Show they are similar. Note that $x_1 < x_2 < x_3$.)

CHAP. 1   Preliminaries

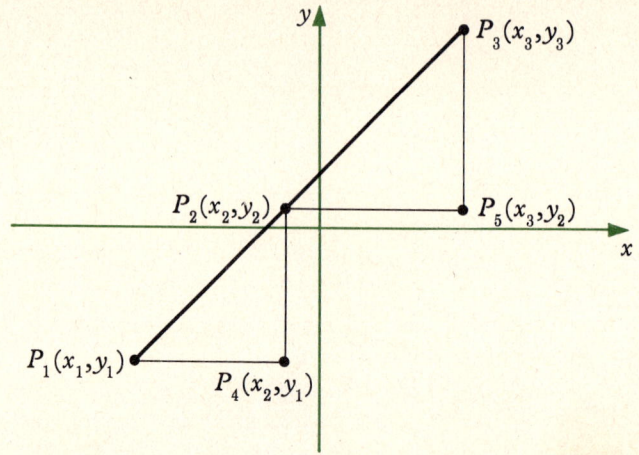

10. Suppose a company has fixed cost equal to $20,000 and has variable cost which depends upon volume. Let the relationship between total cost C and sales volume x be

$$C = \tfrac{1}{2}x + 20{,}000$$

(a) Suppose the company sells $60,000 worth of goods. What is the total cost? What is the fixed cost? What is the profit?
(b) Suppose the company sells $30,000 worth of goods. What is the total cost? What is the profit?
(c) At what point will the company break even? That is, at what value, x, of sales volume, will x equal C the total cost? (*Hint:* Set $x = \tfrac{1}{2}x + 20{,}000 = C$.)

1–4 QUADRATIC FUNCTIONS

The great Italian scientist Galileo Galilei conducted experiments on the study of motion which laid the groundwork for the invention of the calculus. Galileo discovered that all objects regardless of their weight were attracted to the earth by the same force. The following table provides a summary of his findings for a free-falling body.

Time (quarter sec)	0	1	2	3	4	5	...
Distance (ft)	0	1	4	9	16	25	...

Thus the distance s that an object falls is a function of the time t it has fallen, given by $s(t) = t^2$. The graph of $s(t) = t^2$ for all real numbers t is given in Fig. 1-18.

24

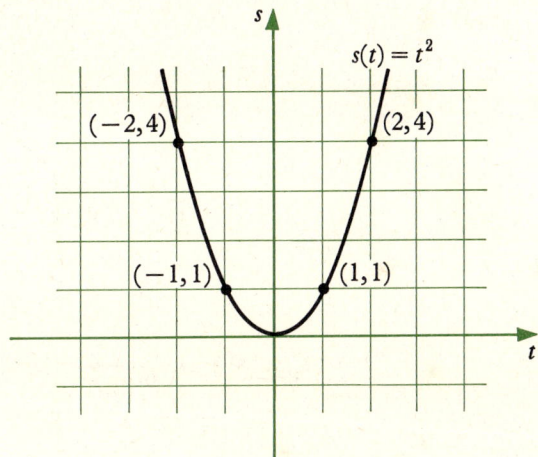

FIGURE 1-18

The function $y = x^2$ is an example of a parabola. Another simple parabola is the function $y = -x^2$, whose graph is given in Fig. 1-19.

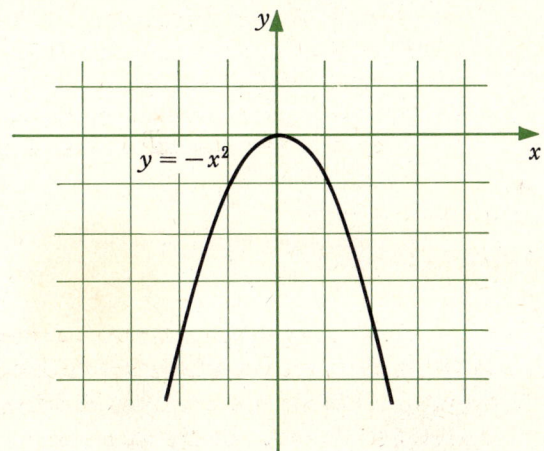

FIGURE 1-19

The graph of $y = x^2$ is a bowl-shaped figure that opens upwards. We say that it is *symmetric* with respect to the y-axis because the portion of the graph to the right of the y-axis is the mirror image of the portion of the graph to the left of the y-axis. Since the point $(0,0)$ is the point where the graph touches the line of symmetry, the y-axis, we say $(0,0)$ is the *vertex*.

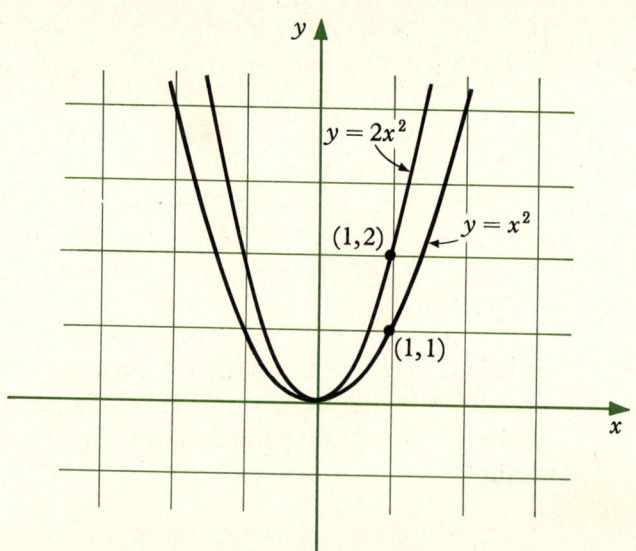

FIGURE 1-20

Note that the graph of the parabola $y = -x^2$ opens downward. It is also symmetric with respect to the y-axis and its vertex is (0,0).

The graphs of the parabolas $y = 2x^2$ and $y = -2x^2$ (Fig. 1-20) are similar to $y = x^2$ and $y = -x^2$ except that they are "narrower".

In general, if $a > 1$, the graphs of $y = ax^2$ and $y = -ax^2$ (Fig. 1-21) are similar to $y = x^2$ and $y = -x^2$ except that they are "narrower". If

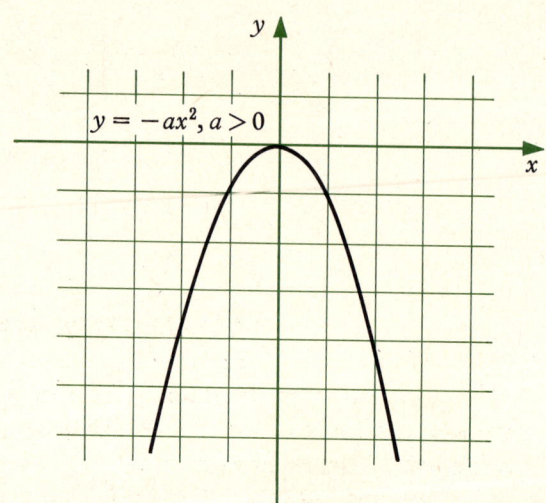

FIGURE 1-21

$0 < a < 1$, then the graphs of $y = ax^2$ and $y = -ax^2$ are "wider" than those of $y = x^2$ and $y = -x^2$.

Consider the graph of $y = x^2 + 4$. To find the functional value y for any x, we first square x and then add 4 to x^2. This has the effect of shifting the graph of $y = x^2$ up four units. Hence the graph of $y = x^2 + 4$ is similar to $y = x^2$ except that it is four units above $y = x^2$. In general, the graphs of $y = x^2 + a$ and $y = -x^2 + a$ are similar to the graphs of $y = x^2$ and $y = -x^2$ except that they are shifted a units up or down depending upon the sign of a.

The graph of the parabola $y = x^2 - 2x + 1 = (x-1)^2$, in Fig. 1-22, is similar to $y = x^2$ except that it is shifted over one unit. Its line of symmetry is $x = 1$ and its vertex is $(1,0)$. In general, the graphs of $y = (x-b)^2$ and $y = -(x-b)^2$ are parabolas with $x = b$ as their line of symmetry and the point $(b,0)$ as their vertex.

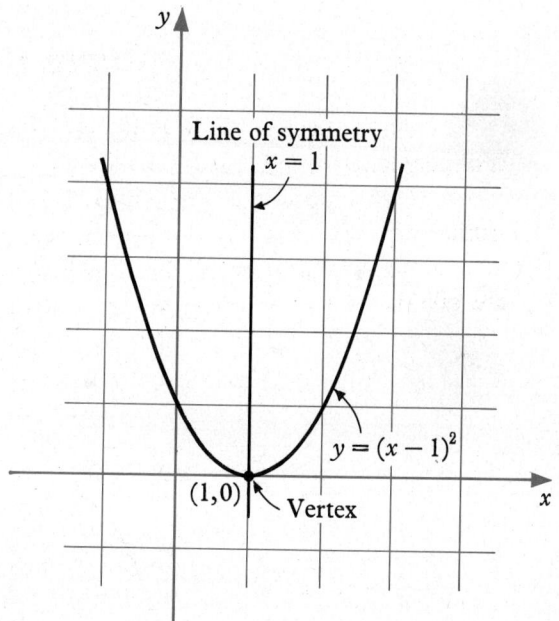

FIGURE 1-22

A quadratic function is one of the form $y = ax^2 + bx + c$. Its graph is similar to $y = x^2$ or $y = -x^2$ depending upon whether a is positive or negative. Its line of symmetry is $x = -b/2a$ and thus the x-coordinate of the vertex is $-b/2a$. To sketch the graph of a quadratic function, we find its line of symmetry and its vertex. We can then simply plot a few more points to the right or left of the line of symmetry, draw a smooth curve through them, and then draw its mirror image through the line of symmetry.

EXAMPLE 1-36

Consider the quadratic function $y = 2x^2 - 4x + 5$. We know its graph is a parabola that opens upward (since $2 > 0$) and its line of symmetry is $x = -(-4)/2 \cdot 2 = 1$ (where $a = 2$, $b = -4$). Its vertex is therefore $(1,3)$ since

$$y(1) = 2 \cdot 1^2 - 4 \cdot 1 + 5 = 3$$

To sketch the graph accurately, we calculate a few additional points to the left of the line $x = 1$. If $x = 0$, then $y(0) = 5$, so $(0,5)$ is on the curve. If $x = -1$, then $y(-1) = 2 - 4(-1) + 5 = 11$. The graph of $y = 2x^2 - 4x + 5$ is given in Fig. 1-23.

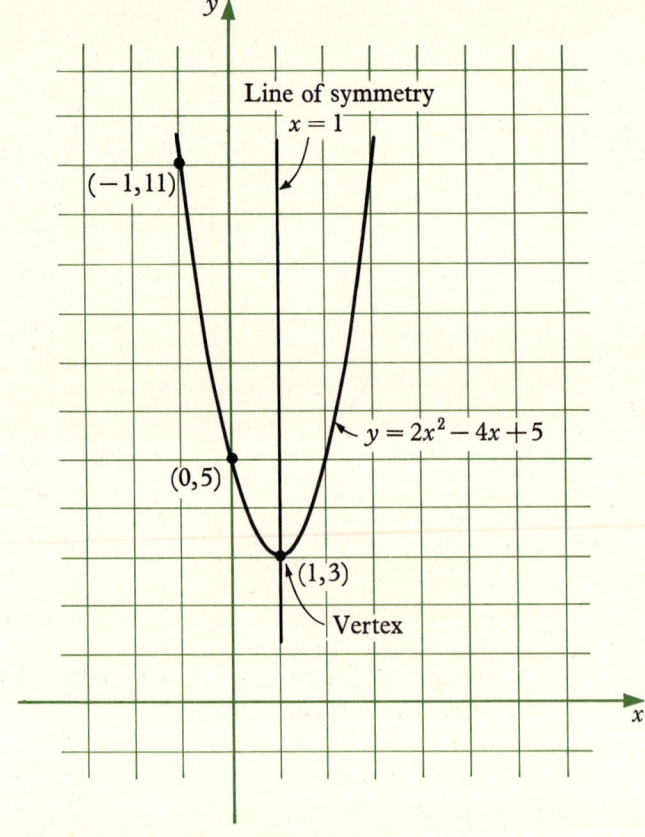

FIGURE 1-23

EXAMPLE 1-37

Consider the quadratic function $y = -3x^2 + 6x + 1$. Its graph is a parabola which opens downward (since $-3 < 0$) and its line of symmetry is $x = -(6)/[2\cdot(-3)] = 1$ (where $a = -3$ and $b = 6$). Its vertex is therefore $(1,4)$ since

$$y(1) = -3(1)^2 + 6(1) + 1 = 4$$

To sketch the graph accurately, we calculate a few additional points to the left of the line $x = 1$ and we find that $(0,1)$ and $(-1,-8)$ are on the parabola. The graph is given in Fig. 1-24.

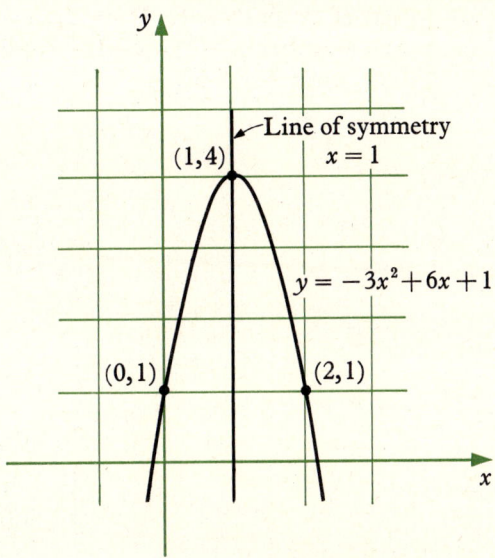

FIGURE 1-24

EXERCISE 1-4

1. Graph the functions by plotting a few points and then drawing a curve through the points.
    - **(a)** $y = 2x^2$
    - **(b)** $y = 3x^2$
    - **(c)** $y = \frac{1}{2}x^2$
    - **(d)** $y = -x^2$
    - **(e)** $y = -2x^2$
    - **(f)** $y = x^2 + 1$
    - **(g)** $y = x^2 + 2$
    - **(h)** $y = x^2 - 1$
    - **(i)** $y = 2x^2 + 1$
    - **(j)** $y = -2x^2 - 1$

2. Graph the functions.
  (a) $y = x^2 + x$
  (b) $y = x^2 + 2x$
  (c) $y = -x^2 + x$
  (d) $y = -2x^2 + 2x$
  (e) $y = 3x^2 - 6x + 1$
  (f) $y = -3x^2 - 12x + x$
  (g) $y = \frac{1}{2}x^2 + 3x - 1$
  (h) $y = -\frac{1}{3}x^2 + 3x - 2$

1-5 SOME ADDITIONAL TYPES OF FUNCTIONS

The function $y = x^2$ is an example of a power function $y = x^n$ for n a positive integer. The graphs of this and an additional example of a power function ($y = x^3$) are given in Fig. 1-25.

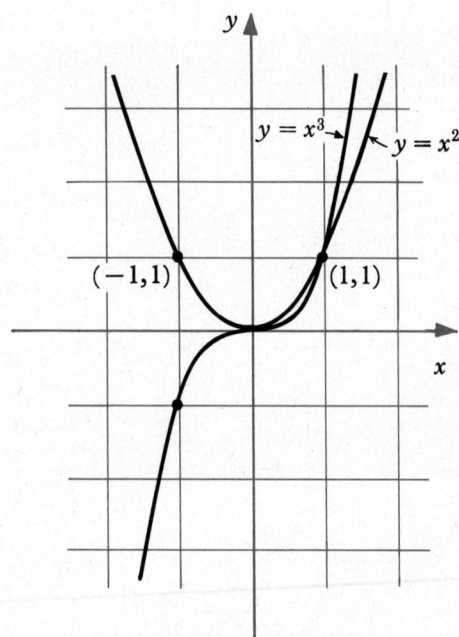

FIGURE 1-25

Let us investigate some other classes of functions and their graphs which we will find useful.

  1. A polynomial function is given by
$$y = a_0 + a_1 x + a_2 x^2 + \cdots + a_n x^n$$
for a_i real numbers. The functions we have already considered are poly-

nomial functions, namely, constant functions $y = c$, linear functions $y = ax + b$, and power functions $y = x^n$. The techniques we shall cover in later chapters will enable us to graph more difficult power and polynomial functions.

2. The function
$$y = \frac{1}{x}$$
is an important function. Its graph is given in Fig. 1-26. Notice that we can consider $f(x) = 1/x$ as the quotient of the two polynomial functions

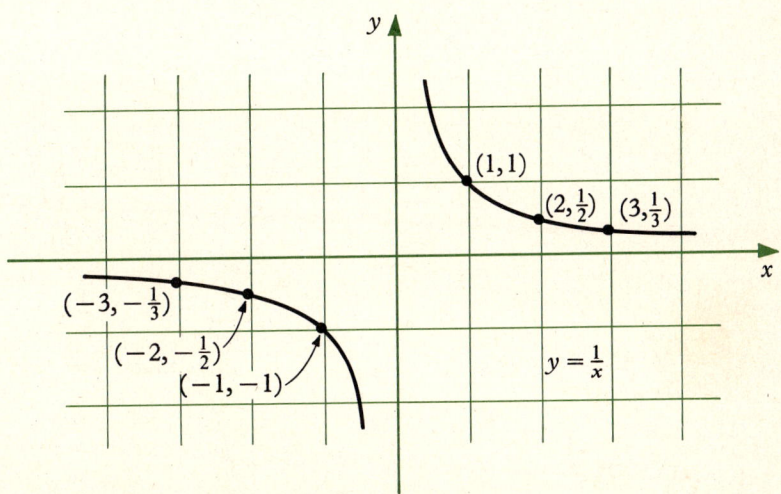

FIGURE 1-26

$g(x) = 1$ and $h(x) = x$. So $f(x) = g(x)/h(x)$. A function that is the quotient of two polynomial functions is called a rational function. Examples of rational functions are
$$f(x) = \frac{x}{x^2 - 4} \quad \text{and} \quad g(x) = \frac{x^3 - 2x - 1}{x^3 - 9}$$

3. The function
$$y = |x|$$
the absolute value function in Fig. 1-27, is important because its graph has a corner at the point (0,0). This fact will be significant in later discussions.

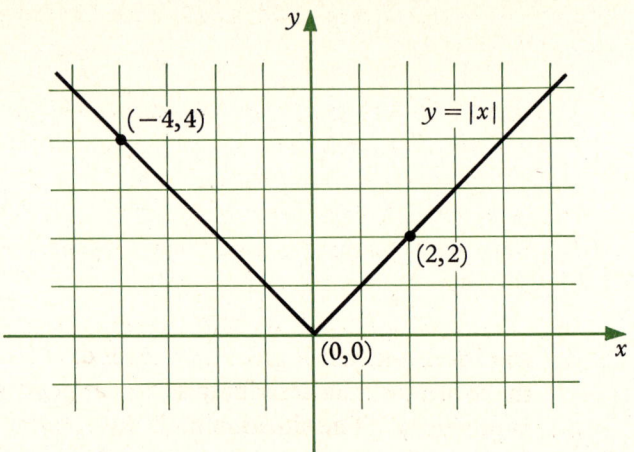

FIGURE 1-27

4. The graph of the function

$$y = \frac{x}{|x|}$$

shown in Fig. 1-28, has a gap at zero; it is not one continuous curve. Note that the function is not defined at zero so the graph does not intersect the y-axis.

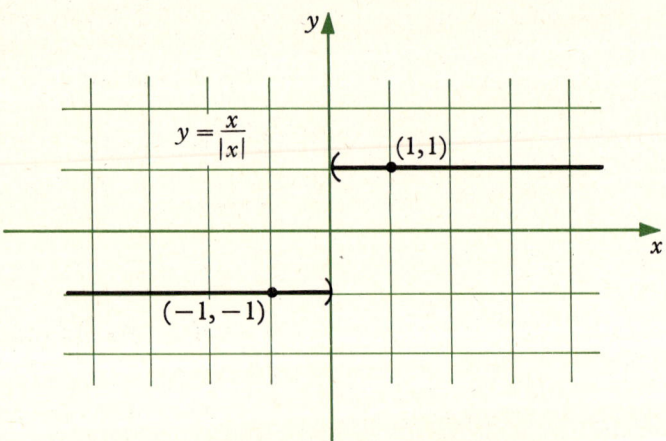

FIGURE 1-28

5. Every real number x can be written in the form

$$x = [\![x]\!] + z$$

where $z \in [0,1)$ and $[\![x]\!]$ is an integer. The number $[\![x]\!]$ is called the greatest integer of x, and the greatest integer function is defined by $f(x) = [\![x]\!]$. The graph of $f(x) = [\![x]\!]$ is given in Fig. 1-29. The greatest integer function is an example of a step function where the "steps" occur at integral values. Examples of step functions appear usually as bar graphs. (See Problems 3, 4, and 6.)

6. Usually in applications the desired function obtained by empirical evidence is given by a set of data presented in a table. To analyze the data a formula describing as best as possible the information in the table is preferred. The situation boils down to the fact that we are given a finite number of points in our function, and hence for our graph. Then a curve is drawn through, or fitted to, these points. This curve is made as smooth as possible so that its formula is easily accessible. This process is called *curve-fitting*. Let us illustrate this procedure with two examples from economics.

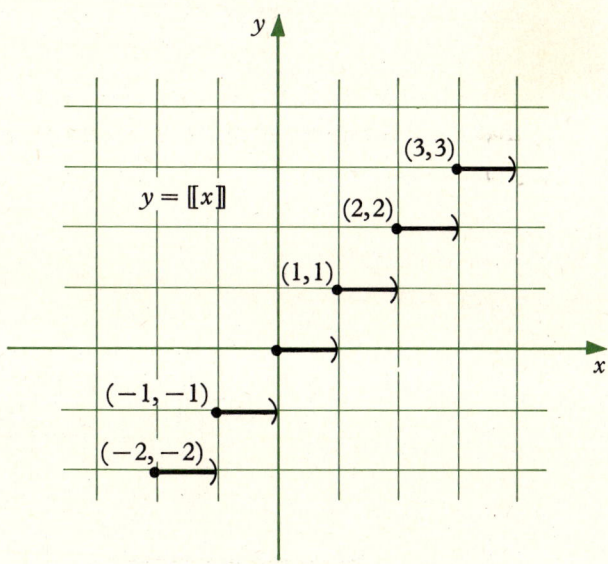

FIGURE 1-29

EXAMPLE 1-38

Demand functions. It is generally true that the quantity of a good that people will buy at any one time depends upon the price charged. The higher the price the less likely you are to buy it, and generally the lower the price the more likely you are to buy it. Thus we see that the quantity demanded of a commodity is a function of the selling price. If p is the selling price and Q is the number of units which can be sold at price p we can write the "demand schedule", or simply "demand function", as

$$Q = f(p)$$

for some formula f in terms of p.

The demand for a good actually depends upon many variables besides the selling price, for instance, the general status of the economy, season of the year, and supply of the item. As a commodity becomes more plentiful, one may purchase more of the good. When water is sparse, only enough to drink is purchased, but if the supply increases more water may be purchased, say, to wash the car, water flowers, and the demand increases. To illustrate our concepts more clearly, we will assume that "everything else is equal" when we discuss demand functions, and that the quantity demanded Q and the price of the good p will be the only variables.

Suppose by empirical evidence we know that quantities of land were sold by a real estate company at prices as shown below:

Price p ($100/acre)	1	2	3	4	5	6	7
Quantity $Q(p)$ (acres)	35	30	25	20	15	10	5

Note that in the table above, 1 means the price was $100 per acre, 2 means the price was $200 per acre, and so on. Thus at the price of $100 per acre ($p = 1$), 35 acres were purchased. Hence $Q(1) = 35$. Also, 30 acres were sold at the price of $200 per acre, so $Q(2) = 30$.

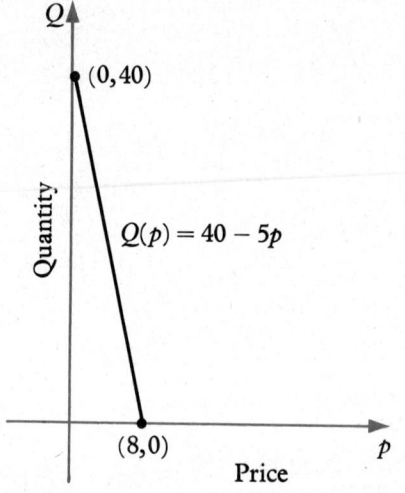

FIGURE 1-30

If we plot these points, namely (1,35), (2,30), ..., we see that they lie on a straight line. The formula thus must be $f(x) = ax + b$ for some real numbers a and b. We know that $35 = a + b$ and $30 = 2a + b$. Solving the equations simultaneously yields $-5 = a$ and $b = 40$. Hence our formula is $f(x) = 40 - 5x$, or equivalently $Q(p) = 40 - 5p$. Clearly the formula cannot be applied for certain values of p, for example $p = -1$, $\sqrt{2}$, 10,000. Usually a specified domain is given. Here we assume $1 \leq p \leq 7$ where p could be taken to be simply integers or perhaps decimals with three nonzero digits. For example, note that $Q(\frac{6}{5}) = 34$ is interpreted as meaning that 34 acres would have been sold at the price of $120 per acre ($p = 1.20 = \frac{6}{5}$).

EXAMPLE 1-39

Supply functions. The amount of goods that a contractor is willing to supply to a market depends upon the price that the market assigns to the goods. The higher the market price the more goods a contractor is usually willing to supply to the market. Suppose that a wheat contractor has determined from previous experience the amount of wheat to contract, depending upon the harvester's price and the predicted market price p:

Price p (dollars/bu)	2	3	4	5
Quantity Q (million bu/mo)	3	9	15	21

By plotting these points we see that the formula required is linear, $f(x) = ax + b$, and so $3 = 2a + b$ and $9 = 3a + b$ imply that $f(x) = 6x - 9$, or $Q(p) = 6p - 9$, is the desired formula.

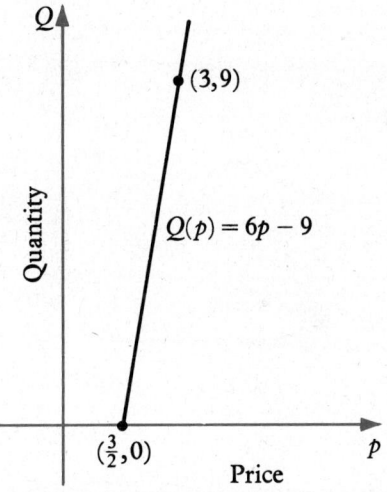

FIGURE 1-31

CHAP. 1   Preliminaries

EXERCISE 1-5

1. Graph the functions.

   (a) $y = x^3 + 1$
   (b) $y = 2x^3$
   (c) $y = -x^3$
   (d) $y = 2(1/x)$
   (e) $y = 1/2x$
   (f) $y = -1/x$
   (g) $y = |2x|$
   (h) $y = 2 - |x|$
   (i) $y = \sqrt{x+1}$
   (j) $y = \sqrt{2x}$
   (k) $y = \sqrt{x} + 1$
   (l) $y = 2\sqrt{x}$
   (m) $y = \dfrac{2x}{|x|}$
   (n) $y = 2[\![x]\!] + 1$

2. Graph the functions.

   (a) $y = \dfrac{x^2}{x}$
   (b) $y = \dfrac{x-1}{x-1}$
   (c) $y = \dfrac{(x-1)^2}{x-1}$

3. The Springfield post office uses the following function to relate the weight and cost of a parcel in order to send it. What type of function is this? Draw a graph of the function.

Weight (oz)	$0 < w \leq 1$	$1 < w \leq 2$	$2 < w \leq 3$	$3 < w \leq 4$
Cost (cents)	8	16	24	32

4. A rent-a-car agency charges $7 a day plus $.10 per mile to rent a car. Let miles travelled per day be the domain and the resultant costs be the range of a function. Graph this function.

5. A landlord who owns a 100-apartment complex makes a profit of $100 each month per occupied apartment but loses $20 each month per vacant apartment. Let x be the number of occupied apartments and let $f(x)$ be the profit per month. Determine a formula for $f(x)$ by a curve-fitting process.

6. With the data for a demand function and a supply function, find by a curve-fitting process a formula reflecting these data.

Price p (dollars/bu)	1	2	3	4	5
Corn demanded (millions of bu)	70	60	50	40	30

Price p (dollars/bu)	1	2	3	4	5
Corn supplied (millions of bu)	23	41	59	77	95

7. A given amount of drug is administered at time $t = 0$, and $f(t)$, the amount of bacteria killed at time t, is given by the following data. Find a formula for $f(t)$.

36

t (hr)	0	1	2	3	4	5
$f(t)$ (millions)	0	1	3	7	15	31

1-6 AN INTRODUCTION TO EXPONENTIAL AND LOGARITHMIC FUNCTIONS

Consider the function
$$f(x) = 2^x$$
Now
$$f(0) = 2^0 = 1 \quad f(1) = 2^1 = 2 \quad f(2) = 2^2 = 4$$
and so on. Also
$$f(\tfrac{1}{2}) = 2^{1/2} = \sqrt{2} \quad f(-1) = 2^{-1} = \tfrac{1}{2}$$
$$f(\tfrac{3}{2}) = 2^{3/2} = \sqrt{8} \quad f(-2) = 2^{-2} = \tfrac{1}{4}$$

If we plot these points and draw a smooth curve through them, we get the graph in Fig. 1-32.

x	$f(x) = 2^x$
-5	$\tfrac{1}{32}$
-4	$\tfrac{1}{16}$
-3	$\tfrac{1}{8}$
-2	$\tfrac{1}{4}$
-1	$\tfrac{1}{2}$
0	1
1	2
2	4
3	8
4	16
5	32

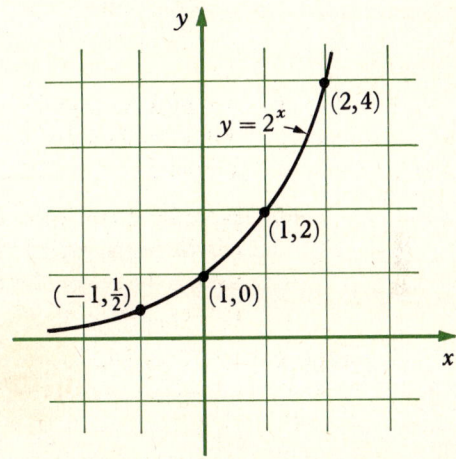

FIGURE 1-32

The alert reader may argue that the function makes no sense for some values of x, say, $\sqrt{2}$, π, or any irrational number. For the present, we will ignore this problem. However, in the near future we will develop the techniques necessary for defining these quantities. The function $y = 2^x$ is an example of an exponential function.

An equation of the form

$$f(x) = b^x$$

where b is a constant called the *base*, defines an exponential function. We must place the restriction on b that b is a positive real number not equal to one. Note that if $b = 1$, then $f(x)$ is simply $f(x) = 1$, and if $b = 0$, then $f(x)$ is simply $f(x) = 0$. If $b < 0$, then $f(x)$ is not defined for many values of x, e.g., if $b = -2$, then $f(1/2) = (-2)^{1/2}$ is not defined. Recall the laws of exponents:

① $b^x b^y = b^{x+y}$

② $\dfrac{b^x}{b^y} = b^{x-y}$

③ $(b^x)^y = b^{xy}$

Hence, if $f(x) = b^x$, then we have the following properties for $f(x)$:

① $f(x) f(y) = f(x+y)$

② $\dfrac{f(x)}{f(y)} = f(x-y)$

③ $(f(x))^y = f(xy)$

Consider again the graph of the exponential function $f(x) = 2^x$ in Fig. 1-32. Notice that as x increases without limit, y increases without limit. As x decreases without limit, y approaches zero. The graph never touches the x-axis, but it comes arbitrarily close to it. If we plot these points and draw

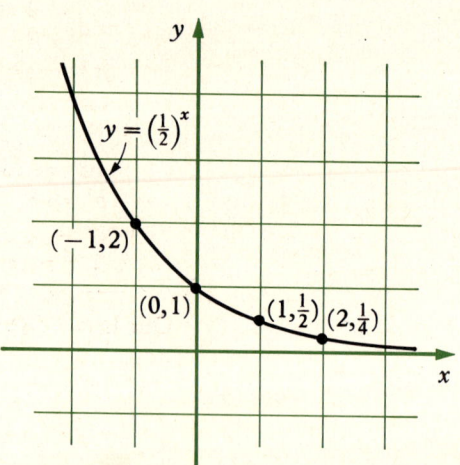

x	$f(x) = \left(\tfrac{1}{2}\right)^x$
-3	8
-2	4
-1	2
0	1
1	$\tfrac{1}{2}$
2	$\tfrac{1}{4}$
3	$\tfrac{1}{8}$
4	$\tfrac{1}{16}$

FIGURE 1-33

a smooth curve between them, we get the graph of $y = 2^x$. This graph is representative of all graphs of exponential functions defined by $f(x) = b^x$ for $b > 1$. If, however, $0 < b < 1$, the graph of $f(x) = b^x$ is altered. For example, consider $f(x) = (\frac{1}{2})^x$. A set of points on the graph of this function and its graph are in Fig. 1-33. Again note that $f(x) > 0$ for all x as before. For $f(x) = (\frac{1}{2})^x$, as x increases without limit, $f(x)$ approaches zero.

Let us now restate the equation $y = b^x$ using what is referred to as logarithmic notation. We define the *logarithm* of the number x to the base b (written $\log_b x$) to be that number y such that $b^y = x$. In symbols we write:

$$\log_b x = y \quad \text{means} \quad b^y = x = b^{\log_b x}$$

Thus a logarithm is nothing more than an exponent. We can now define a logarithmic equation to be one of the form

$$f(x) = \log_b x$$

where the constant b is called the base of the function. Note that we must restrict b to be a positive real number not equal to one, as we mentioned above. Since x takes on precisely those real numbers that the exponential function $x = b^y$ takes on, we see that $x > 0$, and so the domain of any logarithmic function is $(0, \infty)$.

To illustrate the definition of the logarithmic notation, consider the following table:

Exponential form	Logarithmic form
$2^1 = 2$	$\log_2 2 = 1$
$2^3 = 8$	$\log_2 8 = 3$
$2^5 = 32$	$\log_2 32 = 5$
$(\frac{1}{3})^2 = \frac{1}{9}$	$\log_{1/3} \frac{1}{9} = 2$
$10^2 = 100$	$\log_{10} 100 = 2$
$10^4 = 10000$	$\log_{10} 10000 = 4$
$b^1 = b$	$\log_b b = 1$
$b^0 = 1$	$\log_b 1 = 0$
$3^{-2} = \frac{1}{9}$	$\log_3 \frac{1}{9} = -2$

Our laws of exponents can be rewritten in logarithmic form:

① $\log_b x + \log_b y = \log_b xy$

② $\log_b x - \log_b y = \log_b \dfrac{x}{y}$

③ $\log_b x^r = r \log_b x$

Hence, if $f(x) = \log_b x$, we can write

$$f(x) + f(y) = f(xy)$$
$$f(x) - f(y) = f\left(\frac{x}{y}\right)$$
$$f(x^r) = r f(x)$$

Consider the graph of the logarithmic function $f(x) = \log_2 x$. We can plot its points, given in the following table, and then draw a smooth curve through them.

Notice that as x approaches zero, $f(x)$ decreases without limit, and as x increases without limit, so does $f(x)$. The graph of this function is representative of every logarithmic function $f(x) = \log_b x$ where $b > 1$.

Logarithms to the base 10 are an invaluable aid in calculations. Such logarithms are called *common* logarithms. Of supreme importance in mathematics, however, are the so-called *natural* logarithms. The base of the natural logarithms is the number e, which is an irrational number like π and $\sqrt{2}$. Thus it cannot be expressed as the ratio of two integers, but it can be approximated to any desired degree of accuracy. Expressing e to six decimal places we say that e is approximately equal to 2.718281. The graphs of the functions $y = e^x$ and $y = \log_e x$, which is usually written as simply $y = \ln x$, are similar to the graphs of $y = 2^x$ and $y = \log_2 x$ and are given in Fig. 1-35. The importance of these functions is illustrated by the fact that they are called *the* exponential function and *the* logarithmic function.

x	$\log_2 x$
16	4
8	3
4	2
2	1
1	0
$\frac{1}{2}$	-1
$\frac{1}{4}$	-2
$\frac{1}{8}$	-3
$\frac{1}{16}$	-4

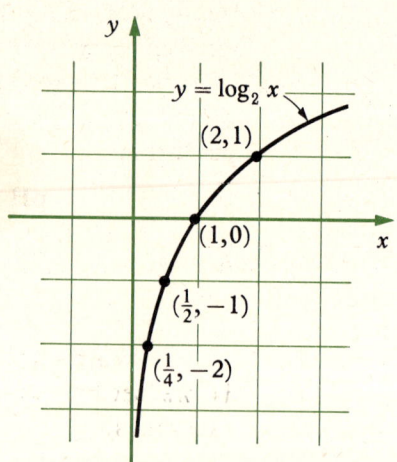

FIGURE 1-34

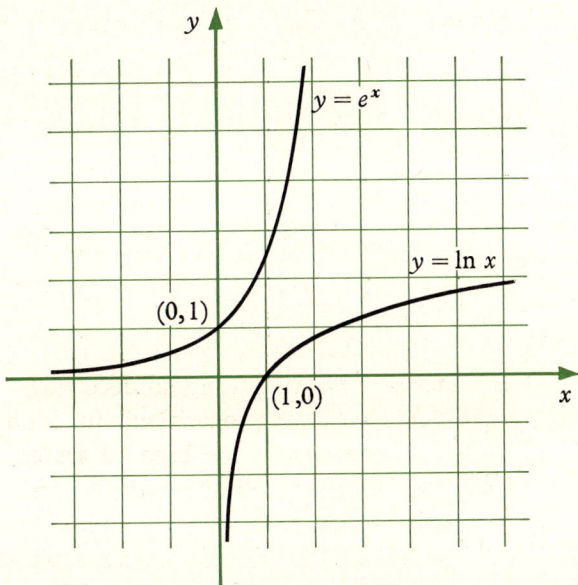

FIGURE 1-35

There are many applications of logarithmic and exponential function in nature. The phenomenon of growth is best described in mathematical models by exponential and logarithmic functions. In living tissue the hydrogen ion concentration (pH) of the cellular fluids is of great biological importance. In nature, the pH can show great variations and it is advantageous to express the pH as the negative of the logarithm function (to the base 10) of the hydrogen ion concentration. Hence the pH for pure distilled water with a hydrogen ion concentration of 1×10^{-7} mole per liter may be expressed $-\log 10^{-7}$, which equals $-(-7)$, or 7 where log means $\log_{10}$. The hydrogen ion concentration of blood is 4×10^{-8} mole per liter so

$$\text{pH} = -\log(4 \times 10^{-8})$$
$$= -(\log 4 + \log 10^{-8})$$
$$= -(0.6021 - 8) = 7.4$$

When one measures sound intensity the unit most frequently used is the decibel (dB), defined in terms of the ratio of two sound-power intensities,

$$\text{dB} = 10 \log\left(\frac{\text{power 1}}{\text{power 2}}\right)$$

Thus the sound intensities of electric instruments, where the power ratio is equal to the square of the voltage ratio, are given by

$$\text{dB} = 10 \log\left(\frac{\text{O power}}{\text{I power}}\right) = 10 \log\left(\frac{\text{O voltage}}{\text{I voltage}}\right) = 20 \log\left(\frac{\text{O voltage}}{\text{I voltage}}\right)$$

where O and I are output and input respectively.

The phenomena of decay processes, those in which the rate of change of a quantity is proportional to the amount of the quantity present, is governed by functions involving $y = e^x$. Some examples of such processes are radioactive decay; biological half-life, the time interval in which the initial amount of material present will be reduced by 50 percent; the amount of light intensity absorbed at a given depth of material; the change of barometric pressure with altitude above the earth. Later when we handle the concept of a differential equation we deal with these processes in more depth.

EXERCISE 1-6

1. Let $f(x) = 3^x$. Find $f(0), f(1), f(2), f(3), f(4), f(\tfrac{1}{2}), f(\tfrac{1}{3}), f(-1), f(-2), f(-3)$.

2. Let $f(x) = 4^x$. Find $f(0), f(1), f(2), f(3), f(4), f(\tfrac{1}{2}), f(\tfrac{1}{3}), f(-1), f(-2), f(-3)$.

3. Graph the functions.
   (a) $y = 3^x$
   (b) $y = 4^x$
   (c) $y = 8^x$
   (d) $y = 10^x$
   (e) $y = (\tfrac{1}{3})^x$
   (f) $y = (\tfrac{1}{8})^x$

4. Let $f(x) = \log_4 x$. Find $f(1), f(2), f(4), f(8), f(16), f(32), f(\tfrac{1}{2}), f(\tfrac{1}{4}), f(\tfrac{1}{32})$.

5. Let $f(x) = \log_{10} x$. Find $f(1), f(10), f(100), f(0.1), f(0.2)$.

6. Graph the functions.
   (a) $y = \log_4 x$
   (b) $y = \log_{10} x$
   (c) $y = \log_8 x$
   (d) $y = \log_{1/2} x$
   (e) $y = \log_{1/8} x$

7. Suppose the following graph was presented to you depicting the rate of growth of a child aged from zero to 15 years. What type of curve do you think is given?

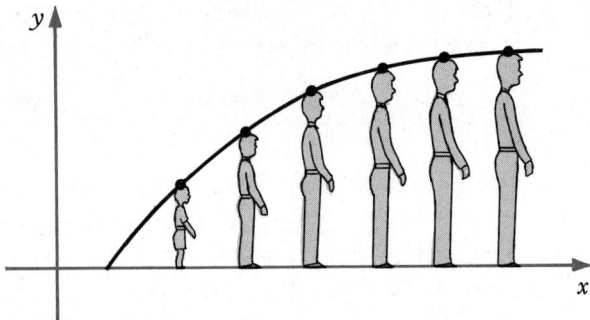

8. Relationships between variable quantities are often written in terms of ratios and proportions. If two quantities x and y are always in the same proportion, we say that one is proportional to or varies directly as the other. Symbolically we write $y = kx$, where the constant k is called the constant of proportionality. For example, the circumference C of a circle varies directly as the radius r, and one writes $C = kr$ (where in this case $k = 2\pi$). Express the following statements as equations.

   (a) *Example:* The cost C of a product varies directly as the output x.
   Solution: $C = kx$.

   (b) The revenue R of a product varies directly as the input t.

   (c) The electrical resistance R of a wire is proportional to its length L.

   (d) Charles' law states that for a perfect gas under constant pressure, the volume V of the gas is proportional to the temperature t.

   (e) Hooke's law states that the force F needed to stretch a spring is proportional to the length l of spring that is stretched.

   (f) Under specific conditions, the natural logarithm $\ln C$ of the current varies directly as the current C.

9. We say that one quantity x varies inversely as another quantity y if $y = k/x$ for the constant of proportionality k. Express the following statements as equations.

   (a) The demand D of a product varies inversely as the price p.

   (b) Boyle's law states that at a given temperature the pressure P of an ideal gas varies inversely as the volume V.

   (c) The gravitational force F of attraction between two bodies varies inversely as the square, d^2, of the distance d between them.

   (d) The illuminance I of a light source varies inversely as the square of the distance d from the source.

10. Find the constant of proportionality and express the following statements as equations.

    (a) *Example:* If $y = kx$ and $y = 2$ when $x = 4$, then $2 = k4$ implies that $k = \frac{1}{2}$. Thus $y = \frac{1}{2}x$.

(b) The profit P of a product varies directly as the output x, and $P = \$10{,}000$ when $x = 100{,}000$ units.
(c) The force F needed to stretch a spring is proportional to the length l of spring that is stretched, and $F = 10$ when $l = 5$.
(d) The illuminance I of a light source varies inversely as the square of the distance d from the source, and $I = 20$ when $d = 5$.

1–7 INVERSE FUNCTIONS AND COMPOSITION OF FUNCTIONS

Consider the functions $f(x) = x^3$ and $g(x) = x^{1/3} = \sqrt[3]{x}$. In a certain sense, one function "undoes" what the other does. That is, $f(x)$ cubes x while $g(x)$ takes the cube root of x. Similarly, the functions $f(x) = 2x - 5$ and $g(x) = (x+5)/2$ undo what the other does. These functions are said to be *inverses* of each other.

To discuss the concept of an inverse function more explicitly we need to mention the composition of functions. It entails finding the "function of a function".

EXAMPLE 1-40

Let $f(x) = x^2 + 1$ and $g(x) = 2x + 1$. Then $g(2) = 5$. Since 5 is in the domain of f, we find by substitution that $f(5) = 26$. These two sentences can be written in the more compact form $f(g(2)) = f(5) = 26$. If a is *any* real number, then

$$f(g(a)) = f(2a+1) = (2a+1)^2 + 1 = 4a^2 + 4a + 2$$

The function

$$f(g(x)) = f(2x+1) = (2x+1)^2 + 1 = 4x^2 + 4x + 2$$

is called the *composition of $f(x)$ with $g(x)$*. Similarly the function

$$g(f(x)) = g(x^2+1) = 2(x^2+1) + 1 = 2x^2 + 2 + 1 = 2x^2 + 3$$

is the *composition of $g(x)$ with $f(x)$*.

Definition If $f(x)$ and $g(x)$ are two functions, then the function $h(x)$ defined by $h(x) = f(g(x))$ is called the *composition of $f(x)$ with $g(x)$*.

To distinguish composition from the product of two functions, often the notation $f \circ g$ is used to denote composition. Thus $(f \circ g)(x) = f(g(x))$. Note from Ex. 1-40 that $f \circ g \neq g \circ f$ in general.

EXAMPLE 1-41

Let $f(x) = 2x - 3$, $g(x) = x^2 + x$, and $h(x) = \sqrt{x+5}$. Then

$$(f \circ g)(x) = f(g(x)) = f(x^2 + x) = 2(x^2 + x) - 3 = 2x^2 + 2x - 3$$

Also

$$g(f(x)) = g(2x - 3) = (2x - 3)^2 + (2x - 3) = 4x^2 - 10x + 6$$
$$f(h(x)) = f(\sqrt{x+5}) = 2\sqrt{x+5} - 3$$
$$h(f(x)) = h(2x - 3) = \sqrt{(2x - 3) + 5} = \sqrt{2x + 2}$$
$$h(g(x)) = h(x^2 + x) = \sqrt{x^2 + x + 5}$$

EXAMPLE 1-42

Let $f(x) = e^x$, $g(x) = \ln x$, and $h(x) = x^2 + 1$. Then

$$(f \circ g)(x) = f(g(x)) = f(\ln x) = e^{\ln x} = x$$

Also,

$$h(g(x)) = h(\ln x) = (\ln x)^2 + 1$$
$$h(f(x)) = h(e^x) = (e^x)^2 + 1 = e^{2x} + 1$$
$$f(h(x)) = f(x^2 + 1) = e^{x^2 + 1}$$
$$g(h(x)) = g(x^2 + 1) = \ln(x^2 + 1)$$

EXAMPLE 1-43

The production of materials in a factory usually consists of many diverse operations. A company which makes beaded necklaces has two basic operations: (1) It manufactures beads and (2) it strings the beads into necklaces. The revenue from the sale of the necklaces depends upon the number of necklaces produced, which in turn depends upon the number of beads produced. Suppose the company determines that the revenue $R(n)$ realized from the production of n necklaces is given by $R(n) = n^{1/2} - 0.2n$. The company also determines that the number of necklaces $n(b)$ produced from a production of b beads is given by $n(b) = 0.09b$. Hence the revenue realized from the production of b beads is

$$R(b) = (0.09b)^{1/2} - 0.2(0.09b) = 0.3b^{1/2} - 0.018b$$

If we take the intuitive approach to the meaning of inverse function, that it undoes what the first function does to x, then when we take the composition of the two functions we should arrive back at where we started, namely x. For example, if $f(x) = x^{1/3}$ and $g(x) = x^3$, then $f(g(x)) = f(x^3) = (x^3)^{1/3} = x$. Also, if $f(x) = 2x - 5$ and $g(x) = (x+5)/2$, then $f(g(x)) = f((x+5)/2) = 2((x+5)/2) - 5 = x$. Note that in each example $g(f(x)) = x$ also.

CHAP. 1   Preliminaries

Definition   If $f(g(x)) = g(f(x)) = x$, then $g(x)$ is said to be the *inverse* of $f(x)$ and is usually denoted by $f^{-1}(x)$, read "f inverse of x".

Later in this section we will see that some functions do not have inverse functions.

EXAMPLE 1-44

Consider again the function $f(x) = 2x - 5$. To find $y = f^{-1}(x)$, we search for a function y such that $f(f^{-1}(x)) = f(y) = x$. But $f(y) = 2y - 5$, so we need y such that $2y - 5 = x$, and hence $2y = x + 5$ and so $y = (x+5)/2$ as we discovered earlier.

EXAMPLE 1-45

If $f(x) = (x^3 + 2)^{1/5}$, then to find $y = f^{-1}(x)$, we search for a function y such that $f(f^{-1}(x)) = f(y) = x$. Hence we consider

$$f(y) = (y^3 + 2)^{1/5} = x$$

and solve for y.

$$(y^3 + 2)^{1/5} = x$$
$$y^3 + 2 = x^5$$
$$y^3 = x^5 - 2$$
$$y = (x^5 - 2)^{1/3}$$

and hence

$$f^{-1}(x) = (x^5 - 2)^{1/3}$$

EXAMPLE 1-46

Consider the function $f(x) = (2x+3)/(4x-5)$. To find $y = f^{-1}(x)$, we solve for y given the equation

$$f(y) = \frac{2y+3}{4y-5} = x$$

$$2y + 3 = x(4y - 5)$$
$$2y + 3 = 4xy - 5x$$
$$5x + 3 = 4xy - 2y$$
$$\frac{5x+3}{4x-2} = y$$

and thus

$$f^{-1}(x) = \frac{5x+3}{4x-2}$$

Suppose $f(a_1) = b$ and $f(a_2) = b$, where $a_1 \neq a_2$. Then if f^{-1} is a function it would follow that $f^{-1}(b) = a_1 \neq a_2 = f^{-1}(b)$. This is a contradiction, so f^{-1} is not a function. For example, consider $f(x) = x^2$. Note that $f(2) = 4 = f(-2)$ and thus f^{-1} is not a function.

A function is said to be *one-to-one* if $f(a_1) = f(a_2)$ implies that $a_1 = a_2$, that is, whenever the situation described in the above paragraph does not happen. Thus, $f(x) = x^2$ is not one-to-one.

EXAMPLE 1-47

Consider the function $f(x) = x^3 - 5$. If $f(a_1) = f(a_2)$, then $a_1^3 - 5 = a_2^3 - 5$ and hence $a_1^3 = a_2^3$ and thus $a_1 = a_2$. Therefore $f(x)$ is one-to-one.

EXAMPLE 1-48

Consider the function $f(x) = (2x-3)/x$. If $f(a_1) = f(a_2)$, then

$$\frac{2a_1 - 3}{a_1} = \frac{2a_2 - 3}{a_2}$$

$$2a_1 a_2 - 3a_2 = 2a_2 a_1 - 3a_1$$

$$-3a_2 = -3a_1$$

$$a_2 = a_1$$

and hence $f(x)$ is one-to-one.

It was noted above that $f(x) = x^2$ is not one-to-one and hence f^{-1} is not a function. If we compute $f(y) = x$ we get $y^2 = x$ and then $y = \pm\sqrt{x}$ which is not a function. However, if we restrict y so that $y \geq 0$, then $y = \sqrt{x}$ and y is a function. Hence, if we restrict the domain of $f(x) = x^2$ to $x \geq 0$, i.e., to $[0, \infty)$, then $f^{-1}(x) = \sqrt{x}$ is a function.

In conclusion, we assert that $f^{-1}(x)$ is a function if and only if $f(x)$ is one-to-one. In addition, if $f(x)$ is not one-to-one, we can restrict the domain of $f(x)$ so that the restricted $f(x)$ is one-to-one and hence $f^{-1}(x)$ will be a function.

EXERCISE 1-7

1. Consider $f(x) = 5x + 3$ and $g(x) = (x-3)/5$. Show that $g(x) = f^{-1}(x)$.
2. Let $f(x) = (3x-5)/(x-7)$. Note that $f(0) = 5/7$, $f(1) = 1/3$ and $f(-1) = 1$. Without computing $f^{-1}(x)$, find $f^{-1}(5/7)$, $f^{-1}(1/3)$, and $f^{-1}(1)$.

3. Find the inverse function $f^{-1}(x)$.
   (a) $f(x) = 2x - 1$
   (b) $f(x) = 3x + 7$
   (c) $f(x) = 2x^3$
   (d) $f(x) = 2x^3 - 5$
   (e) $f(x) = -x^3 + 2$
   (f) $f(x) = 3x^5$
   (g) $f(x) = 5x^5 - 7$
   (h) $f(x) = (x+3)^{1/3}$
   (i) $f(x) = (x^3 - 5)^{1/5}$
   (j) $f(x) = (2x+7)^3$
   (k) $f(x) = (x^{1/3} + 5)^5$
   (l) $f(x) = (7x^3 - 2)^{1/5}$
   (m) $f(x) = (5x-1)/(3-x)$
   (n) $f(x) = (2x+1)/(5x-7)$

4. Determine whether the functions are one-to-one.
   (a) $f(x) = 3x - 7$
   (b) $f(x) = 8x - 1$
   (c) $f(x) = 2x^3 + 2$
   (d) $f(x) = x^3 - 1$
   (e) $f(x) = 1 - 2x^2$
   (f) $f(x) = x^4$
   (g) $f(x) = 2x^5 + 1$
   (h) $f(x) = \sqrt{2x-1}$
   (i) $f(x) = x/(3-x)$
   (j) $f(x) = (2x-1)/(x-7)$

5. The following functions are not one-to-one. Find a restriction of the domain of each so that the resulting function is one-to-one.
   (a) $f(x) = x^2 + 1$
   (b) $f(x) = 3x^2 + 1$
   (c) $f(x) = x^2 + 2x$
   (d) $f(x) = 1 - x^2$
   (e) $f(x) = x^4 + 1$
   (f) $f(x) = \sqrt{1-x^2}$

2

Limits

2-1 THE LIMIT OF A FUNCTION AND CONTINUITY

In the development of the calculus we will encounter two fundamental concepts, the *derivative* and the *integral*. The definition of each of these concepts depends upon the idea of a *limit*. To motivate the definition of the limit of a function, we consider the following four examples.

EXAMPLE 2-1
Consider the function
$$f(x) = \frac{x^2 - 4}{x - 2}$$
We ask the question, What is $f(2)$? If we replace x by 2 in the formula, we get 0/0, which is meaningless. Therefore we conclude that $f(2)$ is not defined. Since $x^2 - 4 = (x+2)(x-2)$, we can express the function in the form
$$f(x) = \frac{(x+2)(x-2)}{x-2}$$
and thus for every $x \neq 2$, we have
$$f(x) = x + 2$$
The graph of this function, given in Fig. 2-1, is identical with the graph of $y = x + 2$, except that there is one point missing, the point (2,4). For every value of x which is close to 2, but not equal to 2, the corresponding functional

value $f(x)$ is close to 4. The last two sentences express intuitively the statement that the limit of $f(x)$ as x approaches 2 is 4, which is written in the form

$$\lim_{x \to 2} f(x) = 4$$

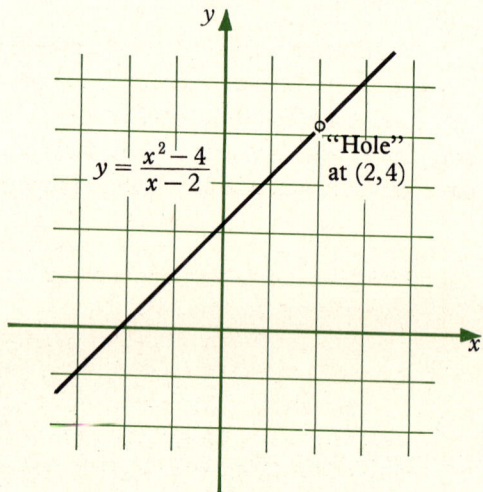

FIGURE 2-1

EXAMPLE 2-2

A wheat contractor determines that the amount of wheat $S(p)$ that he is willing to supply to the market at price p is given by

$$S(p) = \frac{p-2}{\sqrt{p+2}-2}$$

where p is measured in dollars per bushel and $S(p)$ in millions of bushels. Note that where $p = 2$, $S(p)$ is not defined. The value $p = 2$ is a reasonable one, and thus the formula should yield a value for $S(2)$.

One method to resolve the difficulty would be to select first a value of p close to 2 but different from 2, so that $S(p)$ would be defined (why?), and then identify this value for the functional value of $p = 2$. Here we are assuming that if p is close to 2, then $S(p)$ is close to the value that $S(2)$ *should* have.

A better method would be to utilize the following algebraic ploy.

$$S(p) = \left[\frac{p-2}{\sqrt{p+2}-2}\right]\left[\frac{\sqrt{p+2}+2}{\sqrt{p+2}+2}\right]$$

$$= \frac{(p-2)(\sqrt{p+2}+2)}{p-2}$$

Therefore, if $p \neq 2$, then $S(p) = \sqrt{p+2} + 2$. If we let $f(p) = \sqrt{p+2} + 2$, then $f(2) = 4$. As in Ex. 2-1, we write

$$\lim_{p \to 2} S(p) = 4$$

because the graph of $S(p)$ is identical with the graph of $f(p)$ except that the point (2,4) is missing.

In each of the above examples, we described the number $L = \lim_{x \to a} f(x)$ as that number L such that whenever x is close to a but not equal to a, then $f(x)$ is close to L. Let us now describe the case when $\lim_{x \to a} f(x)$ does not exist. You will see that the graph of such a function has a gap at $x = a$, just as $y = x/|x|$ has a gap at $x = 0$.

EXAMPLE 2-3

Biologists define the relative disorder of a system as its *entropy*. The more ordered a system, the lower its entropy, while the more disordered the components of a system, the greater its entropy. If heat is applied to a substance in a solid state, the tightly packed particles of the substance vibrate and there is a steady rise in its entropy. Eventually, as the temperature of the substance increases, enough energy is put into the system so that the chemical bonds holding the solid together are broken, thereby allowing the solid to melt into a liquid state. The critical temperature at which this change takes place is called the *melting point*. When this change of state occurs, a large increase in entropy takes place. The particles of a liquid are in relatively large disorder when compared with the tightly packed particles of a solid.

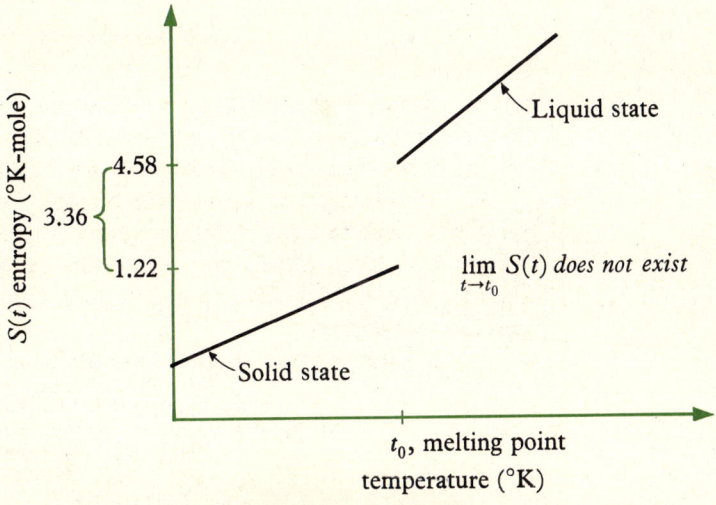

FIGURE 2-2

If we let $S(t)$ denote the measure of entropy of a system, measured in calories per degrees (Kelvin)-mole, at temperature t, measured in degrees Kelvin, then the entropy changes for this substance being heated at constant pressure are as graphed in Fig. 2-2. If t_0 is the melting point, then $\lim_{t \to t_0} S(t)$ does not exist, since, for values of t close to t_0 but not equal to t_0, the corresponding functional values $S(t)$ are not close to any one number. To see this, note that if $t_1 < t_0 < t_2$, then $S(t_1) < S(t_2) - 3.36$, so $S(t_1)$ and $S(t_2)$ are not close for any such values of t_1 and t_2.

EXAMPLE 2-4

Let the function $P(x)$ measure the profits a company realizes if its production is x tons of goods per day, where $P(x)$ is measured in thousands of dollars. Suppose the company determines that $P(x) = 2x - 5$ if it produces $3 \leqslant x \leqslant 10$ tons. If the capacity of the day shift is 10 tons, then to produce more than 10 tons requires the activation of the night shift, which in turn increases the company's fixed cost. Suppose $P(x) = 2x - 7$ for $10 < x \leqslant 17$. Then we can write

$$P(x) = \begin{cases} 2x - 5 \text{ if } x \in [3, 10] \\ 2x - 7 \text{ if } x \in (10, 17] \end{cases}$$

Note that the value $x = 10$ is critical. If the company takes orders requiring production of x tons where x is close to 10 but *less* than 10, then $P(x)$ is close to $2 \cdot 10 - 5 = 15$. But if orders requiring production of x tons where x

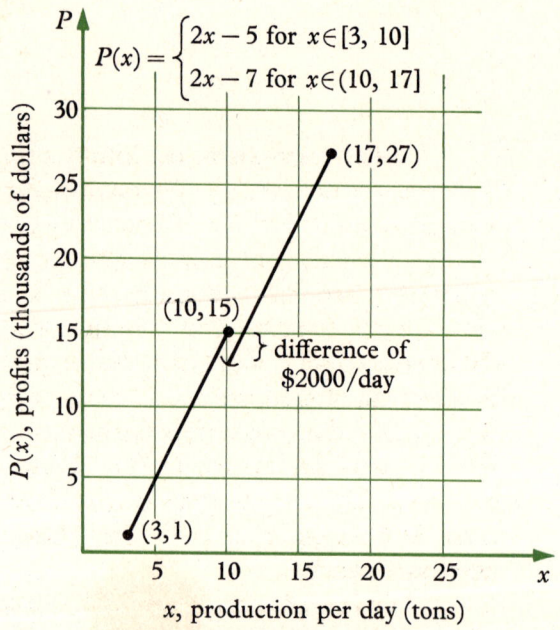

FIGURE 2-3

is close to 10 but *greater* than 10, then $P(x)$ is close to $2 \cdot 10 - 7 = 13$, a difference in profits of almost $2000 per day. One can tell from the graph of $P(x)$ given in Fig. 2-3 that $\lim_{x \to 10} P(x)$ does not exist.

In the above examples we discussed two intuitive interpretations of $\lim_{x \to a} f(x)$. If $f(x)$ is close to the number b whenever x is close to, but not equal to, a, then we say that $\lim_{x \to a} f(x)$. Geometrically, if the graph of $y = f(x)$ is a continuous curve, in the sense that it can be drawn without lifting the pencil from the paper, through the point (a,b), except that the point (a,b) itself may be missing from the graph, then we say $\lim_{x \to a} f(x) = b$. This last statement is equivalent to saying that there is no gap in the graph at $x = a$, but perhaps a hole.

The above descriptions of $\lim_{x \to a} f(x)$ are certainly not mathematically rigorous. However, they do provide us with enough understanding to allow us to compute $\lim_{x \to a} f(x)$ for all the functions $f(x)$ with which we deal.

You have probably observed that we have centered our attention on points where the graph of a function has a hole or a gap. It is at these points that $\lim_{x \to a} f(x)$ may be difficult to evaluate, or may not even exist. For most values a in the domain of $f(x)$, the $\lim_{x \to a} f(x)$ is simply $f(a)$. In this benign case we say that $f(x)$ is *continuous* at a. This is an important definition and we single it out.

Definition   The function $f(x)$ is said to be *continuous at a* if $\lim_{x \to a} f(x) = f(a)$; $f(x)$ is *continuous in an interval* if $f(x)$ is continuous at each point in the interval; $f(x)$ is *continuous* if $f(x)$ is continuous at a for every real number a.

Since their graphs are simply straight lines, one sees that all linear functions are continuous. In fact, all polynomial functions are continuous as we shall see in Ex. 2-8. From the graphs of $y = e^x$ and $y = \ln x$ one can see that each function is continuous and hence $\lim_{x \to a} e^x = e^a$ for all real numbers a and $\lim_{x \to a} \ln x = \ln a$ for all positive real numbers a. Geometrically, a function $f(x)$ is continuous at a if the graph of $f(x)$ is a continuous curve through the point $(a, f(a))$, that is, there is no hole or gap in the graph of the function at the line $x = a$. We've noted in Ex. 2-4 that $\lim_{x \to 10} P(x)$ does not exist, because there is a gap in the graph at $x = 10$, so $P(x)$ is not continuous at 10. From Ex. 2-1, the function $f(x) = (x^2 - 4)/(x - 2)$ is not continuous at 2 since $f(2)$ is not defined, and so there is a hole in the graph at $x = 2$. Both these functions are continuous at every other real number.

To formulate properties of limits we need to define some operations with functions.

Definition For the two functions $f(x)$ and $g(x)$ we define the sum, difference, product, and quotient of $f(x)$ and $g(x)$ by

(I) $(f+g)(x) = f(x) + g(x)$
(II) $(f-g)(x) = f(x) - g(x)$
(III) $(f \cdot g)(x) = f(x)g(x)$
(IV) $(f/g)(x) = f(x)/g(x)$ if $g(x) \neq 0$

EXAMPLE 2-5

If $f(x) = x^2 + 1$ and $g(x) = \sqrt{x+2}$, then

$$(f+g)(x) = x^2 + 1 + \sqrt{x+2}$$

$$(f-g)(x) = x^2 + 1 - \sqrt{x+2} \qquad (g-f)(x) = \sqrt{x+2} - x^2 - 1$$

$$(f \cdot g)(x) = (x^2+1)\sqrt{x+2}$$

$$\frac{f}{g}(x) = \frac{x^2+1}{\sqrt{x+2}} \qquad \frac{g}{f}(x) = \frac{\sqrt{x+2}}{x^2+1}$$

We now can develop the machinery necessary for later chapters. All of these properties will probably seem obvious to you from our intuitive discussion on limits.

Let $\lim_{x \to a} f(x) = A$ and $\lim_{x \to a} g(x) = B$. Then

Property 1 $\quad \lim_{x \to a} [f(x) + g(x)] = A + B = \lim_{x \to a} f(x) + \lim_{x \to a} g(x)$

Property 2 $\quad \lim_{x \to a} [f(x) - g(x)] = A - B = \lim_{x \to a} f(x) - \lim_{x \to a} g(x)$

Property 3 $\quad \lim_{x \to a} f(x)g(x) = A \cdot B = \lim_{x \to a} f(x) \cdot \lim_{x \to a} g(x)$

Property 4 If $B \neq 0$, then $\lim_{x \to a} \dfrac{f(x)}{g(x)} = \dfrac{A}{B} = \dfrac{\lim_{x \to a} f(x)}{\lim_{x \to a} g(x)}$

EXAMPLE 2-6

Let $f(x) = x^2$ and $g(x) = x^3 - 1$. Then $\lim_{x \to 2} f(x) = 4$ and $\lim_{x \to 2} g(x) = 7$. Hence

$$\lim_{x \to 2} f(x) + g(x) = \lim_{x \to 2} (x^2 + x^3 - 1) = 4 + 7 = 11$$

$$\lim_{x \to 2} f(x) - g(x) = \lim_{x \to 2} (x^2 - x^3 + 1) = 4 - 7 = -3$$

$$\lim_{x \to 2} f(x)g(x) = \lim_{x \to 2} x^2(x^3 - 1) = 4 \cdot 7 = 28$$

$$\lim_{x \to 2} \frac{f(x)}{g(x)} = \lim_{x \to 2} \frac{x^2}{x^3 - 1} = \frac{4}{7}$$

Since a constant function $f(x) = c$ is a linear function, we can refer to the above discussion to understand the following property.

Property 5 $\lim_{x \to a} c = c$

In Property 3, if we let $f(x) = c$, a constant function, we get the following:

Property 6 $\lim_{x \to a} cg(x) = c \lim_{x \to a} g(x)$

EXAMPLE 2-7

Let $f(x) = 8x^2$, then
$$\lim_{x \to 2} f(x) = \lim_{x \to 2} 8x^2 = 8 \lim_{x \to 2} x^2 = 8 \cdot 4 = 32$$

Since $f(x) = x$ is a linear function, we see that $\lim_{x \to a} x = a$. From Property 3 we have
$$\lim_{x \to a} x^2 = [\lim_{x \to a} x][\lim_{x \to a} x] = a \cdot a = a^2$$

Similarly, from Property 3,
$$\lim_{x \to a} x^3 = [\lim_{x \to a} x^2][\lim_{x \to a} x] = a^2 \cdot a = a^3$$
and
$$\lim_{x \to a} x^4 = a^4$$

In general, $\lim_{x \to a} x^n = a^n$ is true for any positive integer n. In fact, we can say more.

Property 7 $\lim_{x \to a} x^r = a^r$ for any real number r

EXAMPLE 2-8

$$\lim_{x \to 4} (x^{1/2} - x^{5/2}) = 4^{1/2} - 4^{5/2} = 2 - 2^5 = -30$$

$$\lim_{x \to 1} 3x^2 + 5x - 1 = 7$$

$$\lim_{x \to -1} x^7 - 2x^6 = -3$$

$$\lim_{x \to b} a_n x^n + \cdots + a_1 x + a_0 = a_n b^n + \cdots + a_1 b + a_0$$

We shall often have to find the limit at a of a function of the form $f(x) = h(x)/g(x)$ where $\lim_{x \to a} h(x) = 0 = \lim_{x \to a} g(x)$. Property 4 does not help us in this case, since a hypothesis of Property 4 is that $\lim_{x \to a} g(x) \neq 0$. For example, $\lim_{x \to 0} x^2 = 0 = \lim_{x \to 0} x$, yet $\lim_{x \to 0} x^2/x = 0$. Sometimes when confronted with this situation, we can cancel a factor in the numerator and the denominator, so that the limit of the resulting function is obtainable.

EXAMPLE 2-9

To find $\lim_{x \to 3}(x^2-9)/(x-3)$, we first note that $x^2-9 = (x+3)(x-3)$, and thus

$$\lim_{x \to 3}\frac{x^2-9}{x-3} = \lim_{x \to 3}\frac{(x+3)(x-3)}{x-3} = \lim_{x \to 3}(x+3) = 6$$

In general, if $\lim_{x \to a} f(x) = A$, then

$$\lim_{x \to a}\frac{(x-a)f(x)}{x-a} = \lim_{x \to a} f(x) = A$$

EXAMPLE 2-10

$$\lim_{x \to 1}\frac{(x-3)(x-4)(x-1)}{(x-5)(x-1)} = \lim_{x \to 1}\frac{(x-3)(x-4)}{(x-5)} = \frac{(-2)(-3)}{-4} = -\frac{3}{2}$$

EXAMPLE 2-11

$\lim_{x \to 0} x/x^2 = \lim_{x \to 0} 1/x$, which does not exist.

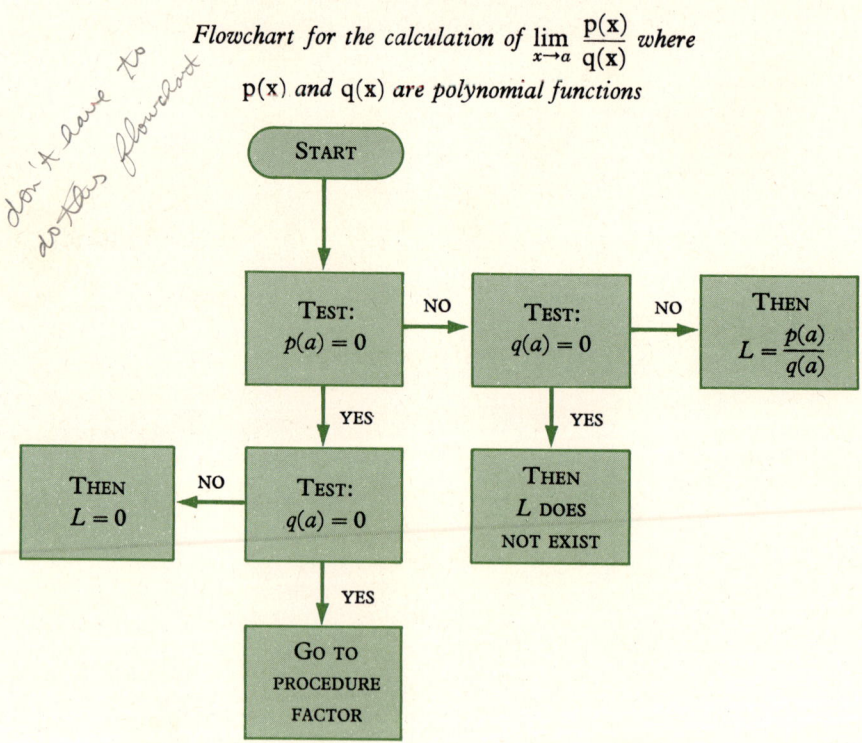

Flowchart for the calculation of $\lim_{x \to a} \dfrac{p(x)}{q(x)}$ *where* $p(x)$ *and* $q(x)$ *are polynomial functions*

PROCEDURE FACTOR: FACTOR $p(x)$ AND $q(x)$, CANCEL ALL COMMON FACTORS, GO TO START.

FIGURE 2-4

Thus we see that if $f(x) = h(x)/g(x)$ and $\lim_{x \to a} h(x) = \lim_{x \to a} g(x) = 0$, we cannot conlude whether or not the limit exists. It may exist, as in Examples 2-9 and 2-10, or it may not, as in Example 2-11. Note that if $\lim_{x \to a} h(x) \neq 0$ and $\lim_{x \to a} g(x) = 0$, then $\lim_{x \to a} h(x)/g(x)$ does not exist.

Much of our ensuing work will entail calculating the limits of various functions. Very often the function in question will be a rational function, so it is imperative that you master the technique of evaluating limits of rational functions. To clarify the procedure we have provided the flowchart, Fig. 2-4, illustrating the possibilities.

EXERCISE 2-1

1. Use our intuitive methods to find $\lim_{x \to a} f(x)$, if it exists, for the following functions from the graphs given in the Chap. 1.

   (a) $f(x) = 2x - 1$, $a = 0$ (e) $f(x) = |x|$, $a = 0$
   (b) $f(x) = 1 - 6x$, $a = 10$ (f) $f(x) = e^x$, $a = 0$
   (c) $f(x) = x^2$, $a = -1$ (g) $f(x) = \ln x$, $a = 1$
   (d) $f(x) = x^2 + 1$, $a = -1$ (h) $f(x) = [\![x]\!]$, $a = 1$

2. Determine whether $\lim_{x \to a} f(x)$ exists for the following graphs and values of a.

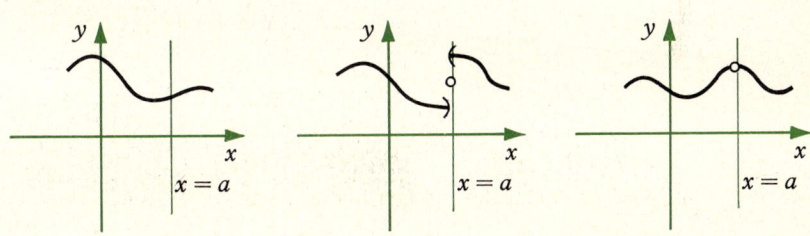

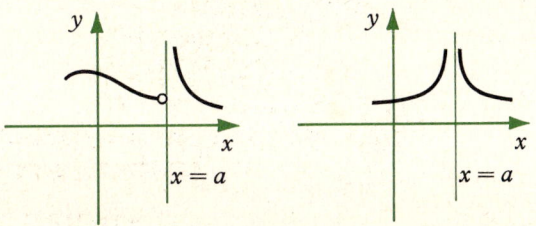

3. Find $\lim_{x \to a} f(x)$, if it exists, for the following functions by first drawing the graph of $y = f(x)$.

(a) $f(x) = \begin{cases} x & \text{if } x < 0 \\ 3x & \text{if } x \geq 0 \end{cases} \quad a = 0$

(b) $f(x) = \begin{cases} x + 2 & \text{if } x < 1 \\ x & \text{if } x \geq 1 \end{cases} \quad a = 1$

(c) $f(x) = \begin{cases} x + 3 & \text{if } x < 2 \\ 3 & \text{if } x = 2 \\ 2 & \text{if } x > 2 \end{cases} \quad a = 2$

In the next four problems we demonstrate the close connection between values in the domain and their corresponding functional values. This connection governs the intrinsic significance of the important concept of the limit of a function. In each, a situation depicting a function whose graph is assumed to be a continuous curve is described. The functional value of one point is found by examining the functional values of points close to it.

4. Suppose you are the editor of a small midwestern newspaper servicing twenty small cities within a 20-mile radius with your town in the geographical center. As you are filling in the temperature table you notice that the surrounding nineteen cities all reported a noon temperature of 90° but your weather bureau's report is lost. What temperature might you fill in for your city?

5. The measurements of the population of a bacteria culture are given by $P(t)$ where t is measured in hours and $P(t)$ in millions of bacteria. Your assistant failed to make a measurement at $t = 5$. What value for $P(t)$ would you fill in?

t	1	2	3	4	5	6	7
$P(t)$	12	14	18	26		74	138

6. Fired into the air a rocket is designed to explode precisely 5 seconds into flight. We want to find out how far it went before it exploded. We have an instrument which measures its distance quite precisely, but exactly at $t = 5$ the instrument cannot measure the distance because of the explosion. The following table gives the distance $s(t)$ travelled after t seconds. How far do you think the rocket travelled?

t	4.9	4.99	4.999
$s(t)$	199	199.9	199.99

7. A laboratory is commissioned to test a new adhesive to see how many pounds it will hold for one minute. Note that the exact maximum weight will never

be known: if the weight falls it was too much, if it doesn't it was too little. Suppose the following data were compiled. What would you guess is the maximum weight?

Weight (lbs)	198	199	199.9	200.1	201	202
Time adhesive held wt.	60	60	60	59.8	59.5	59

8. Give a reason why the following functions are not continuous at 2.

   (a) $f(x) = \begin{cases} x & \text{if } x \leq 2 \\ 3 & \text{if } x > 2 \end{cases}$

   (b) $f(x) = [\![x]\!]$

   (c) $f(x) = \dfrac{x^2-4}{x-2}$

   (d) $f(x) = \dfrac{(x-1)(x-2)}{x-2}$

   (e) $f(x) = \begin{cases} -x & \text{if } x < 2 \\ 5 & \text{if } x = 2 \\ 2 & \text{if } x > 2 \end{cases}$

9. State where each of the following functions is not continuous.

   (a) $f(x) = \begin{cases} 2x-1 & \text{if } x < 3 \\ 5 & \text{if } x \geq 3 \end{cases}$

   (b) $f(x) = \begin{cases} 2x-1 & \text{if } x < 3 \\ 5 & \text{if } x = 3 \\ 6 & \text{if } x > 3 \end{cases}$

   (c) $f(x) = \begin{cases} x^2 & \text{if } x < 0 \\ -1 & \text{if } x = 0 \\ x & \text{if } x > 0 \end{cases}$

   (d) $f(x) = \begin{cases} x & \text{if } x < 0 \\ [\![x]\!] & \text{if } 0 \leq x \leq 3 \\ 2 & \text{if } x > 3 \end{cases}$

   (e) $f(x) = |x|$

10. Which of the following functions are continuous at 2?

    (a) $f(x) = x$

    (b) $f(x) = x + 2$

    (c) $f(x) = 8x^2 - 5$

    (d) $f(x) = \dfrac{1}{x+1}$

    (e) $f(x) = \dfrac{1}{x-2}$

    (f) $f(x) = x$

    (g) $f(x) = \dfrac{x-2}{|x-2|}$

    (h) $f(x) = \sqrt{x}$

    (i) $f(x) = \ln x$

    (j) $f(x) = e^x$

CHAP. 2   Limits

11. Which of the following functions are continuous at $x = 0$?

(a) $f(x) = x + 2$

(b) $f(x) = x^2 + 1$

(c) $f(x) = |x|$

(d) $f(x) = \sqrt{x}$

(e) $f(x) = e^x$

(f) $f(x) = \dfrac{x-2}{x-2}$

(g) $f(x) = \dfrac{1}{x+1}$

(h) $f(x) = \dfrac{x-2}{|x-2|}$

(i) $f(x) = [\![x]\!]$

(j) $f(x) = \begin{cases} x & \text{if } x \leqslant 0 \\ 2x & \text{if } x > 0 \end{cases}$

12. Let $P(t)$ be the production rate of a factory, measured in hundred units per hour, at time t. When all machines are operating, the production rate is 200 units per hour. Suppose all machines operate from 8:00 A.M. ($t = 0$) till noon ($t = 4$), when half the machines are turned off for one hour to give the employees a staggered lunch. Then the machines run from 1:00 P.M. to 4:00 P.M. The function $P(t)$ is expressed by

$$P(t) = \begin{cases} 2 & \text{for } t\ (0,4] \\ 1 & \text{for } t\ (4,5] \\ 2 & \text{for } t\ (5,8] \end{cases}$$

Graph $P(t)$ and determine whether the following limits exist.

(a) $\lim\limits_{t \to 4} P(t)$

(b) $\lim\limits_{t \to 5} P(t)$

(c) $\lim\limits_{t \to 6} P(t)$

13. For the functions $f(x)$ and $g(x)$ find $f(x)+g(x)$, $f(x)-g(x)$, $f(x)g(x)$ and $f(x)/g(x)$.

(a) $f(x) = x$, $g(x) = x + 1$

(b) $f(x) = 3x - 1$, $g(x) = x^2$

(c) $f(x) = x^2 + 1$, $g(x) = 3x^2 - x$

(d) $f(x) = 1/x$, $g(x) = x^2 + 1$

(e) $f(x) = \sqrt{x}$, $g(x) = 3x^2 + 5$

(f) $f(x) = e^x$, $g(x) = x + e^x$

(g) $f(x) = x^2 - 4$, $g(x) = x - 2$

(h) $f(x) = x + \ln x$, $g(x) = x - \ln x$

14. Suppose that $\lim\limits_{x \to 2} f(x) = 5$, $\lim\limits_{x \to 2} g(x) = -3$, and $\lim\limits_{x \to 2} h(x) = 0$. Find the following limits if they exist.

(a) $\lim\limits_{x \to 2} f(x) + g(x)$

(b) $\lim\limits_{x \to 2} f(x)\, h(x)$

(c) $\lim\limits_{x \to 2} h(x) - g(x)$

(d) $\lim\limits_{x \to 2} \dfrac{f(x)}{g(x)}$

(e) $\lim\limits_{x \to 2} \dfrac{h(x)}{f(x)}$

(f) $\lim\limits_{x \to 2} \dfrac{f(x)}{h(x)}$

15. Find the following limits.
    (a) $\lim_{x \to 3} x^2 - 2x + 1$
    (b) $\lim_{x \to -1} x^5 + 5x^4 - 10$
    (c) $\lim_{x \to 0} x^6 - 10x^5 + 25$
    (d) $\lim_{x \to 1} x^{1/2} - x$
    (e) $\lim_{x \to 4} x^{3/2} - \frac{1}{x}$
    (f) $\lim_{x \to 1/2} \frac{x-1}{x^2-2}$

16. Is it possible to find functions $f(x)$ and $g(x)$ and a point a such that:
    (a) $\lim_{x \to a} f(x)$ and $\lim_{x \to a} g(x)$ exist, but $\lim_{x \to a} f(x) + g(x)$ does not exist?
    (b) $\lim_{x \to a} f(x)$ and $\lim_{x \to a} g(x)$ do not exist, but $\lim_{x \to a} f(x) + g(x)$ does exist? (*Hint:* Let $f(x) = x/|x|$ and $g(x) = -x/|x|$.)

17. Calculate the following limits if they exist.
    (a) $\lim_{x \to 1}(x^2 + 2x - 3)$
    (b) $\lim_{x \to -2} (3x^3 - 8x)$
    (c) $\lim_{x \to 2}(8x^5 + 5x - 2)$
    (d) $\lim_{x \to 2} e^x$
    (e) $\lim_{x \to 5} e^x$
    (f) $\lim_{x \to 2} \ln x$
    (g) $\lim_{x \to 3} \ln x$
    (h) $\lim_{x \to 2} \frac{(3x-2) \ln x}{x-1}$
    (i) $\lim_{x \to 5} \frac{(x-5)e^x}{3x}$
    (j) $\lim_{x \to 1} \frac{8x^2 - x - 3}{e^x(x^3 - 4)}$
    (k) $\lim_{x \to 3} \frac{x^2 - 9}{x - 3}$
    (l) $\lim_{x \to 4} \frac{x^2 - 16}{x - 4}$
    (m) $\lim_{x \to 1} \frac{x^2 - 1}{x - 1}$
    (n) $\lim_{x \to 2} \frac{x^2 - 4}{x - 2}$
    (o) $\lim_{x \to -3} \frac{x^2 - 9}{x + 3}$
    (p) $\lim_{x \to 0} \frac{x(x - 3)}{x}$
    (q) $\lim_{x \to 1} \frac{x^2 - 3x + 2}{x - 1}$
    (r) $\lim_{h \to 0} \frac{2h - 3h^2 + h^3}{h}$

18. One intuitively defines the *right-hand limit of $f(x)$ as x approaches a*, written
$$\lim_{x \to a^+} f(x)$$
to be that number L such that $f(x)$ is arbitrarily close to L if x is arbitrarily close to a and $x > a$. For example,
$$\lim_{x \to 0^+} \frac{x}{|x|} = 1$$

One similarly defines the *left-hand limit of $f(x)$ as x approaches a*, written

$$\lim_{x \to a^-} f(x)$$

for $x < a$. For example,

$$\lim_{x \to 0^-} \frac{x}{|x|} = -1$$

Find $\lim_{x \to 2^+} f(x)$ and $\lim_{x \to 2^-} f(x)$ for the functions in Problems 8 and 10.

19. The number e was first introduced in Sec. 1-6. The number e is an irrational number whose decimal equivalent to six decimal places is 2.718281. In Chapter 7, once we have developed the machinery of the calculus, we present a rigorous definition of the function $y = \ln x$ and then define e as that number which satisfies $\ln x = 1$. We can alternately define e as a limit as follows:

$$e = \lim_{x \to 0} (1+x)^{1/x} \quad \text{where } x > 0$$

Let $f(x) = (1+x)^{1/x}$. Then $f(1) = (1+1)^1 = 2$, $f(\frac{1}{2}) = (1+\frac{1}{2})^2 = 2.25$, $f(\frac{1}{3}) = (1+\frac{1}{3})^3 = (\frac{4}{3})^3 = \frac{64}{27} \cong 2.37$. Find $f(\frac{1}{4})$, $f(\frac{1}{5})$, and $f(\frac{1}{10})$.

3

The Derivative

3-1 DEFINITION OF THE DERIVATIVE

The *derivative* of a function $y = f(x)$ is another function derived from $y = f(x)$. This new "derived" function will help us to understand more fully the properties of $y = f(x)$. One uses the derivative to study topics concerning *rates of change*. One example of the use of the derivative is the concept of the tangent line to a curve. In geometry, a tangent line to a circle is defined to be a line intersecting the circle at one and only one point. In extending this notion to tangent lines of arbitrary curves, a bit of caution is in order. Most people have a good intuitive idea of a tangent line. In Fig. 3-1 we would say that l_1 and l_2 are tangent to $y = f(x)$ but the other lines are not.

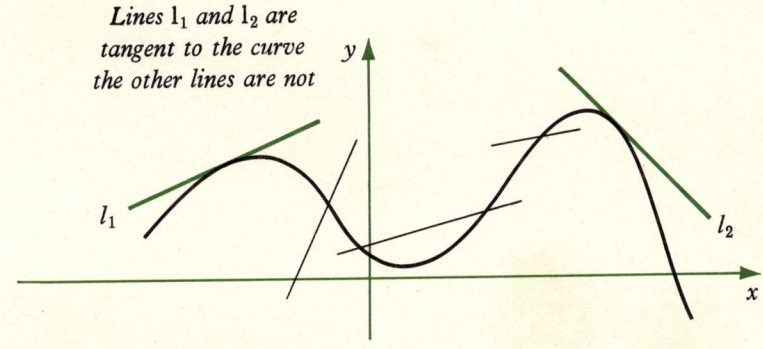

FIGURE 3-1

CHAP. 3   The Derivative

Suppose an airplane travelling along the path of the curve $y = x^2$ shot a flare directly in front of it. The flare would not continue on the same path along $y = x^2$, but would, according to Newton's third law of motion, rush along the tangent line to the curve at that point, as in Fig. 3-2. We now want to calculate the equation of the tangent line to the curve $f(x) = x^2$ at the point $(2,4)$. In order to accomplish this, we consider the

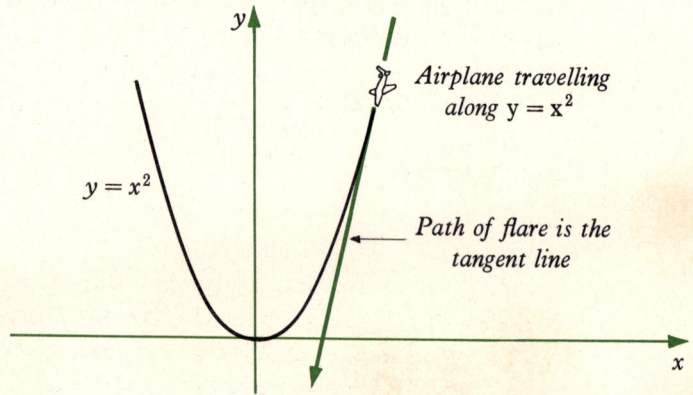

FIGURE 3-2

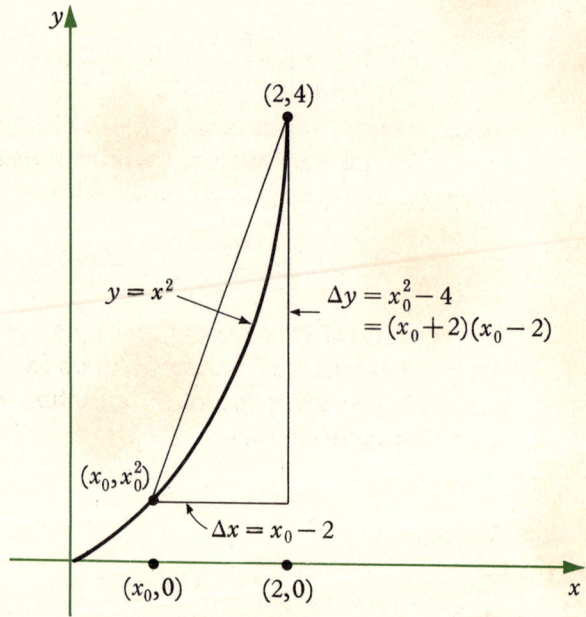

FIGURE 3-3

lines that go through $(2,4)$ and (x_0, x_0^2), which is close to $(2,4)$, as seen in Fig. 3-3. The slope of this type of line is

$$\frac{\Delta y}{\Delta x} = \frac{f(x_0) - f(2)}{x_0 - 2} = \frac{x_0^2 - 4}{x_0 - 2} = \frac{(x_0 + 2)(x_0 - 2)}{x_0 - 2} = x_0 + 2$$

Intuitively it is clear that the closer x_0 is to 2, the closer the slope of the line through $(2,4)$ and (x_0, x_0^2) is to the slope of the tangent line. Indeed, this is the language used in talking intuitively about limits, and if we let $f'(2)$ be the slope of the tangent line to $f(x) = x^2$ at $x = 2$, then we are led to write

$$f'(2) = \lim_{x_0 \to 2} \frac{f(x_0) - f(2)}{x_0 - 2} = \lim_{x_0 \to 2} (x_0 + 2) = 4$$

The expression

$$\frac{f(x_0) - f(2)}{x_0 - 2}$$

is called the *difference quotient* of $f(x)$ at $x = 2$, and $f'(2)$ is called the *derivative* of $f(x)$ at $x = 2$. Recall that the point-slope form of the straight line through the point (a,b) with slope m is given by $y - b = m(x - a)$. We now can say that the tangent line to $f(x) = x^2$ at $x = 2$ is $y - 4 = 4(x - 2)$.

Using this same procedure, we can calculate the equation of the tangent line to the curve $f(x) = x^2$ at $(3,9)$. We find the derivative $f'(3)$ of $f(x) = x^2$ at $x = 3$ by considering

$$f'(3) = \lim_{x_0 \to 3} \frac{f(x_0) - f(3)}{x_0 - 3} = \lim_{x_0 \to 3} \frac{x_0^2 - 9}{x - 3} = \lim_{x_0 \to 3} (x_0 + 3) = 6$$

Hence the slope of the tangent line is 6 and the equation is $y - 9 = 6(x - 3)$.

Notice that we can carry this reasoning further. At any point (a, a^2) the slope of $f(x) = x^2$ is given by

$$f'(a) = \lim_{x \to a} \frac{f(x) - f(a)}{x - a} = \lim_{x \to a} \frac{x^2 - a^2}{x - a} = \lim_{x \to a} x + a = 2a$$

The equation of the tangent line to $f(x) = x^2$ at $x = a$ is $y - f(a) = 2a(x - a)$.

Apart from any geometrical consideration, this limit of the difference quotient has wide application for many functions. We are thus led to present the following definition.

Definition The *derivative of $f(x)$ at $x = a$*, if it exists, is denoted by $f'(a)$. We define

$$f'(a) = \lim_{x \to a} \frac{f(x) - f(a)}{x - a}$$

provided that this limit exists.

If $f'(a)$ exists, we say that $f(x)$ is *differentiable* at $x = a$. Also if $f'(a)$ exists, we define the slope of the tangent line to be $f'(a)$ and say that the equation of the tangent line to $f(x)$ at $(a, f(a))$ is $y - f(a) = f'(a)(x - a)$, which is what our intuition led us to believe.

We have noted that when $f(x) = x^2$ then $f'(a) = 2a$ for any real number a. We can simply say that $f'(x) = 2x$, i.e., the derivative of $f(x)$ can be regarded as a function whose value for any real number a is $2a$. Hence taking the derivative of a function $f(x)$ gives rise to another function $f'(x)$ called the *derivative of $f(x)$*.

EXAMPLE 3-1

Let $f(x) = x$. For each a, we have

$$f'(a) = \lim_{x \to a} \frac{x - a}{x - a} = \lim_{x \to a} 1 = 1$$

Hence $f'(x) = 1$.

EXAMPLE 3-2

Let $f(x) = x^3$. For each a, we have

$$f'(a) = \lim_{x \to a} \frac{x^3 - a^3}{x - a} = \lim_{x \to a} \frac{(x - a)(x^2 + ax + a^2)}{x - a} = 3a^2$$

Hence $f'(x) = 3x^2$.

EXAMPLE 3-3

Let $f(x) = 4x^2$. For each a, we have

$$f'(a) = \lim_{x \to a} \frac{4x^2 - 4a^2}{x - a} = \lim_{x \to a} \frac{4(x - a)(x + a)}{x - a} = 4 \cdot 2a = 8a$$

Hence $f'(x) = 8x$.

EXAMPLE 3-4

Let $f(x) = \sqrt{x}$. For each a, we have

$$f'(a) = \lim_{x \to a} \frac{\sqrt{x} - \sqrt{a}}{x - a}$$

$$= \lim_{x \to a} \frac{\sqrt{x} - \sqrt{a}}{x - a} \cdot \frac{\sqrt{x} + \sqrt{a}}{\sqrt{x} + \sqrt{a}}$$

$$= \lim_{x \to a} \frac{x - a}{(x - a)(\sqrt{x} + \sqrt{a})}$$

$$= \lim_{x \to a} \frac{1}{\sqrt{x} + \sqrt{a}} = \frac{1}{2\sqrt{a}}$$

Notice that this limit does not exist when $a = 0$. Why? We therefore say that $f(x) = \sqrt{x}$ is not differentiable at $x = 0$. We can say that $f'(x) = 1/(2\sqrt{x})$, so $f'(x)$ exists for all positive real numbers.

There are many other notations used for the derivative. In place of $f'(x)$ we will sometimes use $D_x f(x)$. Thus from Ex. 3-4,

$$D_x x = 1 \qquad D_x x^2 = 2x \qquad D_x x^3 = 3x^2 \qquad D_x x^{1/2} = \tfrac{1}{2}x^{-1/2}$$

Another common notation is $df(x)/dx$. Thus

$$\frac{d}{dx}x^2 = \frac{dx^2}{dx} = 2x \quad \text{and} \quad \frac{d}{dx}x^3 = \frac{dx^3}{dx} = 3x^2$$

As we have just seen in Ex. 3-4, there do exist functions which are not differentiable at certain points. In other words, the limit of the difference quotient does not exist at these points. For instance, let us consider $y = |x|$ and the point $x = 0$. Then the difference quotient at zero is simply

$$\frac{|x| - |0|}{x - 0} = \frac{|x|}{x}$$

But we have previously seen that $\lim_{x \to 0} |x|/x$ does not exist. Intuitively, we can see that the tangent line to $f(x) = |x|$ does not exist at $x = 0$ because, on the one hand, if $x > 0$, $f(x) = x$ and the tangent line should be $y = x$, and on the other, if $x < 0$, $f(x) = -x$ and the tangent line should be $y = -x$.

Any function that has a "corner" at $x = a$, i.e., that abruptly changes direction at $x = a$, as $y = |x|$ does at $x = 0$, is not differentiable at $x = a$. This is so for the same reason that $y = |x|$ is not differentiable at $x = 0$. Geometrically, the tangent line to the curve at $x = a$ appears to be two different lines when one considers the behavior of the function to the right and to the left of $x = a$. Mathematically, the limit of the difference quotient does not exist. The functions whose graphs are in Fig. 3-4 are not differentiable at $x = a$.

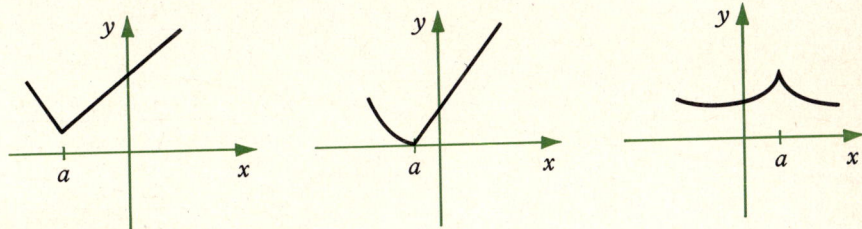

Functions that have a "corner" at **a** *are not differentiable at* **a**.

FIGURE 3-4

The derivative measures the rate of change of a variable quantity, usually with respect to time in physical applications. Thus when one studies the price of a stock, one is concerned with the rate of increase or decrease of the price with respect to time. Physicians study the rate of increase of blood pressure. Statisticians and economists study the rate of increase or decrease of the unemployment rates. Ecologists are concerned with the rate of increase of a species occupying a given territory. Let us study two examples in a little more depth. Additional applications will be considered in the exercises.

EXAMPLE 3-5

The Spread of a Disease Model.

Let $p(t)$ denote the proportion of the people in a country who have been infected by a communicable disease prior to time t. If there are no forces present to check the spread of the epidemic, then $p(t)$ will increase as t increases, so that either $p(t) = 1$ at some time t or $p(t)$ approaches the value 1 without actually reaching 1. If the epidemic starts at time $t = 0$, then the domain of $p(t)$ is the set $T = \{t \mid t \geq 0\}$, with $0 \leq p(t) \leq 1$. For a finite population of N people, $p(t)$ will not be continuous but must increase in jumps of integral multiples of N. Thus $p(t)$ is a step function. See Exercise 1-5, Problems 4 and 6, for examples of such functions. For large populations, we can assume that $p(t)$ is continuous. This epidemic model assumes that infection spreads by contact between members of a community in which there is no removal from circulation by death, recovery, or isolation during the main part of the epidemic. The common cold is such an epidemic. In order for medical authorities to handle such an epidemic, it is necessary to ascertain the rate of increase of $p(t)$ with respect to t. Suppose authorities have determined that an epidemic is spreading according to $p(t) = \frac{1}{1000}t^2$, and they wish to know what the instantaneous rate of increase of $p(t)$ is when $t = 2$. To solve we calculate

$$p'(2) = \lim_{t \to 2} \frac{p(t) - p(2)}{t - 2}$$

$$= \lim_{t \to 2} \frac{\frac{1}{1000}(t^2 - 4)}{t - 2}$$

$$= \frac{1}{1000} \lim_{t \to 2} t + 2$$

$$= \frac{4}{1000}, \text{ or } 0.004$$

Such information will help determine when a vaccine should be made available to the public. In Chap. 4 we investigate this model in greater depth to determine the significance of $p'(t)$.

EXAMPLE 3-6
A Marginal Analysis Model.

In its attempt to improve profit, management is frequently confronted by decisions about whether to change input. For example, would it increase profit if a new machine or a new salesman were added, or if inventory or advertising were increased? Thus management is seeking the rate of change of a quantity, say, cost, with respect to another, say, units produced. If the units produced were increased, how would cost be affected? Economists give the name *marginal analysis* to this type of analysis of a problem whose solution depends upon determining the rate of change of the quantity in question.

For example, suppose that a firm has determined that the cost $C(x)$ of making x sets of a hundred units per hour is given by

$$C(x) = 20 + 0.5x - 0.1x^2$$

where $C(x)$ is measured in thousands of dollars and $x \in [0, 2.5]$. The rate of increase of $C(x)$ when $x = 2$ is called the *marginal cost* at $x = 2$ and is given by

$$\begin{aligned}
C'(2) &= \lim_{x \to 2} \frac{C(x) - C(2)}{x - 2} \\
&= \lim_{x \to 2} \frac{20 + 0.5x - 0.1x^2 - 20 - 1 + 0.4}{x - 2} \\
&= \lim_{x \to 2} \frac{0.5(x - 2) - 0.1(x^2 - 4)}{x - 2} \\
&= \lim_{x \to 2} \frac{(x - 2)(0.5 - 0.1x - 0.2)}{x - 2} \\
&= 0.1
\end{aligned}$$

The economist would then refer to this value of 0.1 as the *extra* cost of making an *extra* unit. Let us assume that the production capacity of the firm is $x = 2.5$ hundred units per hour. From the graph of the cost function $C(x)$ given in Fig. 3-5, we see that as the company increases production, cost also increases, but the rate of increase of $C(x)$, i.e., the derivative $C'(x)$, which is measured by the slope of the tangent line, is decreasing. For instance, using the definition of the derivative, one can determine that $C'(1) = 0.3$ whereas $C'(2) = 0.1$. Thus the tangent line at $x = 1$ has a larger slope than at $x = 2$, so the function $C(x)$ is increasing at a slower rate at $x = 2$ than at $x = 1$. Thus the firm is operating more efficiently when it is producing $x = 2$ hundred units per hour rather than $x = 1$ hundred units per hour. Is the company operating even more efficiently at $x = 2.5$? This model reflects the rationale for mass production.

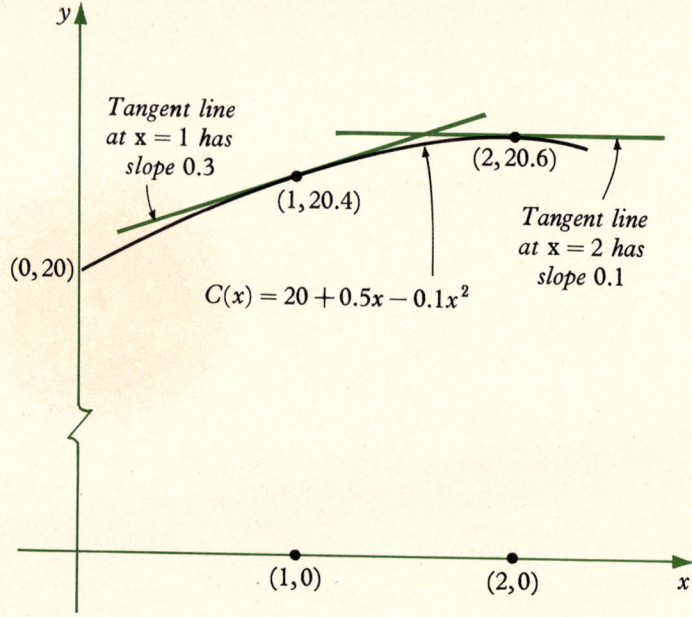

FIGURE 3-5

It will sometimes be helpful for us to use an alternate definition of $f'(a)$. Notice that if we set $h = x - a$, as x approaches a, then h approaches 0. Also note that $f(x) = f(a+h)$. Hence we can rewrite the difference quotient as

$$\frac{f(a+h) - f(a)}{h}$$

and the definition of $f'(a)$ by

$$f'(a) = \lim_{h \to 0} \frac{f(a+h) - f(a)}{h}$$

For example, we use this alternate definition of $f'(x)$ show that $D_x 3x = 3$. If we let $f(x) = 3x$ then

$$f'(a) = \lim_{h \to 0} \frac{f(a+h) - f(a)}{h}$$
$$= \lim_{h \to 0} \frac{3(a+h) - 3a}{h}$$

$$= \lim_{h \to 0} \frac{3h}{h}$$

$$= 3$$

EXERCISE 3-1

1. Sketch the curves and draw the tangent line to the given point $x = a$. Guess as to what the slope of this line may be.
   - (a) $f(x) = x - 2$, $a = 2$
   - (b) $f(x) = 2x + 1$, $a = 1$
   - (c) $f(x) = x^2 + 1$, $a = 2$
   - (d) $f(x) = 2x^2$, $a = -1$
   - (e) $f(x) = -x^2 + 1$, $a = \frac{1}{2}$
   - (f) $f(x) = -3x^2 - 5$, $a = 1$
   - (g) $f(x) = -x^2 + 5$, $a = 2$
   - (h) $f(x) = x^2 - x$, $a = 0$

2. Compute the derivative of the given function at the given point and find the equation of the tangent line at the point. Then compute $f'(x)$.
   - (a) $f(x) = 3x - 5$, $a = 1$
   - (b) $f(x) = -\frac{1}{2}x + 1$, $a = 2$
   - (c) $f(x) = x^2 + 1$, $a = 2$
   - (d) $f(x) = 2x^2$, $a = 2$
   - (e) $f(x) = 3x^2$, $a = 2$
   - (f) $f(x) = x^3$, $a = 2$
   - (g) $f(x) = 1/x$, $a = 1$
   - (h) $f(x) = 2x^3$, $a = 2$

3. At what point do the functions fail to have a derivative?
   - (a) $f(x) = |x|$
   - (b) $f(x) = 2 + |x|$
   - (c) $f(x) = \begin{cases} x & \text{if } x \in (-\infty, 0] \\ 0 & \text{if } x \in [0, \infty) \end{cases}$
   - (d) $f(x) = x/|x|$
   - (e) $f(x) = \begin{cases} x^2 & \text{if } x \in (-\infty, 1] \\ 1 & \text{if } x \in [1, \infty) \end{cases}$

4. The medical authorities of a community have determined that the proportion $p(t)$ of the people who have been infected by a communicable disease prior to time t is $p(t) = \frac{1}{1000}(t^2 + t)$. What is the instantaneous rate of increase of $p(t)$ with respect to t when $t = 2$? when $t = 3$?

5. A firm has determined that the cost $C(x)$ of producing x units of goods is $C(x) = 10 + 3x$. Find the marginal cost when $x = 2$, $x = 3$, $x = 4$.

6. The amount of goods demanded $D(x)$ for a product at a price x is given by $D(x) = 7 - 2x$. The *marginal demand* when $x = a$ is defined to be $D'(a)$. Find the marginal demand when $x = 1$, $x = 2$, $x = 3$.

7. Suppose that the amount of a goods that a producer is willing to supply to the market $S(x)$ at a price x is given by $S(x) = 1 + 3x$. The *marginal supply* at $x = a$ is defined to be $S'(a)$. Find the marginal supply when $x = 1$, $x = 2$, $x = 3$.

8. The revenue $R(x)$ that a company takes in when x units are produced is given

by $R(x) = 6x - x^2$. The *marginal revenue* at $x = a$ is defined by $R'(a)$. Find the marginal revenue when $x = 2$, $x = 3$.

3-2 FINDING DERIVATIVES

It will be very useful for us to develop tools for finding derivatives so that we will not have to use the somewhat cumbersome definition each time we desire the derivative of a function.

Earlier we noted the following differentiation formulas:

$$D_x x = 1 \qquad D_x x^2 = 2x \qquad D_x x^3 = 3x^2$$

and

$$D_x x^{1/2} = \tfrac{1}{2} x^{-1/2} \qquad \text{where } x^{1/2} = \sqrt{x}$$

This pattern suggests an important formula called the *power rule*.

Rule D1 $D_x x^r = r x^{r-1}$ *for every real number r*

This formula is actually a theorem which requires proof. In Exercise 3-2 we indicate how to prove the theorem when r is a natural number.

We have also seen that $D_x 4x^2 = 8x = 4 \cdot 2x = 4 D_x x^2$. If you work out a few more examples you will convince yourself that "the derivative of a constant times a function is equal to the constant times the derivative of the function".

Rule D2 $D_x a f(x) = a D_x f(x)$ *for any real number a*

The proofs of this formula and the next are not hard and are included at the end of this section. We first describe their utility.

Rule D3 $D_x [f(x) + g(x)] = D_x f(x) + D_x g(x)$

This formula states, "The derivative of a sum is the sum of the derivatives". The rule states that the derivative of the sum of two functions is the sum of the derivatives, but one sees readily that the rule is valid for the sum of any number of functions.

We can now find the derivative of any polynomial function. For example,

$$D_x(4x^5 - 8x^3 + 3x^2 + 5) = D_x 4x^5 + D_x(-8x^3) + D_x 3x^2 + D_x 5$$
$$\text{(Rule D3)}$$
$$= 4 D_x x^5 + (-8) D_x x^3 + 3 D_x x^2 + 5 D_x x^0$$
$$\text{(Rule D2)}$$

$$= 4(5x^4) - 8(3x^2) + 3(2x) + 5(0 \cdot x^{-1})$$
(Rule D1)

$$= 20x^4 - 24x^2 + 6x$$

In general we can write, where $a_n, a_{n-1}, \ldots, a_0$ are constants,

$$D_x(a_n x^n + a_{n-1} x^{n-1} + \cdots + a_1 x + a_0)$$
$$= D_x a_n x^n + D_x a_{n-1} x^{n-1} + \cdots + D_x a_1 x + D_x a_0$$
$$= a_n D_x x^n + a_{n-1} D_x x^{n-1} + \cdots + a_1 D_x x + 0$$
$$= n a_n x^{n-1} + (n-1) a_{n-1} x^{n-2} + \cdots + a_1$$

Let us look at one more application of our first three rules.

$$D_x(8 + 2x^{-1} + 2x^{-5} + 2x^{5/3}) = D_x 8 + D_x 2x^{-1} + D_x 2x^{-5} + D_x 2x^{5/3}$$
(Rule D3)

$$= 0 + 2 D_x x^{-1} + 2 D_x x^{-5} + 2 D_x x^{5/3}$$
(Rule D2)

$$= 2(-1)x^{-2} + 2(-5)x^{-6} + (2)(\tfrac{5}{3})x^{2/3}$$
(Rule D1)

$$= -2x^{-2} - 10x^{-6} + \tfrac{10}{3} x^{2/3}$$

Let us again consider the derivative of the function $y = f(x)$ as the instantaneous rate of change of $f(x)$ with respect to x. As we have seen before (in Sec. 1-4) Galileo discovered that the relationship between distance d and time t of a free falling object is $d(t) = -16t^2$, where $d(t)$ is measured in feet and t in seconds. A negative sign appears in the expression because a positively directed distance is measured upward. Our object is falling and hence is moving in the negative direction. The average velocity from time t_0 to t_1 is just

$$\frac{d(t_1) - d(t_0)}{t_1 - t_0}$$

which is our difference quotient. The average velocity is the average rate of change of $d(t)$ with respect to t. To find the instantaneous velocity at $t = t_0$, i.e., how fast the object is falling at $t = t_0$, we let t_1 approach t_0 and find the limit of the difference quotient at $t = t_0$:

$$\lim_{t \to t_0} \frac{d(t) - d(t_0)}{t - t_0}$$

But this is simply the derivative $d'(t_0)$ at t_0, which we know is equal to $-32 t_0$. Hence we say that the instantaneous velocity of a free falling object is $v(t) = d'(t) = -32t$. Similarly, the instantaneous acceleration is given by $a(t) = v'(t) = -32$.

EXAMPLE 3-7

Suppose that an object is thrown upward and the position of the object above ground is given by $d(t) = -16t^2 + 64t$. Then the velocity is given by $v(t) = d'(t) = -32t + 64$, and the acceleration is $a(t) = v'(t) = -32$. The object will reach its maximum height when $v(t) = 0$. (Why?) Letting $v(t) = 0$ yields $-32t + 64 = 0$, which implies that $t = 2$. Hence the object travelled $d(2)$, or 64 units, upward.

We now present two more differentiation formulas. Their proofs are presented in Ch. 7.

Rule D4 $D_x e^x = e^x$

Rule D5 $D_x \ln x = \dfrac{1}{x}$

EXAMPLE 3-8

$$D_x(3e^x + 4\ln x) = 3D_x e^x + 4D_x \ln x = 3e^x + 4\left(\dfrac{1}{x}\right)$$

EXAMPLE 3-9

In the study of medicine the reaction of the body to a dose of a drug depends upon both the amount of drug administered and the concentration of the drug already in the body, or the so-called initial concentration. For our purposes, we may assume that the drug concentration in the body at the time we administer more of the drug is insignificant. If we let R be the strength of the reaction, we measure R in units—the type of units depending upon what test is required. That is, if R measures the change in blood pressure, the units are millimeters of mercury; if R is the change in temperature, the units are degrees. If x is the amount of the drug, then R is a function of x, $R = R(x)$.

The *sensitivity* of a dose of drug is described as the change in the reaction $R(x)$, given a "small" change in the size of the dose x. Thus the term *sensitivity* measures the rate of change of $R(x)$ with respect to x. Since the derivative of a function measures the rate of change of a function, the *sensitivity* of a dose of drug x_0 is defined to be the derivative $R'(x_0)$.

Medical authorities generally wish to administer that dose of drug for which the sensitivity of the body is greatest. The problem therefore resolves to finding that dose such that $R'(x)$ is a maximum.

If M is the largest dose of a drug which can safely be administered, it is often assumed that the reaction $R(x)$ is given by the formula

$$R(x) = \dfrac{x^2}{6}(3M - 2x) = \dfrac{3Mx^2}{6} - \dfrac{2x^3}{6} = \dfrac{Mx^2}{2} - \dfrac{x^3}{3}$$

Thus
$$R'(x) = \frac{2Mx}{2} - \frac{3x^2}{3} = Mx - x^2$$

In Chap. 4 we provide the necessary machinery to determine from this information that $R'(x)$ is a maximum when $x = M/2$.

EXAMPLE 3-10

In the study of optics one considers the subjective brightness which the viewer senses from a source. If x is a measure of the actual brightness, then the subjective brightness, S, is a function of x. Such a function can often be described by a power function

$$S(x) = kx^n$$

where k and n are constants depending upon the light intensity to which the participants in the experiment were subjected. The constant k is greater for people who were in relative darkness and n is greater for people who were in light previous to the experiment. An important problem to be considered by such an experiment is that which considers the ranges of sensitivity to changes in the strength of the light, which is greater for those who were previously in darkness and those who were in brighter light. Hence one investigates $S'(x) = nkx^{n-1}$. Suppose that for those in darkness before the experiment the subjective brightness $S_1(x)$ is given by

$$S_1(x) = 0.001x^{1/4}$$

and for those in light

$$S_2(x) = 0.002x^{1/2}$$

Hence

$$S_1'(x) = (0.001)\tfrac{1}{4}x^{-3/4} = (0.00025)x^{-3/4}$$

and

$$S_2'(x) = 0.001x^{-1/2}$$

To solve for the value of x where both types of subjects have equal sensitivity we get

$$0.001x^{-1/2} = 0.00025x^{-3/4}$$

and so $x = (0.25)^4$. If x is smaller than this value, people who were in the dark have greater sensitivity, and if x is greater, the opposite is true.

PROOF OF RULE D2 Let $g(x) = af(x)$ for some real number a. We need to prove that $g'(x) = D_x g(x) = D_x af(x) = a D_x f(x) = af'(x)$. We calculate

$$g'(x) = \lim_{h \to 0} \frac{g(x+h) - g(x)}{x}$$

$$= \lim_{h \to 0} \frac{af(x+h) - af(x)}{h}$$

$$= \lim_{h \to 0} a\left[\frac{f(x+h) - f(x)}{h}\right]$$

$$= a \lim_{h \to 0} \frac{f(x+h) - f(x)}{h}$$

$$= af'(x)$$

PROOF OF RULE D3 Let $k(x) = f(x) + g(x)$. We need to show that $k'(x) = f'(x) + g'(x)$. We calculate

$$k'(x) = \lim_{h \to 0} \frac{k(x+h) - k(x)}{h}$$

$$= \lim_{h \to 0} \frac{f(x+h) + g(x+h) - f(x) - g(x)}{h}$$

$$= \lim_{h \to 0}\left[\frac{f(x+h) - f(x)}{h} + \frac{g(x+h) - g(x)}{h}\right]$$

$$= \lim_{h \to 0}\frac{f(x+h) - f(x)}{h} + \lim_{h \to 0}\frac{g(x+h) - g(x)}{h}$$

$$= f'(x) + g'(x)$$

EXERCISE 3-2

1. Find $f'(x)$ for the given functions.
   - (a) $f(x) = 2$
   - (b) $f(x) = 5x - 1$
   - (c) $f(x) = 8x^2 - 4x + 1$
   - (d) $f(x) = 6x^3 - 5x^2 + 5$
   - (e) $f(x) = 9x^4 - \frac{3}{2}x^3 + 5x^2 - x + 3$
   - (f) $f(x) = 2x^5 + \frac{6}{7}x^4 - x^2 + 5$
   - (g) $f(x) = x^{-2} + 3x^{-4}$
   - (h) $f(x) = x^{3/2} + 3x^{1/4} + 5x^{-1/8}$
   - (i) $f(x) = x^2 - x^{-1/2}$
   - (j) $f(x) = x^2 - x - x^{5/8}$
   - (k) $f(x) = x + e^x$
   - (l) $f(x) = 5e^x$
   - (m) $f(x) = x^2 + \ln x$
   - (n) $f(x) = 3x^2 + 5\ln x$
   - (o) $f(x) = e^x - \ln x$
   - (p) $f(x) = 3e^x - 8\ln x$

2. An object is thrown upward to the position given by $d(t) = -16t^2 + 32t + 6$. Find the velocity $v(t)$ and acceleration $a(t)$. How far up will the object travel?

3. Find the equations of the tangent lines to the functions in Problem 1 at the points $x = 1$ and $x = 2$.

4. The medical authorities of a community have determined that the proportion $p(t)$ of the people who have been infected by a communicable disease prior to time t is $p(t) = \frac{1}{500}(15t^2 - t^3)$. What is the instantaneous rate of increase of $p(t)$ with respect to t when $t = 4$? $t = 9$?

5. A firm has determined that the cost $C(x)$ of producing x units of goods is $C(x) = 20 + x^{2/3} + x$. Find the marginal cost when $x = 9$.

6. The revenue a company takes in when x units are produced is given by $R(x) = x^3 - \frac{9}{2}x^2 - 30x$. Find the marginal revenue when $x = 3, 4, 5, 6$.

7. Suppose a firm determines that $C(x) = 10 + x - 0.1x^2$ and $R(x) = x^3 - \frac{15}{2}x^2 - 6x$. Then the profit the firm realizes from selling x units is $P(x) = R(x) - C(x)$. Find $P(x)$ and $P'(x)$.

Recall that

$$(a+h)^n = a^n + na^{n-1}h + \frac{n(n-1)}{2}a^{n-2}h^2 + \cdots + nah^{n-1} + h^n$$

for any real numbers a and h and for any positive integer n. Use this binomial expansion in the following problem.

8. Let $f(x) = x^n$ for any positive integer n. Prove that

$$f'(a) = \lim_{h \to 0} \frac{f(a+h) - f(a)}{h} = na^{n-1}$$

using the binomial expansion.

9. Using Rule D3 twice show that
$$D_x[f(x) + g(x) + h(x)] = D_x f(x) + D_x g(x) + D_x h(x).$$

10. Using the definition of the derivative, prove that Rule D2 is true for any real number a.

11. Using the definition of the derivative and Property 1, which says the limit of a sum is the sum of the limits (Sec. 2-1), prove Rule D3.

3-3 PRODUCT AND QUOTIENT RULES

The formulas developed up to this point allow us to find the derivatives of polynomial functions and functions that are sums of functions of the form ax^n where n is not necessarily a natural number. Often we deal with functions whose derivatives cannot be obtained from these rules, for instance,

$$f(x) = \frac{x}{x+1} \quad \text{or} \quad g(x) = \frac{x^2 - 2x}{x^3 - 4x^2} \quad \text{or} \quad h(x) = x^2 e^x$$

We shall first discuss the derivative of the product of two functions. One might expect that the derivative of the product of two functions is the product of their derivatives. A quick inspection of simple functions shows that this is not true. For example, if we let $f(x) = g(x)k(x)$ where $g(x) = 2$ and $k(x) = x$, so that $f(x) = 2x$, then $g'(x) = 0$ and $k'(x) = 1$, and so $g'(x)k'(x) = 0$, but $f'(x) = 2$. Thus the derivative of the product, $f'(x)$ is not equal to the product of the derivatives, $g'(x)k'(x)$. The expression for the derivative of a product is a little more complicated:

☆ **Rule D6** *If $f(x) = g(x)k(x)$, then $f'(x) = g'(x)k(x) + g(x)k'(x)$.*

Rule D6 is read: "The derivative of a product is the derivative of the first times the second plus the first times the derivative of the second."

EXAMPLE 3-11

Let $f(x) = (x^2+1)^2$, that is, $f(x) = (x^2+1)(x^2+1)$. Then
$$f(x) = g(x) \cdot k(x)$$
where $g(x) = k(x) = x^2+1$. So $g'(x) = k'(x) = 2x$ and
$$f'(x) = g'(x)k(x) + g(x)k'(x)$$
$$= (2x)(x^2+1) + (x^2+1)(2x)$$
$$= 4x(x^2+1)$$

EXAMPLE 3-12

Let $f(x) = (x^2+1)^3$. You might expect to be able simply to employ the power formula and get $f'(x) = 3(x^2+1)^2$, but this is not quite correct, as we shall see. Let $g(x) = (x^2+1)^2$ and $k(x) = x^2+1$. Then
$$g'(x) = 4x(x^2+1) \quad \text{and} \quad k'(x) = 2x$$
Thus
$$f'(x) = 4x(x^2+1) \cdot (x^2+1) + (x^2+1)^2(2x)$$
$$= 6x(x^2+1)^2$$

Note that both these functions can be expanded,
$$(x^2+1)^2 = x^4 + 2x^2 + 1$$
and
$$(x^2+1)^3 = x^6 + 3x^4 + 3x^2 + 1$$
and their derivatives obtained from these expressions. Thus
$$D_x(x^4 + 2x^2 + 1) = 4x^3 + 4x = 4x(x^2+1)$$
and
$$D_x(x^6 + 3x^4 + 3x^2 + 1) = 6x^5 + 12x^3 + 6x$$
$$= 6x(x^4 + 2x^2 + 1)$$
$$= 6x(x^2+1)^2$$

In the following example we encounter functions for which such expansions are impossible and we have to use the product formula.

EXAMPLE 3-13

Using Rules D4, D5, and D6 we have

$$D_x x^2 e^x = x^2 D_x e^x + e^x D x^2 \qquad \text{(Rule D6)}$$
$$= x^2 e^x + e^x 2x$$
$$= e^x(x^2 + 2x)$$

and

$$D_x e^x \ln x = e^x D_x \ln x + \ln x\, D_x e^x \qquad \text{(Rule D6)}$$
$$= e^x \frac{1}{x} + \ln x (e^x) \qquad \text{(Rules D5 and D4)}$$
$$= e^x \frac{1}{x} + e^x \ln x$$
$$= e^x \left(\frac{1}{x} + \ln x \right)$$

Rule D6 is not as readily predictable from the limit theorems as Rules D2 and D3, which govern the derivative of a constant times a function and the derivative of a sum. Likewise, its proof is not as straightforward. Since the inquisitive reader desires validation of Rule D6, as well as of Rule D7, their proofs are included.

PROOF OF RULE D6 If $f(x) = g(x)k(x)$, then

$$f'(x) = \lim_{h \to 0} \frac{f(a+h) - f(a)}{h}$$
$$= \lim_{h \to 0} \frac{g(a+h)k(a+h) - g(a)k(a)}{h}$$

Now we add and subtract the quantity $g(a+h)k(a)$.

$$f'(x) = \lim_{h \to 0} \frac{1}{h}[g(a+h)k(a+h) - g(a+h)k(a) + g(a+h)k(a) - g(a)k(a)]$$
$$= \lim_{h \to 0} \left[\frac{g(a+h)k(a+h) - g(a+h)k(a)}{h} \right]$$
$$\qquad + \lim_{h \to 0} \left[\frac{g(a+h)k(a) - g(a)k(a)}{h} \right]$$

CHAP. 3   The Derivative

$$= \lim_{h \to 0} g(a+h) \left[\frac{k(a+h) - k(a)}{h} \right] + \lim_{h \to 0} \left[\frac{g(a+h) - g(a)}{h} \right] k(a)$$

$$= \lim_{h \to 0} g(a+h) \cdot \lim_{h \to 0} \frac{k(a+h) - k(a)}{h} + \lim_{h \to 0} \frac{g(a+h) - g(a)}{h} \cdot \lim_{h \to 0} k(a)$$

$$= g(a) k'(a) + g'(a) k(a)$$

We now give the quotient formula, which is the rule governing the derivative of the quotient of two functions.

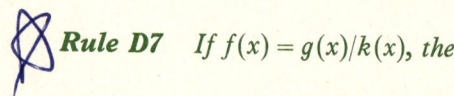**Rule D7**   If $f(x) = g(x)/k(x)$, then

$$f'(x) = \frac{g'(x) k(x) - g(x) k'(x)}{[k(x)]^2}$$

PROOF   If $f(x) = g(x)/k(x)$, then

$$f'(a) = \lim_{h \to 0} \frac{1}{h} \left[\frac{g(a+h)}{k(a+h)} - \frac{g(a)}{k(a)} \right]$$

$$= \lim_{h \to 0} \frac{1}{h} \left[\frac{g(a+h) k(a) - k(a+h) g(a)}{k(a+h) k(a)} \right]$$

$$= \lim_{h \to 0} \frac{1}{h} \left[\frac{g(a+h) k(a) - g(a) k(a) + g(a) k(a) - k(a+h) g(a)}{k(a+h) k(a)} \right]$$

$$= \lim_{h \to 0} \left[\frac{k(a)}{k(a+h) k(a)} \right] \left[\frac{g(a+h) - g(a)}{h} \right]$$

$$- \lim_{h \to 0} \left[\frac{g(a)}{k(a+h) k(a)} \right] \left[\frac{k(a+h) - k(a)}{h} \right]$$

$$= \lim_{h \to 0} \frac{k(a)}{k(a+h) k(a)} \cdot \lim_{h \to 0} \frac{g(a+h) - g(a)}{h}$$

$$- \lim_{h \to 0} \frac{g(a)}{k(a+h) k(a)} \cdot \lim_{h \to 0} \frac{k(a+h) - k(a)}{h}$$

$$= \frac{k(a) g'(a) - g(a) k'(a)}{[k(a)]^2}$$

EXAMPLE 3-14

Let $f(x) = x^2/(x+1)$. Then

$$g(x) = x^2 \quad \text{and} \quad k(x) = x + 1$$

so that

$$g'(x) = 2x \quad \text{and} \quad k'(x) = 1$$

Hence

$$f'(x) = \frac{2x(x+1) - x^2(1)}{(x+1)^2} = \frac{x^2 + 2x}{(x+1)^2}$$

EXAMPLE 3-15

Let $f(x) = (x^2 + 1)/(x^3 + x - 1)$. Then

$$g(x) = x^2 + 1 \quad \text{and} \quad k(x) = x^3 + x - 1$$

so that

$$g'(x) = 2x \quad \text{and} \quad k'(x) = 3x^2 + 1$$

Hence

$$f'(x) = \frac{(2x)(x^3 + x - 1) - (x^2 + 1)(3x^2 + 1)}{(x^3 + x - 1)^2} = -\frac{x^4 + 2x^2 + 2x + 1}{(x^3 + x - 1)^2}$$

EXAMPLE 3-16

Let $f(x) = (x^{1/2} + 1)/(x^2 + x^{3/4})$. Then

$$f'(x) = \frac{\frac{1}{2}x^{-1/2}(x^2 + x^{3/4}) - (x^{1/2} + 1)(2x + x^{-1/4})}{(x^2 + x^{3/4})^2}$$

EXAMPLE 3-17

Using Rules D4 and D5 we have

$$D_x\left[\frac{e^x}{x^2 + x}\right] = \frac{D_x e^x (x^2 + x) - e^x D_x(x^2 + x)}{(x^2 + x)^2} \quad \text{(Rule D7)}$$

$$= \frac{(x^2 + x)e^x - e^x(2x + 1)}{(x^2 + x)^2} \quad \text{(Rules D4 and D1)}$$

$$= \frac{e^x(x^2 - x - 1)}{(x^2 + x)^2}$$

and

$$D_x\left[\frac{x + \ln x}{e^x}\right] = \frac{[D_x(x + \ln x)]e^x - (x + \ln x)D_x e^x}{(e^x)^2} \quad \text{(Rule D7)}$$

$$= \frac{(1 + 1/x)e^x - (x + \ln x)e^x}{e^{2x}} \quad \text{(Rules D5 and D4)}$$

$$= \frac{e^x(1 + 1/x - x - \ln x)}{e^{2x}}$$

$$= \frac{1 - x + 1/x - \ln x}{e^x}$$

EXERCISE 3-3

1. Find $D_x f(x)$ for the given $f(x)$.

   (a) $f(x) = (x+1)^3$

   (b) $f(x) = (2x+1)^3$

   (c) $f(x) = (5x+3)^3$

   (d) $f(x) = (5x+3)^4$

   (e) $f(x) = (x^2+1)^2$

   (f) $f(x) = (x^2+1) \ln x$

   (g) $f(x) = (x^2+1) e^x$

   (h) $f(x) = (x^3+1)(e^x + \ln x)$

   (i) $f(x) = (x^3+2x-1)^2$

   (j) $f(x) = (2x^3 - 5x + 1) e^x$

   (k) $f(x) = (x^4 - 3x^{1/2})^2$

   (l) $f(x) = \dfrac{1}{x+1}$

   (m) $f(x) = \dfrac{e^x}{2x-1}$

   (n) $f(x) = \dfrac{x}{x+1}$

   (o) $f(x) = \dfrac{2x^3}{3x+2}$

   (p) $f(x) = \dfrac{\ln x}{x^2-1}$

   (q) $f(x) = \dfrac{e^x + \ln x}{2x^2 + x}$

   (r) $f(x) = \dfrac{2x-1}{x^2+x}$

2. Find the equation of the tangent line to the given curve at the given point.

   (a) $f(x) = (3x-2)^2 e^x$, $a = 1$

   (b) $f(x) = (2x^2+1)^2$, $a = 2$

   (c) $f(x) = (x^{1/2} + x)^2$, $a = 4$

3. The cost function $C(x) = x + (x^2+1)^2$ and the revenue function $R(x) = (x^{1/2} + x)^3$ for a certain firm are given. Find the marginal cost function and the marginal revenue function.

4. A stream is being polluted by waste such that at time t the quantity of waste in the stream is $Q(t) = (t^{1/3} + 1)^2$. Find how fast the amount of waste in the stream is increasing and the instantaneous rate of increase of $Q(t)$ with respect to t when $t = 8$.

3-4 COMPOSITION AND THE CHAIN RULE

The rules for differentiation so far developed do not allow the differentiation of more complicated functions such as

$$f(x) = (x^3 + x)^{1/2} \qquad g(x) = e^{x^2} \qquad h(x) = \ln(x^2 + 1)$$

In this section we shall present the indispensable tool for calculating derivatives called the *Chain Rule*. Our first task is to express more complicated functions, such as those above, in terms of simpler ones. This process is called *composition* of functions and was first mentioned in Sec. 1-6.

EXAMPLE 3-18

Let $f(x) = 2x+3$ and $g(x) = 1-x^2$. Then $f(g(x)) = f(1-x^2) = 2(1-x^2) + 3 = 5-2x^2$ and $g(f(x)) = g(2x+3) = 1-(2x+3)^2 = 1-4x^2-12x-9 = -4x^2-12x-8$.

EXAMPLE 3-19

Let $f(x) = e^x$, $g(x) = \ln x$, and $h(x) = x^2+1$. Then

$$(f \circ g)(x) = f(g(x)) = f(\ln x) = e^{\ln x} = x$$

Also,

$$h(g(x)) = h(\ln x) = (\ln x)^2 + 1$$

$$h(f(x)) = h(e^x) = (e^x)^2 + 1 = e^{2x} + 1$$

$$f(h(x)) = f(x^2+1) = e^{x^2+1}$$

$$g(h(x)) = g(x^2+1) = \ln(x^2+1)$$

EXAMPLE 3-20

The production of materials in a factory usually consists of many diverse operations. A company which makes beaded necklaces has two basic operations: (1) It manufactures beads and (2) it strings the beads into necklaces. The revenue from the sale of the necklaces depends upon the number of necklaces produced, which in turn depends upon the number of beads produced. Suppose the company determines that the revenue $R(n)$ realized from the production of n necklaces is given by $R(n) = n^{1/2} - 0.2n$. The company also determines that the number of necklaces $n(b)$ produced from a production of b beads is given by $n(b) = 0.09b$. Hence the revenue realized from the production of b beads is

$$R(b) = (0.09b)^{1/2} - 0.2(0.09b) = 0.3b^{1/2} - 0.018b$$

We now present an argument meant to convince you of the validity of the Chain Rule. This discussion is not a rigorous proof of the Chain Rule since there are a few difficult points which need to be clarified. For a rigorous proof the reader is invited to see a more advanced calculus text.

Let $h(x) = f(g(x))$. Then

$$h'(a) = \lim_{h \to 0} \frac{f(g(a+h)) - f(g(a))}{h}$$

If we first multiply numerator and denominator by $g(a+h) - g(a)$,

$$h'(a) = \lim_{h \to 0} \frac{f(g(a+h)) - f(g(a))}{h} \cdot \frac{g(a+h) - g(a)}{g(a+h) - g(a)}$$

which may be written also as

$$h'(a) = \lim_{h \to 0} \frac{f(g(a+h)) - f(g(a))}{g(a+h) - g(a)} \cdot \frac{g(a+h) - g(a)}{h}$$

then make the substitutions y for $g(a+h)$ and b for $g(a)$, we have

$$h'(a) = \lim_{h \to 0} \frac{f(y) - f(b)}{y - b} \cdot \lim_{h \to 0} \frac{g(a+h) - g(a)}{h}$$

Now let's consider the left-hand factor of the product first. Remembering that $y = g(a+h)$ and $b = g(a)$, as $h \to 0$, so does $g(a+h) \to g(a)$, which shows that $y \to b$. Hence

$$\lim_{h \to 0} \frac{f(y) - f(b)}{y - b} = \lim_{y \to b} \frac{f(y) - f(b)}{y - b} = f'(b)$$

Also we have

$$\lim_{h \to 0} \frac{g(a+h) - g(a)}{h} = g'(a)$$

so that

$$h'(a) = f'(b) g'(a) = f'(g(a)) g'(a)$$

We are therefore led to conclude that

Theorem 3-2   *Chain Rule*
If $h(x) = f(g(x))$, then $h'(x) = f'(g(x)) g'(x)$.

EXAMPLE 3-21

Consider the function $h(x) = e^{x^2}$. We can set $f(x) = e^x$ and $g(x) = x^2$ so that $h(x) = f(g(x))$. Then $f'(x) = e^x$ and $f'(g(x)) = e^{x^2}$ while $g'(x) = 2x$. From the Chain Rule we then conclude that $h'(x) = 2xe^{x^2}$.

We will use the Chain Rule in three basic forms given in the following three rules.

Rule D8 $D_x f(x)^n = n f(x)^{n-1} D_x f(x)$

Rule D9 $D_x e^{f(x)} = e^{f(x)} D_x f(x)$

Rule D10 $D_x \ln f(x) = \dfrac{1}{f(x)} D_x f(x)$

Rules D8, D9, and D10 can be derived directly from the Chain Rule. We will give the proof of Rule D8 and leave the remaining two for Exercise 3-4.

PROOF OF RULE D8   Let $h(x) = f(x)^n$. If we let $g(x) = x^n$, then $h(x) = g(f(x))$. Applying the Chain Rule with

$$g'(x) = nx^{n-1} \quad \text{and} \quad g'(f(x)) = nf(x)^{n-1}$$

we have
$$h'(x) = g'(f(x))f'(x) = nf(x)^{n-1}f'(x)$$

EXAMPLE 3-22

Consider the function $h(x) = (x^2+1)^{1/2}$. If we let $f(x) = x^2+1$ and apply Rule D8, we have
$$D_x h(x) = \tfrac{1}{2}(x^2+1)^{-1/2} D_x(x^2+1)$$
$$= \tfrac{1}{2}(x^2+1)^{-1/2}(2x)$$
$$= x(x^2+1)^{-1/2}$$

EXAMPLE 3-23

Consider the function $h(x) = e^{3x^2}$. If we let $f(x) = 3x^2$ and apply Rule D9, we have
$$D_x h(x) = e^{3x^2} D_x 3x^2$$
$$= 6xe^{3x^2}$$

EXAMPLE 3-24

Consider the function $h(x) = \ln(x^2+1)$. If we let $f(x) = x^2+1$ and apply Rule D10 we have
$$D_x h(x) = \frac{1}{x^2+1} D_x(x^2+1)$$
$$= \frac{2x}{x^2+1}$$

EXAMPLE 3-25

Often one has to combine the Chain Rule with the product and quotient rules to calculate a derivative. Consider the following illustration:

$$\begin{aligned}
D_x x^2 e^{3x} &= x^2 D_x e^{3x} + e^{3x} D_x x^2 & \text{(Rule D6)} \\
&= x^2(e^{3x} D_x 3x) + e^{3x}(2x) & \text{(Rule D9)} \\
&= 3x^2 e^{3x} + 2xe^{3x} \\
&= e^{3x}(3x^2 + 2x)
\end{aligned}$$

EXAMPLE 3-26

The quality control of a company's product is usually of prime importance. The quality level of a product often depends upon production level. If too little production takes place, morale and concentration of the working force may sharply reduce quality, while too large a production rate may cause faulty control and hence reduce quality. The usual assembly line is such an example. Often quality $Q(x)$ can be measured in terms of output x, where $Q(x)$ is measured on a scale from 0 to 100 (100 is a "perfect" product) and x is measured in 100 units per day. Suppose a company determines that

$$Q(x) = 90 - 0.4(x-10)^2 = 50 + 8x - 0.4x^2 \quad x \in (0, 18)$$

whose graph is given in Fig. 3-6.

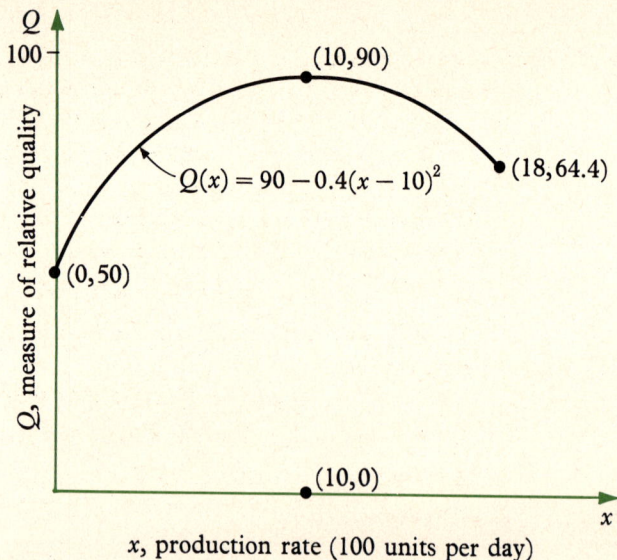

FIGURE 3-6

In other words, $Q(x)$ is the (relative) quality of the product produced if the production rate is $100x$ units per day. Production, in turn, depends upon time of day, t, measured in hours, and given by

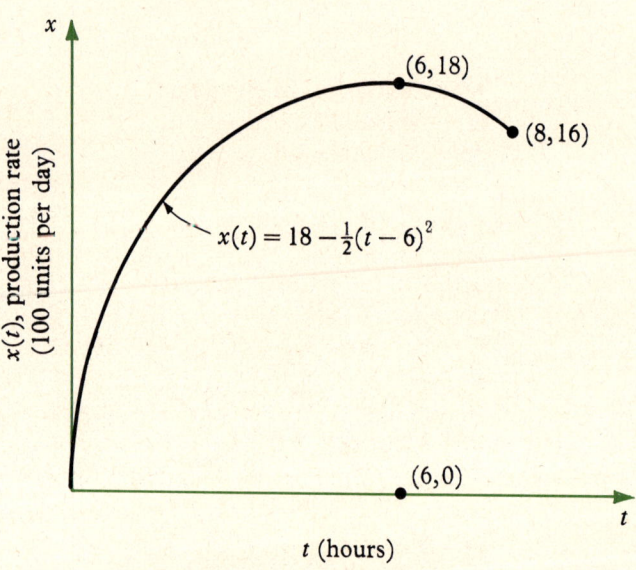

FIGURE 3-7

$$x(t) = 18 - \tfrac{1}{2}(t-6)^2 = 6t - \tfrac{1}{2}t^2 \qquad t \in (0, 8)$$

Thus, at time of day t, the factory is producing goods at a rate such that if that particular rate were held constant all day, the factory would produce $100x(t)$ units. The graph of $x(t)$ is given in Fig. 3-7.

We see that early in the day the production rate is slow. It increases to a maximum at $t = 6$ (6 hours into the working day) and then slows again towards the completion of the day, at $t = 8$. The rate of change of the quality $Q(x)$ with respect to x is

$$Q'(x) = -0.8(x-10) = 8 - 0.8x$$

Since Q is a function of x and x is a function of t, we see that Q is a function of t. In fact,

$$\begin{aligned} Q(t) = Q(x(t)) &= Q(6t - \tfrac{1}{2}t^2) \\ &= 50 + 8(6t - \tfrac{1}{2}t^2) - 0.4(6t - \tfrac{1}{2}t^2)^2 \end{aligned}$$

To find the rate of change of Q with respect to t, we could differentiate the above formula directly or use the Chain Rule. Using the Chain Rule, we have

$$\begin{aligned} Q'(t) &= Q'(x(t)) x'(t) \\ &= [8 - 0.8x(t)] x'(t) \\ &= [8 - 0.8(6t - \tfrac{1}{2}t^2)](6 - t) \\ &= (8 - 4.8t + 0.4t^2)(6 - t) \\ &= 48 - 36.8t + 7.2t^2 - 0.4t^3 \end{aligned}$$

EXERCISE 3-4

1. Find $D_x f(x)$ for the given $f(x)$.
   - (a) $f(x) = (3x+1)^2$
   - (b) $f(x) = (3x+1)^3$
   - (c) $f(x) = (3x+1)^{10}$
   - (d) $f(x) = (3x+1)^{-1}$
   - (e) $f(x) = (3x+1)^{-10}$
   - (f) $f(x) = (2x^2+1)^4$
   - (g) $f(x) = (8x^2-x)^{-1}$
   - (h) $f(x) = (x^{1/2}+x)^{-2}$
   - (i) $f(x) = \sqrt{2x+1}$
   - (j) $f(x) = \sqrt{x^2+2x+1}$
   - (k) $f(x) = (3x-5)^{1/3}$
   - (l) $f(x) = (x^2+1)^{2/5}$
   - (m) $f(x) = (x^{1/2}+2)^{-1/3}$
   - (n) $f(x) = \sqrt{2x-1} + \sqrt{5x+1}$
   - (o) $f(x) = \sqrt{3x-1} + (5x+1)^3$
   - (p) $f(x) = 2x + \sqrt{3x+1}$
   - (q) $f(x) = e^{2x^2}$
   - (r) $f(x) = e^{x^3}$
   - (s) $f(x) = e^{x^2+x}$
   - (t) $f(x) = \ln(x^3+x^2)$
   - (u) $f(x) = \ln(x+e^x)$

2. Find $D_x f(x)$ by using the product, quotient, and chain rules for the given $f(x)$.
   - (a) $f(x) = (x+1)(3x-1)^{-1}$
   - (b) $f(x) = (x^2+1)\sqrt{8x-1}$
   - (c) $f(x) = (x^3-x+1)\sqrt{x^2+x}$

(d) $f(x) = (3x-1)^{1/3}(8x-1)^{5/2}$
(e) $f(x) = (6x+x^{1/2})^{-2}(x^{1/3}+2)^{-3}$
(f) $f(x) = \dfrac{x^{1/2}+1}{(2x+1)^{2/5}}$
(g) $f(x) = (3x-1)(8x+1)\sqrt{6x+5}$
(h) $f(x) = e^{x^2}\ln(x^2+x)$
(i) $f(x) = \ln(x+e^{x^2})$
(j) $f(x) = \dfrac{x^2+e^{x^2}}{e^{x^3}+\ln x}$

3. Find the equation of the tangent lines to the given functions at the given points.
   (a) $f(x) = (3x+1)\sqrt{3x-2}$, $a = 1$
   (b) $f(x) = \dfrac{(2x^2-1)^{1/2}}{x^2+3x-2}$, $a = 1$

3–5 HIGHER DERIVATIVES

Consider the function $f(x) = x^3$. Then $f'(x) = 3x^2$. If we let $g(x) = 3x^2$, then $g'(x) = 6x$. One immediately sees that there is a close connection between the functions $h(x) = 6x$ and $f(x) = x^3$, that is, $h(x) = 6x$ is the derivative of the derivative of $f(x) = x^3$. Thus $(f'(x))' = 6x$. We simplify this notation by writing $f''(x) = 6x$ and we say the *second derivative* of $f(x)$ is $6x$.

EXAMPLE 3-27

If $f(x) = x^4 - 7x^3 - 3x + 1$, then $f'(x) = 4x^3 - 21x^2 - 3$ and hence the second derivative $f''(x)$ is given by

$$f''(x) = 12x^2 - 42x$$

EXAMPLE 3-28

If $g(x) = 12x^2 - 42x$, then $g'(x) = 24x - 42$ and the second derivative $g''(x)$ is given by

$$g''(x) = 24$$

If you study the above two examples carefully, you'll note that $g(x) = 12x^2 - 42x = f''(x)$. Hence $g'(x) = 24x - 42$ is the *third* derivative of $f(x)$, that is, the derivative of the derivative of the derivative of $f(x)$. Also $g''(x) = 24$ is the fourth derivative of $f(x)$. We write $f'''(x) = 24x - 24$ and $f^{(4)}(x) = 24$. The latter notation is used to avoid confusing powers of $f(x)$ with the higher derivatives of $f(x)$ as well as avoiding a confusing amount of primes.

There are various notations used for higher derivatives. If $y = f(x)$, the first, second, third and nth derivatives can be expressed as follows:

First derivative: y', f', $D_x y$, $\dfrac{dy}{dx}$, $\dfrac{df}{dx}$

Second derivative: y'', f'', $D_x^2 y$, $\dfrac{d^2 y}{dx^2}$, $\dfrac{d^2 f}{dx^2}$

Third derivative: y''', f''', $D_x^3 y$, $\dfrac{d^3 y}{dx^3}$, $\dfrac{d^3 f}{dx^3}$

nth derivative: $y^{(n)}$, $f^{(n)}$, $D_x^n y$, $\dfrac{d^n y}{dx^n}$, $\dfrac{d^n f}{dx^n}$

It often takes some ingenuity to find an expression for the general nth derivative of a function. Sometimes such an expression does not even exist.

EXAMPLE 3-29

To find the nth derivative of $f(x) = x^{-1}$, we first find the first four derivatives and then determine the pattern so formed.

$$f'(x) = -x^{-2}$$
$$f''(x) = (-2)(-1)x^{-3} = 2x^{-3}$$
$$f'''(x) = (-3)(-2)(-1)x^{-4} = -6x^{-4}$$
$$f^{(4)}(x) = (-4)(-3)(-2)(-1)x^{-5} = 24x^{-5}$$

In the expression for $f^{(n)}(x)$, we see there are three distinct entities. There is a power of x, a constant, and, something a little less obvious, the derivatives alternate in sign. The power of x is negative and its absolute value is one greater than the number of the derivative, that is, $x^{-(n+1)}$. The constant is $1 \cdot 2 \cdot 3 \ldots \cdot n = n!$. To express the fact that the signs alternate, we write $(-1)^n$. That is, if n is odd, $(-1)^n$ is negative one, if n is even, $(-1)^n$ is plus one. Hence

$$f^{(n)}(x) = (-1)^n n! \, x^{-(n+1)}$$

Note that this formula agrees with our first four derivatives. Substituting $n = 5$ into the formula yields

$$f^{(5)}(x) = (-1)^5 (5!) x^{-(5+1)}$$
$$= -120 x^{-6}$$

which is the same expression that one gets by differentiating $f^{(4)}(x)$.

We shall see in the next chapter that the geometrical significance of the second derivative is of fundamental importance. Intuitively, the first derivative measures the slope or steepness of a curve, and in addition, it determines if the curve is going up or down as x increases. The second derivative determines whether the curve is bending up or down.

EXERCISE 3-5

1. Find the second derivative.

    (a) $y = x^4 + 3x^2 + x$
    (b) $y = x^5 - 2x^3$
    (c) $y = 3x^2 - 4x^{1/2}$
    (d) $y = 5x^2 + 4x^{1/2} - x$
    (e) $y = \frac{1}{2}x^4 - x^{1/5}$
    (f) $y = x^{-1} + 2x^{-3}$
    (g) $y = x^{1/2} - 3x^{-2}$
    (h) $y = x^{2/3} - 5x^{-3}$
    (i) $y = (2x+1)^{-1}$
    (j) $y = (x^2+1)^{-3}$
    (k) $y = (x^{-1}+x)^{1/2}$
    (l) $y = (2x+1)/(x-1)$

2. Find the third derivative.

    (a) $y = x^5 - 8x^2$
    (b) $y = x^4 - 2x^{1/2}$
    (c) $y = x^5 + x^{-2}$
    (d) $y = (2x+1)^{-1}$
    (e) $y = (x^2+1)^{-3}$
    (f) $y = (1+x^{-1})^{1/2}$

3. Find the fourth derivative.

    (a) $y = x^4 + x$
    (b) $y = x^6 - x$
    (c) $y = x^2 + 3$
    (d) $y = x^{-2}$
    (e) $y = 3x^{1/2}$
    (f) $y = (x+1)^{-1}$

4. Find the nth derivative.

    (a) $y = x^2$
    (b) $y = x^{-1/2}$
    (c) $y = (x+1)^{-1}$
    (d) $y = (2x+1)^{-1}$
    (e) $y = e^x$
    (f) $y = e^{2x}$

5. If $f(x) = g(x) \cdot h(x)$, find $f''(x)$.

6. Show that the function $y = x + \sqrt{1+x^2}$ satisfies the equation

$$(1+x^2)y'' + xy' = y$$

4

Applications of the Derivative

4-1 MAXIMA AND MINIMA

Very often when one studies the rate of change of one variable quantity with respect to another, the problem entails maximizing or minimizing the quantity. Familiar examples from various fields are abundant: an engineer seeks to minimize the stress in beams, a manufacturer wants to maximize profits, a doctor needs to know the maximum amount of a drug which can be administered safely, a physiologist studies the minimum rate of stimulus needed to achieve a given response from an organism. The derivative plays a fundamental role in such problems.

We need to first define a few concepts. To illustrate their geometrical significance, consider the graph in Fig. 4-1.

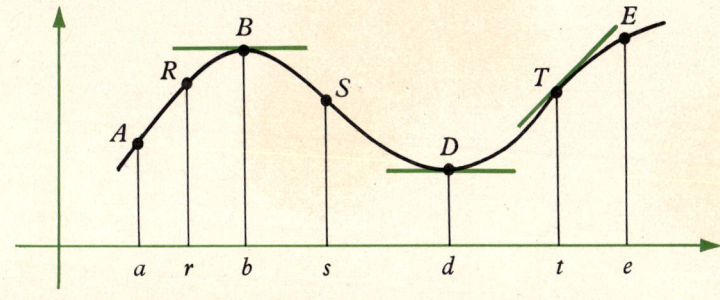

FIGURE 4-1

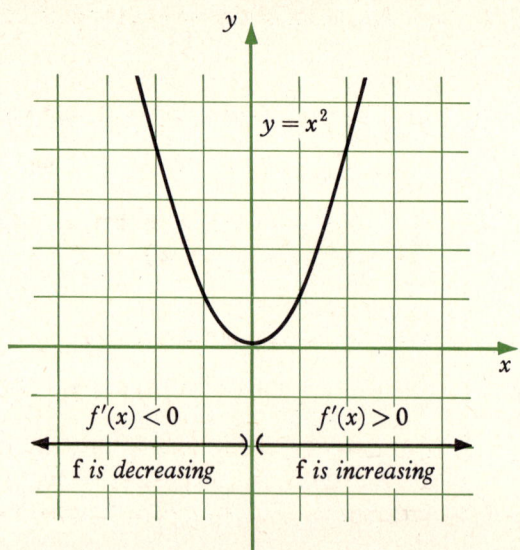

FIGURE 4-2

The curve in Fig. 4-1 is increasing, or rising, in the intervals (a, b), (d, e) and decreasing in the interval (b, d). The tangent line at the points B and D are horizontal; at R and T the slope of the tangent line is positive as the curve is increasing, and is negative at S where the curve is decreasing.

The first concept we wish to define is that of an increasing (decreasing) function. If you've ever perused the stock market averages, this idea is not new to you.

Definition A function $f(x)$ is said to be *increasing* [*decreasing*] at x_0 if for every x_1 and x_2 sufficiently close to x_0 with $x_1 < x_0 < x_2$, the function
$$f(x_1) < f(x_0) < f(x_2) \quad [f(x_1) > f(x_0) > f(x_2)]$$
We say that $f(x)$ is *increasing on the interval* (a, b) if $f(x)$ is increasing at every $x_0 \in (a, b)$. One similarly defines *decreasing on an interval* (a, b).

Try this simple test. Draw a line segment through the point $(a, f(a))$ with slope 1. Suppose this is the tangent line to the curve at a. What must the curve look like? One thing you can say for sure is that the curve must be increasing at a. This geometric test should help convince you of the validity of the following important rule.

Rule M-1 *If $f'(x) > 0$ for every x in some interval I, then $f(x)$ is increasing in I. Similarly, if $f'(x) < 0$ for every x in some interval I, then $f(x)$ is decreasing in I.*

The proof of this rule is given in Sec. 4-5.

EXAMPLE 4-1

Let $f(x) = x^2$. Then $f'(x) = 2x$, and thus

$$f'(x) > 0 \text{ if } x > 0 \quad \text{and} \quad f'(x) < 0 \text{ if } x < 0$$

By Rule M-1 we conclude that $f(x)$ is increasing in $(0, \infty)$ and decreasing in $(-\infty, 0)$. (See Fig. 4-2.)

EXAMPLE 4-2

Let $f(x) = x^2 - 4x + 3 = (x-1)(x-3)$. Then

$$f'(x) = 2x - 4 = 2(x-2)$$

and thus $f'(x) > 0$ when $x - 2 > 0$, i.e., in $(2, \infty)$. Similarly $f'(x) < 0$ when $x - 2 < 0$, i.e., in $(-\infty, 2)$. Hence $f(x)$ is increasing in $(2, \infty)$ and decreasing in $(-\infty, 2)$. What happens to $f'(x)$ and the graph of $f(x)$ when $x = 2$?

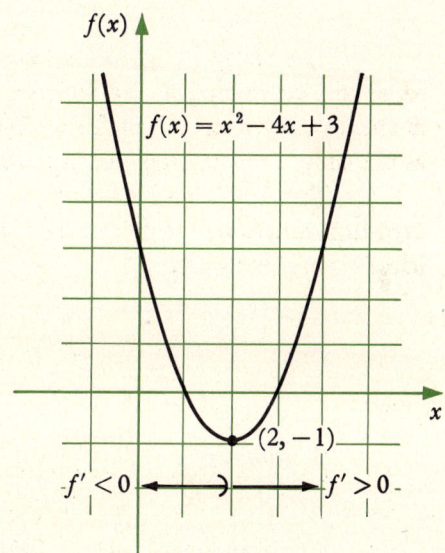

FIGURE 4-3

EXAMPLE 4-3

Let $f(x) = 2x^3 - 9x^2 + 12x$. Then

$$f'(x) = 6x^2 - 18x + 12 = 6(x^2 - 3x + 2) = 6(x-1)(x-2)$$

To determine where $f(x)$ is increasing we set $f'(x) > 0$. There are two cases:

(1) $$x - 1 > 0 \quad \text{and} \quad x - 2 > 0$$
$$x > 1 \quad \text{and} \quad x > 2$$

so $$x \in (2, \infty)$$

(2) $$x - 1 < 0 \quad \text{and} \quad x - 2 < 0$$
$$x < 1 \quad \text{and} \quad x < 2$$

so $$x \in (-\infty, 1)$$

Hence $f(x)$ is increasing in $(-\infty, 1)$ and $(2, \infty)$. By setting $f'(x) < 0$ we find that $f(x)$ is decreasing in $(1, 2)$. Given this information we can give an accurate sketch of the graph of this function by locating just a few points. Now $f(1) = 5$ and $f(2) = 4$. Hence the graph of $f(x)$ must be similar to that in Fig. 4-3. The methods of the next few sections will enable us to be more exact when sketching curves.

The roles played by the points at $x = 1$ and $x = 2$ are very important. Note that the functional values of points close to $x = 1$ are all less than $f(1) = 5$. Thus $f(x)$ has a relative maximum value of 5 at $x = 1$. We say "relative" because there do exist functional values which are greater than 5, for example $f(3) = 9$. Similarly $f(2) = 4$ is a relative minimum value.

Definition A function $y = f(x)$ is said to have a *relative maximum* [*relative minimum*] value at x_0 if $f(x_0)$ is greater than [smaller than] all values of $f(x)$ where x is sufficiently close to x_0, i.e., if there exists an interval (a, b) containing x_0 such that whenever $x \in (a, b)$, $x \neq x_0$, then

$$f(x) \leq f(x_0) \qquad [f(x) \geq f(x_0)]$$

A function $y = f(x)$ has an *absolute maximum* [*minimum*] value at x_0 if $f(x_0) \geq f(x) [f(x_0) \leq f(x)]$ for all x. Thus a relative maximum [minimum] is an absolute maximum [minimum] on some interval. For this reason we will henceforth refer to a relative maximum [minimum] as simply a maximum [minimum].

A value or a point which is a maximum or a minimum is sometimes called an *extremum* value or point.

Although Fig. 4-1, 4-2, 4-3, and 4-4 indicate that a maximum or minimum point on a curve occurs where the tangent is parallel to the x-axis, one can see from Fig. 4-5 that a function can have an extremum where the tangent does not exist, i.e., where the graph has a corner or an endpoint.

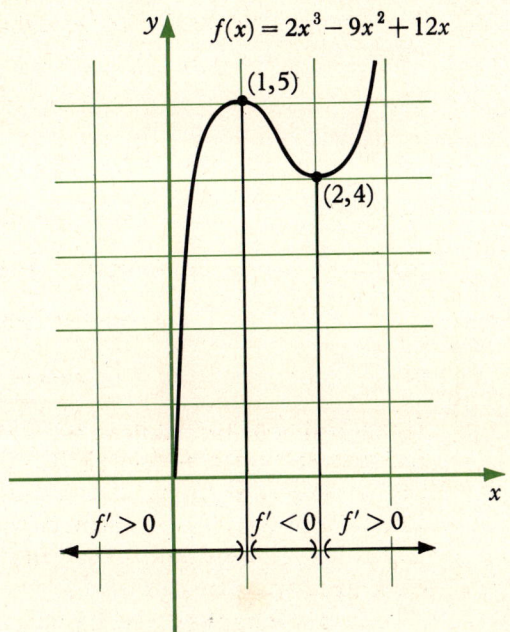

FIGURE 4-4

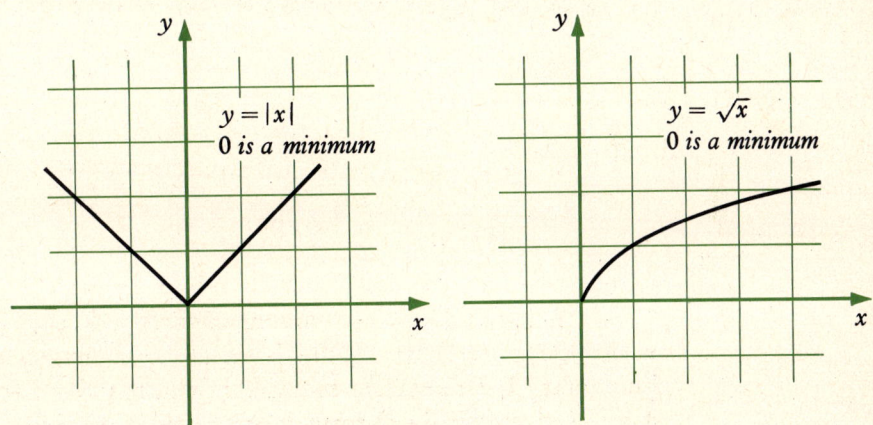

FIGURE 4-5

Fortunately, most applications involving extrema give rise to functions that are differentiable in the desired interval.

Rule M-2 *If a function $f(x)$ is differentiable in the open interval (a,b) and if $f(x)$ has a relative maximum or minimum at $x_0 \in (a,b)$, then $f'(x_0) = 0$.*

Our task of devising techniques to solve extrema problems thus boils down to determining those points x_0 where $f'(x_0) = 0$. However, a function $f(x)$ may have a point x_0 such that $f'(x_0) = 0$ but have no extremum at x_0, as illustrated by Ex. 4-4.

EXAMPLE 4-4

Let $f(x) = x^3$. Then $f'(x) = 3x^2$, and $f'(x) = 0$ when $x = 0$. It is clear from the graph of $f(x) = x^3$ that the function does not have an extremum at zero.

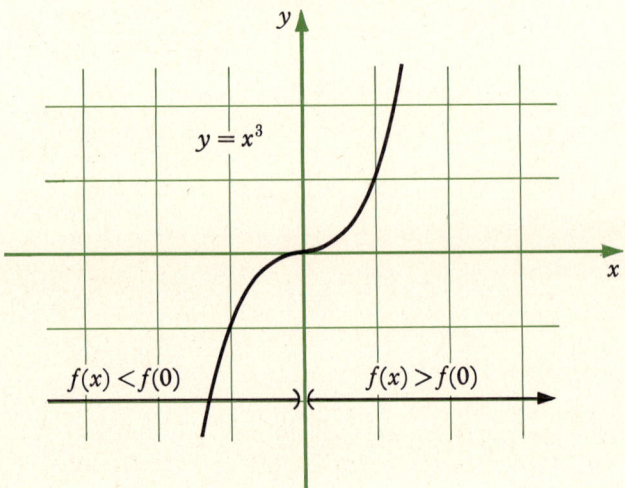

FIGURE 4-6

The values of x for which $f'(x) = 0$ are often called the *critical values* for the function, and the corresponding points $(x, f(x))$ of the graph are called the *critical points* of the curve.

We are now about to present our first procedure for solving extrema problems, called the *first derivative test*. If $f(x)$ is a differentiable function, then for $f(x)$ to have a maximum at x_0, it is clear that $f(x)$ must be increasing immediately to the left of x_0 and decreasing immediately to the right of x_0. Similarly, for $f(x)$ to have a minimum at x_0, $f(x)$ must be decreasing to the left of x_0 and increasing to the right of x_0. Recall that in order to determine where $f(x)$ is increasing or decreasing, we must determine the sign of $f'(x)$. We say that $f'(x)$ is positive (negative) to the left of x_0 if there exists an interval $(x_0 - h, x_0)$ in which $f'(x)$ is positive (negative). Similarly, we say that $f'(x)$ is positive (negative) to the right of x_0 if there exists an interval $(x_0, x_0 + h)$ in which $f'(x)$ is positive (negative).

SEC. 4-1 Maxima and Minima

Flowchart for determining the extrema of a differentiable function using the First Derivative Test

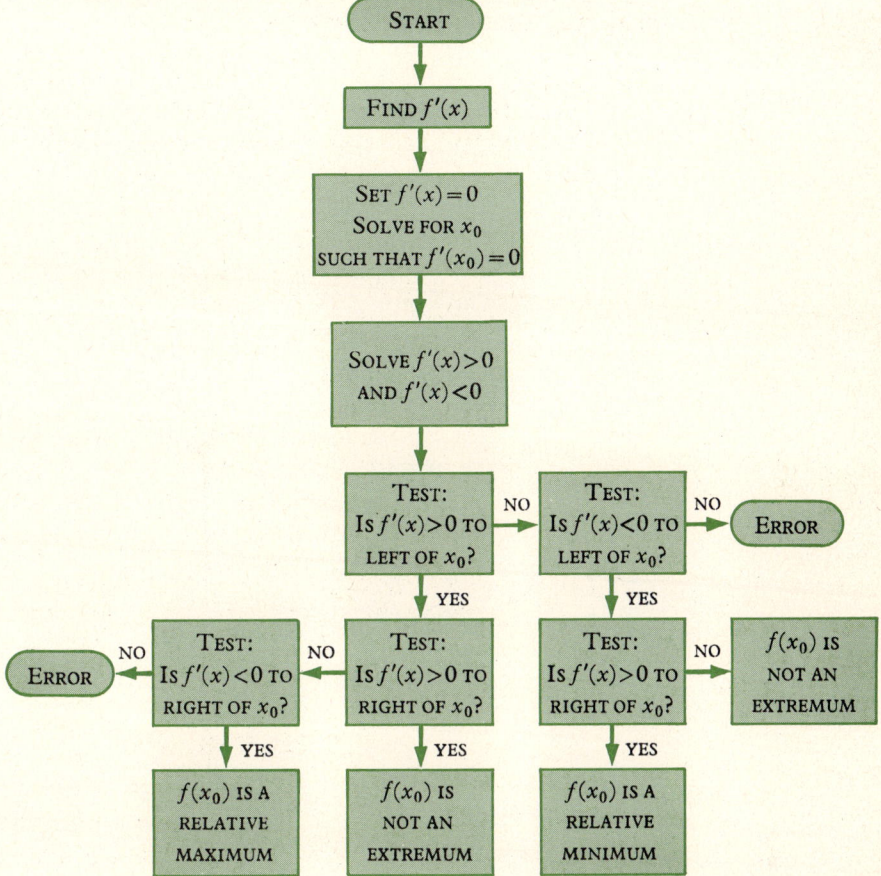

FIGURE 4-7

First Derivative Test

Suppose $f(x)$ is differentiable. To find the relative maxima and minima of $f(x)$, proceed as follows (and see Fig. 4-7):
1. Find the critical values by setting $f'(x) = 0$.
2. Determine the intervals where $f'(x)$ is positive and where $f'(x)$ is negative.
3. If x_0 is a critical value, then:
   (i) $f(x_0)$ is a maximum if $f'(x)$ is positive to the left of x_0 and negative to the right

97

(ii) $f(x_0)$ is a minimum if $f'(x)$ is negative to the left of x_0 and positive to the right

(iii) $f(x_0)$ is neither a maximum nor a minimum if $f'(x)$ has the same sign to the left and right of x_0.

EXAMPLE 4-5

Let $f(x) = (x-2)^2(x-5)$. Then $f'(x) = 3(x-2)(x-4)$.
1. Let $f'(x) = 0$. The critical values are $x = 2, 4$.
2. Let $f'(x) > 0$. Then
   (i) $x - 2 > 0$ and $x - 4 > 0$ implies that $x > 4$.
   (ii) $x - 2 < 0$ and $x - 4 < 0$ implies that $x < 2$ so that $f'(x)$ is positive and hence $f(x)$ is increasing in $(-\infty, 2) \cup (4, \infty)$. Similarly $f'(x) < 0$ in $(2, 4)$, and thus $f(x)$ is decreasing in $(2, 4)$.
3. (i) $f(x)$ has a maximum at 2 since $f(x)$ is increasing to the left of 2 and decreasing to the right.
   (ii) $f(x)$ has a minimum at 4 since $f(x)$ is decreasing to the left of 4 and increasing to the right.

We get an accurate sketch by graphing the two points $(2, 0)$ and $(4, -4)$ and utilizing the above information. (See Fig. 4-8.)

EXAMPLE 4-6

Let $f(x) = x^4 - 8x^3 + 22x^2 - 24x + 9$. Then

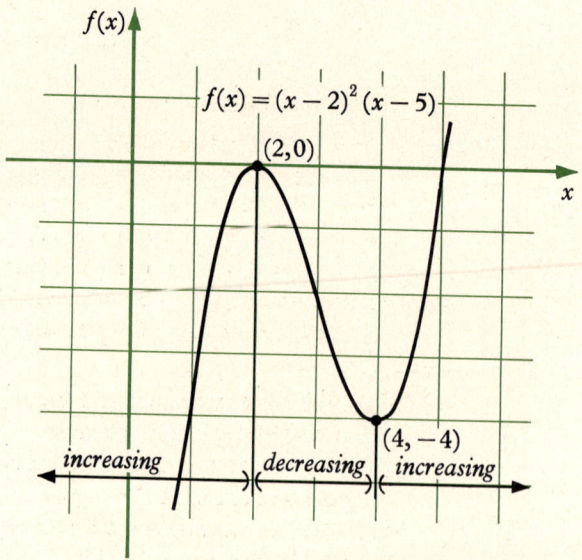

FIGURE 4-8

$$f'(x) = 4x^3 - 24x^2 + 44x - 24$$
$$= 4(x^3 - 6x^2 + 11x - 6)$$
$$= 4(x-1)(x-2)(x-3)$$

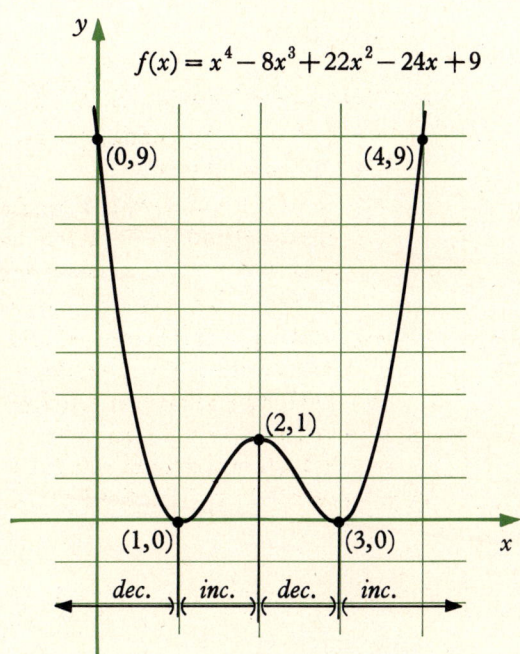

FIGURE 4-9

1. The critical values are $x = 1, 2, 3$.
2. If we set $f'(x) > 0$, we have:
   (i) $x-1 > 0$, $x-2 > 0$, $x-3 > 0$, which implies $x > 3$.
   (ii) $x-1 > 0$, $x-2 < 0$, $x-3 < 0$, which implies $1 < x < 2$.
   (iii) $x-1 < 0$, $x-2 < 0$, $x-3 > 0$, which is impossible.
   (iv) $x-1 < 0$, $x-2 > 0$, $x-3 < 0$, which is impossible.

Hence $f'(x) > 0$, and thus $f(x)$ is increasing, in $(1, 2) \cup (3, \infty)$.

Similarly we see that $f'(x) < 0$, and thus $f(x)$ is decreasing in $(-\infty, 1) \cup (2, 3)$.

3. (i) $f(x)$ has a minimum at 1 because $f(x)$ is decreasing to the left of 1 and increasing to the right.
   (ii) $f(x)$ has a maximum at 2 because $f(x)$ is increasing to the left of 2 and decreasing to the right.
   (iii) $f(x)$ has a minimum at 3 because $f(x)$ is decreasing to the left of 3 and increasing to the right.

We can graph the three points $(1,0)$, $(2,1)$, $(3,0)$ and sketch an accurate graph.

EXAMPLE 4-7

Consider the following problem: Two cars A and B, travelling on perpendicular roads, start out at the same time and travel toward the intersection at the respective rates of 60 mph and 30 mph with A situated 300 miles and B 100 miles from the intersection. How soon after they start will the distance between them be a minimum? (Generalizations of this simplified model are fundamental problems in celestial mechanics, e.g., determination of when the distance between two planets will be a minimum.) We set up the coordinate system so that the axes correspond to the two roads. Let A and B (Fig. 4-10) represent the positions of cars A and B, and d the distance

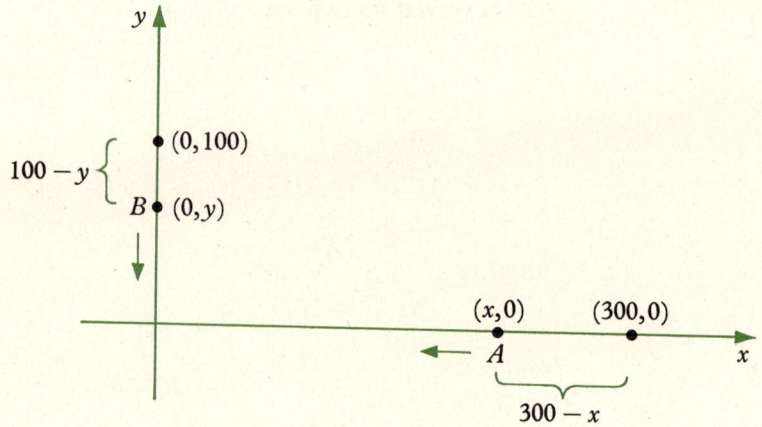

FIGURE 4-10

between them. Then $d = \sqrt{x^2 + y^2}$. At time t, we have $60t = 300 - x$ and $30t = 100 - y$. Thus

$$\frac{100-y}{30} = \frac{300-x}{60} = t$$

which implies that $x = 100 + 2y$ and so $d = \sqrt{10^4 + 400y + 5y^2}$. Hence

$$d'(y) = \frac{400 + 10y}{2(10^4 + 400y + 5y^2)^{1/2}}$$

Letting $d' = 0$ yields $y = -40$. The first derivative test shows that d has a minimum when $y = -40$, or when $t = \frac{14}{3}$ hr.

EXAMPLE 4-8

An important concept considered by biologists in the study of the central nervous system is the excitation of a nerve fiber by means of a stimulus. The stronger the stimulus, S, the higher is the frequency of the nerve impulses, which is often referred to as the intensity of excitation E. Suppose for a given organ the relation between the excitation and the stimulus is given by $E = a(S^2 - 2Sh + a)$ for two constants a and h, and where a negative reading for $E'(S)$ means that the stimulus was not strong enough to excite the nerve fiber. It follows that $E'(S) = 2aS - 2ah$. Letting $E' = 0$ yields $S = h$. By the first derivative test we see that E has a minimum at h. The value h is called the *threshold* since S must exceed this value in order that excitation may be released at all.

EXAMPLE 4-9

The following problem is related to Snell's empirical law of refraction in geometrical optics, which states that a ray of light will minimize the time necessary to travel from one point to another point in a different medium. We want to find a point P on a line L such that the time it takes for a particle

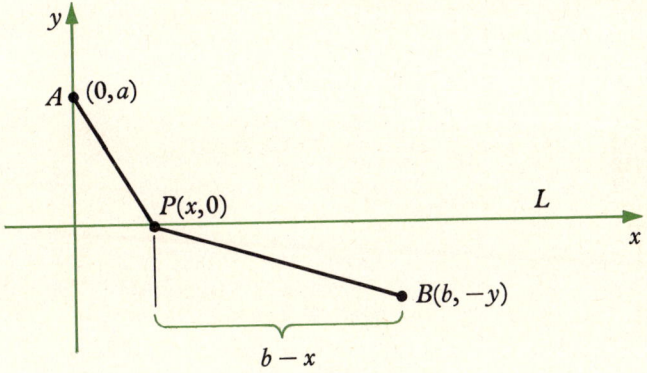

FIGURE 4-11

to travel in a straight line from a point A to P and then in another straight line to a point B on the other side of L is a minimum. We assume that the velocity in the region above L is v, and w in the region below. L. We choose the x-axis to be L and let A be on the y-axis as in Fig. 4-11. If T_1 is the time it takes the particle to go from A to P, then $T_1 v = \sqrt{x^2 + a^2}$, and if T_2 is the time it takes to go from P to B, then $T_2 w = \sqrt{(b-x)^2 + y^2}$. Hence the total time T is given by

$$T = T_1 + T_2 = \frac{\sqrt{x^2 + a^2}}{v} + \frac{\sqrt{(b-x)^2 + y^2}}{w}$$

Hence

$$T'(x) = \frac{x}{v\sqrt{x^2 + a^2}} + \frac{x - b}{w\sqrt{(b-x)^2 + y^2}}$$

If we set $T'(x) = 0$ and solve for x, a very difficult problem in itself, we have the desired point P. The reader who is familiar with trigonometry is invited to manipulate the above equation into a more accessible form.

EXAMPLE 4-10

Ecology, the study of interactions between organisms and their environment, is a field of study which is abundant with concrete examples of constructive mathematics, expecially calculus. The effects of predation in a steady-state population can be graphically illustrated by the methods we've employed to solve extrema problems. The function that is to be minimized is the difference between the death distribution of predator and prey. For the population to thrive, both a maximum steady yield for the predator and a life-span enabling the population of the prey to withstand predation are desired. An ecologist can use empirical evidence to determine the death rate $D_1 = D_1(t)$ (a derivative itself) for the predator and the death rate for the prey $D_2 = D_2(t)$. Then the function to be minimized is $D_1(t) - D_2(t) = F(t)$. Thus we set $F'(t) = 0$, i.e.,

$$\frac{d}{dt}(D_1 - D_2) = 0$$

Nature provides that animals will not be taken as prey at too tender an age. The prey population must be allowed to multiply sufficiently. On the other hand, the prey cannot be taken at too old an age because then the causes of death other than predation will accumulate and there will be too small an available prey population for the predator. Hence the predator will feed on the largest and weakest animals it can find. In nature then, it would be expected that both predator and prey would be allowed to evolve to that time t_0 at which animals taken as yield are old enough so that they are threatened by death from causes other than predation. In other words, predation on animals about to die from nonpredatory causes minimizes the difference between the two death rates $F(t)$.

EXERCISE 4-1

1. Sketch the functions and determine the maxima and minima.

   (a) $f(x) = x^2 - 4$
   (b) $f(x) = x^2 + 4x - 2$
   (c) $f(x) = x^2 + x$
   (d) $f(x) = x(x^2 - 1)$
   (e) $f(x) = (x-1)(x+3)^2$
   (f) $f(x) = x^3 + 3x^2 + 3x + 1$
   (g) $f(x) = (x-2)^2(x+1)^2$
   (h) $f(x) = 1 + x^2 - x^6$
   (i) $f(x) = \dfrac{1}{x-2}$
   (j) $f(x) = \tfrac{1}{3}x^3 + \tfrac{1}{2}x^2 - 6x + 10$
   (k) $f(x) = x^4 - 4x + 3$
   (l) $f(x) = x^2 - 3x + 1$
   (m) $f(x) = 3x^2 - 6x + 5$
   (n) $f(x) = 20 + 0.2x - 0.1x^2$
   (o) $f(x) = 200 + 0.8x - 0.1x^2$
   (p) $f(x) = ax^2 + bx + c$

2. Examine the functions for extrema in the given interval only.

   (a) $f(x) = -x^3$, $(-3, 3)$
   (b) $f(x) = (x-3)(x+2)^2$, $(0, 5)$

(c) $f(x) = x + \dfrac{1}{x}$, $(0, 5)$

(d) $f(x) = 1 - \dfrac{1}{x} + \dfrac{1}{x^2}$, $(0, 3)$

3. Divide the number 300 into two parts such that the product of one part and the square of the other is a maximum.

4. A box with no top and a square bottom is to have a volume of 8 cu in. If the material of which the sides are to be constructed costs 4 cents per sq in. and the material of which the bottom is to be constructed costs 8 cents per sq in., find the dimensions of the box which costs the least to make.

5. Find the maximum total revenue of a product when the price per x hundred is $2000/(400 + x^2)$ where the price is measured in thousands of dollars.

6. At 12:00 P.M. ship A was 60 miles due north of ship B. If ship B sails due east at 10 mph and A sails due south at 15 mph when will they be nearest each other and how near?

7. Show that among all rectangles having a fixed perimeter, the square has the maximum area.

8. Show that among all rectangles having a fixed area, the square has the minimum perimeter.

9. A farmer wishes to enclose with a fence a rectangular field which is adjacent to a river, where no fencing is required along the river. What are the dimensions of the field if 4000 ft of fencing is to be used and the area of the field is to be a maximum?

10. Suppose the farmer in the above exercise wishes to enclose another field with 1000 ft of fencing while utilizing part of an existing 600-ft stone wall as one side. Find the dimensions of the field if the area of the field is to be a maximum.

11. An apple orchard contains 50 trees per acre and the average yield is 400 apples per tree. The farmer wishes to plant more trees to maximize his yield per acre. From empirical evidence, as an application of the law of diminishing returns, the farmer has determined that for each additional tree planted per acre, the yield per tree is reduced by 1.25%, or 5 apples. Find the number of additional trees per acre which should be planted to maximize the yield per acre.

12. The fixed cost of running a truck is $15 per hr and the cost of fuel is proportional to the square of the speed and is $20 per hr for a speed of 20 mph. Find the speed which minimizes the cost per mile.

13. A manufacturing company offers the following discount schedule of prices:

$20 for orders of 500 or less, with the cost being reduced 2 cents for each unit above 500. Find the order that will maximize the company's revenue. revenue.

14. A man is on an island 10 miles from shore. He wishes to get to a point 12 miles up the coast. If he can row at the rate of 2 mph and walk at the rate of 4 mph, to what point on the shore should he row?

15. Suppose that a manufacturing company wants to package its product in a closed can in the shape of a right circular cylinder. The merchandizing department has determined that the volume of the can should be 16 cu in. What must be the dimensions of the can so that the total surface area shall be a minimum? (*Hint:* $V = 16 = \pi r^2 h$ and $S = 2\pi r^2 + 2\pi rh$.)

16. Suppose that a manufacturing company wants to package its product in a rectangular box with a square base with a volume of 32 cu in. The cost of the material used for the top and bottom is 3 cents per sq in., while the cost of the material used for the sides is 6 cents per sq in. What must the dimensions be for the total cost to be a minimum?

17. A manufacturing company wants to find the dimensions of a box with a square base of minimum cost: the volume is 64 cu in.; the cost of the base is 1 cent per sq in., the cost of the top is 3 cents per sq in., and the cost of the sides is 2 cents per sq in.

18. An open box is to be constructed from a 5×6 inch rectangular piece of cardboard by cutting out equal squares at each corner and then folding up the sides. What must the length of the side of the square be in order that the box will have the maximum volume?

19. If $s(t)$ is the position of a point moving on a straight line at time t, then $s'(t)$ is the velocity of the point at time t and $s''(t)$ is the acceleration at time t. For the following functions find the velocity and acceleration, the intervals where $s(t)$ is increasing and decreasing, and the extrema of $s(t)$.
   (a) $s(t) = t^2 + 2t$
   (b) $s(t) = t^3 - 2t^2 + 1$
   (c) $s(t) = \dfrac{t}{t+1}$ for $t \in (0, \infty)$
   (d) $s(t) = e^{t^2}$
   (e) $s(t) = t + \ln t$ for $t \in (0, \infty)$

20. A useful principle in economics states that maximum profits are obtained when the marginal revenue equals the marginal cost. If $P(x)$ is the profit realized with the production of x amount of goods and $R(x)$ and $C(x)$ are the revenue and cost, respectively, when x goods are produced, then $P(x) = R(x) - C(x)$. Given this formula, explain the above principle.

21. A manufacturing company determines that the revenue function $R(x)$ is

given by $R(x) = 60x - 3x^2$ for $x \in (1, 10)$ and that the cost function $C(x)$ is given by $C(x) = 200 + 0.6x$. Find the output x at which the profits are a maximum.

22. The techniques of calculus are often used in the area of inventory control. One critical phase of inventory control is procurement lead time, the determination of the optimum time to reorder a product. This decision is based on a number of factors: the cost of ordering a unit of goods, the cost of holding that unit in inventory, fluctuations in the demand for the good. Suppose a company that reorders a certain product monthly determines from empirical evidence that if they order at time t, measured in days, $0 < t \leq 30$, the revenue $R(t)$ from the sale of the product is $R(t) = 10t^2 + 34t - 10$ and the cost $C(t)$ is $C(t) = 11t^2 + 4t + 51$. At what value of t is the profit a maximum?

23. A number of biological investigators in Britain and the United States have established that the propagation of the nerve impulse is essentially an electrical process. The nerve impulse is a wavelike variation of potential difference across the nerve-axon membrane. The membrane potential E is related to the current I_m across any segment of the membrane and the current I_i of ionic elements by

$$\frac{dE}{dt} = C(I_m - I_i)$$

where C is a constant and t is time. Suppose that for a given segment of membrane it is determined that $I_i = 4 + 2t^2$ where I is in amperes and t in seconds while $I_m = 3t^2 - 7t$. At what time t is E a minimum?

4-2 THE SECOND DERIVATIVE TEST

Consider the graphs of the two functions in Fig. 4-12. Each function is increasing, yet one is bending upward ($y = x_2$) and the other is bending downward ($y = \sqrt{x}$) in $(0, \infty)$.

We are led to present the following definition.

Definition A function $f(x)$ is *concave up* [*down*] in an interval if for every a, b, and c in the interval such that $a < b < c$, the point $(b, f(b))$ lies below [above] the secant line of the curve connecting $(a, f(a))$ and $(c, f(c))$.

If you draw a few graphs of functions that are concave up and concave down and investigate the behavior of the derivative of these functions you will readily convince yourself of the validity of the following theorem, whose proof we omit.

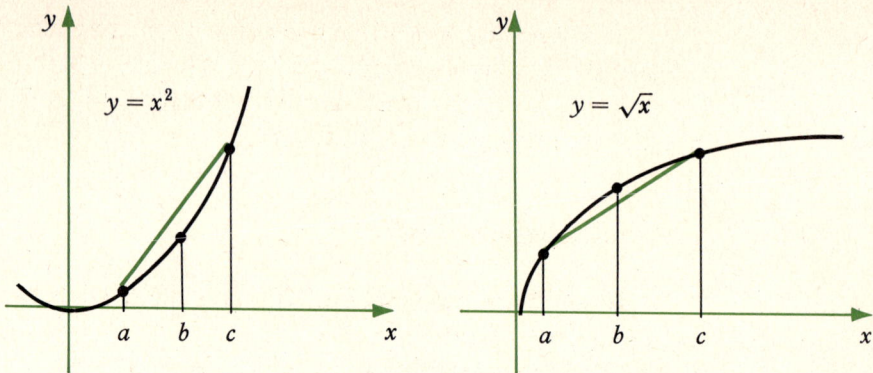

FIGURE 4-12

Theorem 4-1 *If f' is increasing [decreasing] in an interval, then f is concave up [down] in that interval.*

If the derivative of $f'(x)$ exists, it is called the second derivative of $f(x)$, written $f''(x)$. If $f(x) = x^2$ then $f'(x) = 2x$ and $f''(x) = 2$. If $g(x) = (x+1)^{1/2}$, then $g''(x) = -\frac{1}{4}(x+1)^{-3/2}$. One similarly defines the third derivative of $f(x)$ and higher derivatives.

Recall Rule M-1: If $f' > 0$ in an interval, then f is increasing in the interval. If we replace f by f' in this rule, we obtain the following.

Rule M-3 *If $f'' > 0$ for every x in some interval I, then $f(x)$ is concave up in I. Similarly, if $f'' < 0$ for every x in some interval I, then $f(x)$ is concave down in I.*

EXAMPLE 4-11

> Let $f(x) = x^3$. Then $f'(x) = 3x^2$ and $f''(x) = 6x$. Hence $f'' < 0$ in $(-\infty, 0)$ and thus f is concave down in $(-\infty, 0)$. Similarly, $f'' > 0$ in $(0, \infty)$ and thus f is concave up in $(0, \infty)$.

In Ex. 4-11 note that at $x = 0$ the function $f(x)$ changes from being concave down to the left of zero to concave up to the right. A function $f(x)$ is said to have a *point of inflection* $(a, f(a))$ at a if $f(x)$ is concave up (or down) to the left of a and concave down (or up) to the right of a. Hence $(0,0)$ is a point of inflection for $f(x) = x^3$.

If you know that $f(x)$ is concave down in an interval I and you also know that $f(x)$ has a point $a \in I$ such that $f'(a) = 0$, then $f(x)$ must have a maximum at a. Similarly if $f(x)$ is concave up in I and $f'(a) = 0$, for $a \in I$, then $f(x)$ has a minimum at a.

Second Derivative Test

Suppose $f(x)$ is differentiable. To find the relative maxima and minima of $f(x)$, proceed as follows:
1. Find the critical values by setting $f'(x) = 0$.
2. Find $f''(x)$.
3. If x_0 is a critical value, then:
   (i) $f(x_0)$ is a maximum if $f''(x_0) < 0$.
   (ii) $f(x_0)$ is a minimum if $f''(x_0) > 0$.
   (iii) $f(x_0)$ may or may not be an extremum if $f''(x_0) = 0$, i.e., the test fails if $f''(x_0) = 0$.

It is interesting to note that if $f''(a) = 0$ and if $(a, f(a))$ is a point of inflection (neither of which implies the other), then $y = f'(x)$ has a maximum or a minimum at a. In fact, in many applications a function $y = f(x)$ is given and the problem to be solved involves finding a point of inflection of f which is a maximum or minimum of f'. Examples 4-13 and 4-14 illustrate this.

EXAMPLE 4-12

Consider $f(x) = 4x^3 - 9x^2 + 6x$. Then
$$f'(x) = 12x^2 - 18x + 6 = 6(x-1)(2x-1)$$
and so the critical values are $x = 1, \frac{1}{2}$. We apply the second derivative test with $f''(x) = 24x - 18$ and get $f''(1) = 6$, while $f''(\frac{1}{2}) = -6$ so that $f(x)$ has a minimum at $x = 1$ and a maximum at $\frac{1}{2}$. Note further that $f'' > 0$ and hence f is concave up in $(\frac{3}{4}, \infty)$, and that $f'' < 0$ and hence f is concave down in $(-\infty, \frac{3}{4})$. Hence $f(x)$ has a point of inflection at $x = \frac{3}{4}$. We can present this information in Fig. 4-13.

In Ex. 4-12 we are impressed with the importance of the use of derivatives in finding extrema. If the normal procedure of grinding out points on a curve were used here, plotting points corresponding to $x = 0, 1, 2, \ldots$, the maximum at $x = \frac{1}{2}$ would be passed over.

x	f	f'	f''
0	0	Pos.	Neg.
$\frac{1}{2}$	Pos.	0	Neg.
$\frac{3}{4}$	Pos.	Neg.	0
1	Pos.	0	Pos.
2	Pos.	Pos.	Pos.

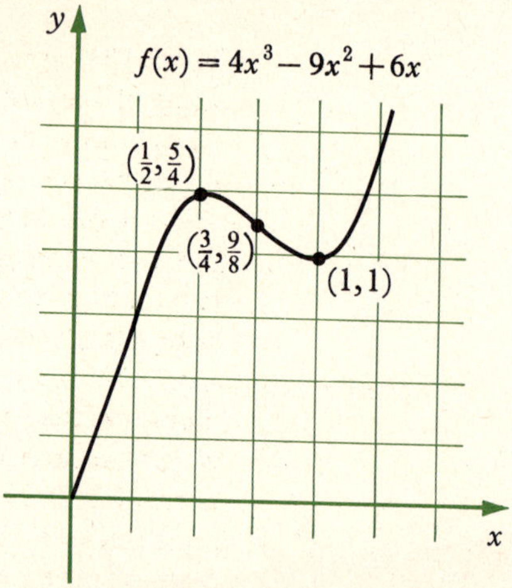

[Graph of $f(x) = 4x^3 - 9x^2 + 6x$ showing points $(\frac{1}{2}, \frac{5}{4})$, $(\frac{3}{4}, \frac{9}{8})$, and $(1,1)$.]

FIGURE 4-13

EXAMPLE 4-13

When combating the spread of a disease it is important for health officials to know, for many reasons including minimizing cost, when the rate of increase of the disease begins to slow, i.e., when the rate of increase has reached a maximum. Hence if $y = P(t)$ is the function giving the percentage $P(t)$ of a population which has contracted a disease at time t, the health officials want to find that value t_0 such that $P'(t)$ has a maximum, i.e., when $P''(t_0) = 0$ for some t_0. At that point t_0, $P(t)$ will have a point of inflection (see Fig. 4-14).

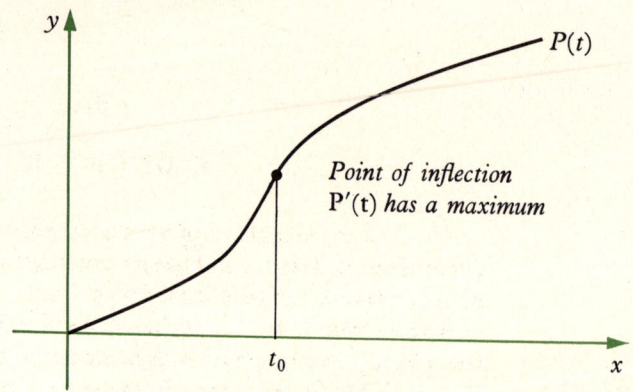

FIGURE 4-14

EXAMPLE 4-14

In psychology there has recently been a vast amount of research on the study of learning. Calculus has proved to be a valuable tool in this field of study. In such a mathematical model, a learning curve gives a measure of performance P as a function of training time or trials, where P is assumed to be a differentiable function of time t. Thus the learning curve is a function derived from assumptions about the learning process. When a subject is given a new task, whether it be a worker on a production line, a salesman in the field, or perhaps a duffer correcting a golf swing, it is generally true that although the rate of the measure of success is slow at first, it increases as the subject gains experience with the task, and then it slows again as performance of the task is perfected. Mathematically, we mean that the derivative P' of $P(t)$ increases slowly, then sharply and finally slows as t increases (see Fig. 4-15). Now the point in time of maximum increased-learning rate, i.e., the value of t where P' is a maximum, is of fundamental importance. Generally a great deal of effort is exerted to allow the subject to reach this point, since effort can then be reduced considerably. What is desired therefore is a value t_0 such that $P'(t_0)$ is a maximum, i.e., we need $P''(t_0) = 0$. Thus at t_0, $P(t)$ has a point of inflection.

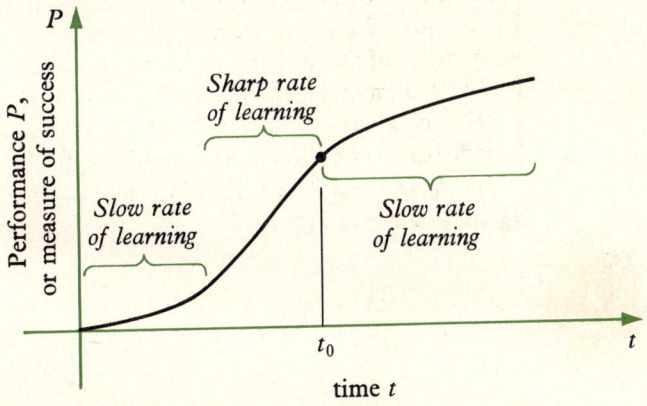

FIGURE 4-15

As an example, suppose a psychologist determines that the learning curve associated with a task group experiment is given by $f(t) = 3t^2 - 0.2t^3$ where $f(t)$ is a measure of the learning which a group has acquired, measured on a scale from 0 to 100, at time t, measured in minutes. The domain of $f(t)$ is $[0,10]$, that is, ten minutes is the time it takes for a group to accomplish the task. The learning rate, the measure of how fast a group is learning the ins and outs of the task, is given by $f'(t) = 6t - 0.6t^2$. To determine when the learning rate is a maximum we find the derivative of the learning rate, that is, the second derivative $f''(t)$ of $f(t)$, which is $f''(t) = 6 - 1.2t$, and set $f''(t)$ equal to zero and get $6 - 1.2t = 0$, $t = 5$. Thus at $t = 5$ the learning rate is a maximum. Note that at $t = 5$, the learning curve $f(t)$ has a point of inflection. The graph of $f(t)$ is given in Fig. 4-16. Note that $f(t)$ is concave up in (0,5) since $f''(t) > 0$ in that interval and $f(t)$ is concave down in (5,10) since $f''(t) < 0$ in that interval.

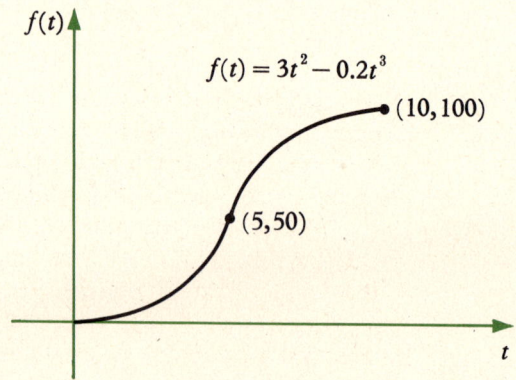

FIGURE 4-16

EXAMPLE 4-15

Biologists often assume that a given organism has the optimal possible design with respect to economy of material used, energy expended, and other necessary functions. One such organism that merits a great deal of scrutiny is the aorta. It is generally accepted that the size of the aorta could be smaller in most animals if it were not for the possibility of turbulences in the blood flow which are connected with loss of energy through stress and strain. It is known from hydrodynamics that the flow of a fluid becomes turbulent when the Reynolds number, R, exceeds a certain critical value, where R depends upon the specific gravity σ, the viscosity η, the velocity v of the fluid, and the radius r of the tube through which the fluid flows.

Empirical evidence shows that for certain organisms, where we assume σ, η, v, and the total volume of blood, C, flowing through the aorta are constant, the values for R for which nonturbulent flow is maintained is given by

$$R(r) = A \ln r - Br$$

where $A = C\sigma/\pi\eta$ and B is a constant depending upon the particular fluid and organism in question.

To find the maximum value that R can attain for nonturbulent flow, we calculate $R' = A/r - B$ and $R'' = -A/r^2$. Letting $R' = 0$ yields

$r = A/B$. Since $A > 0$, we have $R''(A/B) < 0$, and thus R has a maximum at $r = A/B$. For example, for the vascular system of a dog weighing 13 kg, the cardiac output is given as $0 = 40 \text{ cm}^3/\text{sec}$, $\eta/\sigma = 0.027 \text{ cm}^2/\text{sec}$, $B = 1100$, and thus the maximum value of R necessary to maintain nonturbulent blood flow is

$$r = \frac{A}{B} = \frac{C\sigma}{B\eta} \approx 0.43 \text{ cm}$$

EXAMPLE 4-16

A firm determines that its profit P is a function of input x given by the formula $P(x) = 0.1x^2 + 0.2x$ where P is measured in thousands of dollars per month and x in hundred of units, and the production capacity of the firm is restricted to utilization of 1000 input units per month, that is, the domain of $P(x)$ is $[0,10]$. To determine that value of input which maximizes profit, we can graph $P(x)$. Thus we find $P'(x) = 0.2x + 0.2$ and in $[0,10]$, $P'(x) > 0$ so that $P(x)$ is increasing in $[0,10]$. The graph of $P(x)$ is given in Fig. 4-17. It is clear that profit is maximized at the endpoint $x = 10$ and the maximum profit is $P(10) = 0.1(10)^2 + 0.2(10) = 12$ thousand dollars.

From the above example it is apparent that if the domain of a function contains one or more endpoints, the value at each endpoint must be investigated for possible extrema. Note that the derivative of the function does not exist at an endpoint, even though a relative maximum or minimum will usually occur at an endpoint.

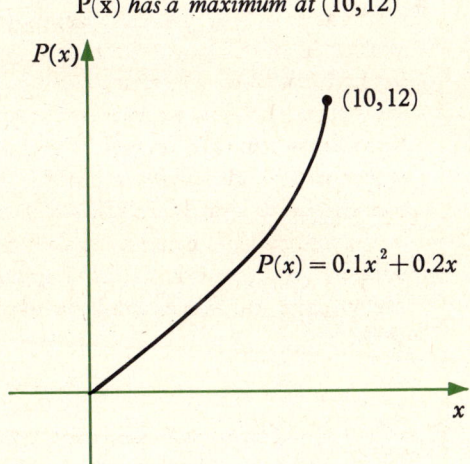

FIGURE 4-17

In the same manner, if the domain of a function contains a point where the derivative does not exist, the function may have an extremum at that point. For example, $y = |x|$ has a corner at $x = 0$ and $y'(0)$ does not exist, but the function takes on the minimum value of 0 at $x = 0$.

One of three possibilities may occur at a point of a function $y = f(x)$ where $f'(a)$ does not exist because $f(x)$ has a corner at $x = a$. In Fig. 4-18, the function $y = f(x)$ has a minimum at $x = a$, $y = g(x)$ has a maximum, and $y = h(x)$ does not have an extremum.

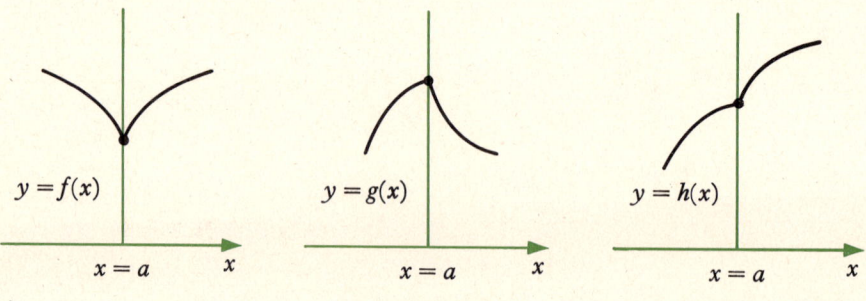

FIGURE 4-18

EXAMPLE 4-17

To graph the function

$$f(x) = \begin{cases} x & \text{if } x \leq 1 \\ x^2 - 4x + 4 & \text{if } x \geq 1 \end{cases}$$

we note that if $x < 1$, then $f'(x) = 1$; whereas if $x > 1$, $f'(x) = 2x - 4$. Hence there is no point to the left of $x = 1$ such that $f'(x) = 0$. By setting $2x - 4 = 0$ we get $x = 2$, so that $f'(2) = 0$. Thus $(2,0)$ is a critical point. By using the Second Derivative Test we see that $f''(x) = 2$ and hence $f''(2) = 2$, and therefore $(2,0)$ is a minimum. Now let us consider the function at $x = 1$. By considering points to the left of $x = 1$, one sees that the derivative at $x = 1$ should be 1. By considering points to the right of 1 where $f'(x) = 2x - 4$, one sees that the derivative at $x = 1$ should be -2. Therefore the derivative at $x = 1$ does not exist. We now use the First Derivative Test to investigate whether $f(x)$ has an extremum at $x = 1$. For $x < 1$, $f'(x) = 1$ so that $f(x)$ is increasing to the left of $x = 1$. For $x > 1$, $f'(x) = 2x - 4$ is less than zero in $(1,2)$ so that $f(x)$ is decreasing in $(1,2)$. Therefore $f(x)$ has a maximum at $x = 1$. The graph of $f(x)$ is given in Fig. 4-19.

SEC. 4-2 The Second Derivative Test

f(x) *has a minimum at* (2,0) *and a maximum at*
(1,1) *where* f'(x) *does not exist*

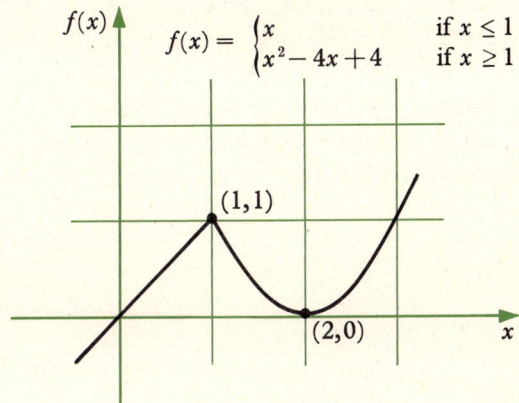

$$f(x) = \begin{cases} x & \text{if } x \leq 1 \\ x^2 - 4x + 4 & \text{if } x \geq 1 \end{cases}$$

FIGURE 4-19

EXERCISE 4-2

1. Sketch the functions and determine the maxima, minima, and points of inflection.
   (a) $f(x) = x^2 - 2x + 1$
   (b) $f(x) = x^3 - 4$
   (c) $f(x) = x(x^2 - 4)$
   (d) $f(x) = x^3 - 3x^2 + 3x - 1$
   (e) $f(x) = x^2(x - 5)$
   (f) $f(x) = x(x - 2)(x - 3)$
   (g) $f(x) = (x - 3)^2(x - 5)^2$
   (h) $f(x) = x^2(x^2 - 3x + 2)$
   (i) $f(x) = 2x^4 + x - 5$
   (j) $f(x) = 1 + 6x + x^6$
   (k) $f(x) = (x - 1)^2(x - 2)$
   (l) $f(x) = (x - 2)^2(x + 1)$
   (m) $f(x) = (x + 1)^2(x - 3)$
   (n) $f(x) = (x - 3)^2(x - 1)$

2. In the second derivative test it is asserted that no conclusion can be made about the critical value x_0 if $f''(x_0) = 0$. If $f''(x_0) = 0$ for a critical value x_0, any of the following may happen: $f(x_0)$ is a maximum, $f(x_0)$ is a minimum, or $f(x_0)$ is not an extremum. Give examples of functions that illustrate each of the three possibilities. (*Hint:* Consider $y = x^4$, $y = -x^4$, $y = x^3$.)

3. Divide the number 120 into two parts such that the product of one part and the cube of the other is a maximum.

4. A box with no top and a square bottom is to have a volume of 10 cu in. Find the dimensions of the box costing the least to make if the material of which the

sides are to be constructed costs one-third per square inch as much as the material of which the bottom is to be constructed.

5. At 1:00 P.M. plane A is 650 miles due west of plane B. If plane A flies at 100 mph due south and plane B flies at 150 mph due west, when will they be nearest each other and how near?

6. An electric company wishes to lay a cable from point A on an island 10 miles offshore to point B on the coast 20 miles from the point, C, on the shore nearest the island. If it costs $5 per mile to run the cable under water and $7 per mile to run the cable on land, find the point, D (between B and C), to which the cable should be run from A and then to B in order to minimize the cost.

7. A sheet of metal 60 ft long and 20 ft wide is to be made into a trough by turning up equal sides at right angles. Assume that the ends are to be sealed in another operation. Find the dimensions that yield the maximum volume.

8. Let C be the cost of operating a boat and v its speed in still water. Suppose $C = Rv^3$ for some constant R. If the velocity of the stream is 20 mph, find the most economical speed to operate the boat upstream.

9. A railroad company offers the following discount rates for chartered trips of at least 300 passengers: The fare is $10 per person for exactly 300 passengers, while the fare per person decreases by one cent for each passenger over 300. Find the number of passengers which will yield the maximum revenue.

10. In Problem 22 of Exercise 4-1 we discussed inventory control. Another fundamental facet of this theory is the economic lot size, or that quantity to order at given intervals in order to minimize inventory cost. Suppose that total inventory cost $T(x)$ of buying and selling x units of the product is a function of (1) the cost of ordering x units, $O(x)$, which is proportional to the number of orders placed, (2) the cost of holding x units in inventory, $H(x)$, which is proportional to the quantity ordered, and (3) the cost of the purchase, $C(x)$. Then $T(x) = O(x) + H(x) + C(x)$. Therefore, were a company to determine that for a given product $O(x) = 1080/x$, $H(x) = 3x$, and $C(x) = 27x$, what value of x would minimize $T(x)$?

11. A manufacturing company has determined from empirical data that an average employee placed on a new job learns the job according to the schedule $P(t) = 15t^2 - t^3$ where $P(t)$ is the average proportion of the standard job hourly quota completed by the employee and where t is measured in days, for $t \in (0, 9)$ for practical reasons. Find the time such that $P'(t_0)$ is a maximum. The management would expend the maximum effort and cost until t_0 and then gradually decrease the help given the employee.

12. Suppose that health officials predict that the percentage $P(t)$ of a population which contracts a disease is following the pattern given by $P(t) = (0.01)(18t^2 - t^3)$ where t is measured in days for $t \in [0,15]$. On what day will the rate that the percentage is increasing be a maximum?

13. Suppose that empirical evidence shows that for certain organisms, where we assume $\eta/\sigma = 0.011 \text{ cm}^2/\text{sec}$, $C = 20 \text{ cm}^3/\text{sec}$, and $B = 900$, the relationship between the radius of the aorta and the Reynolds number, R, is given by

$$R(r) = \frac{C\sigma \ln r}{\eta} - Br$$

Find the maximum value that R can attain for nonturbulent flow.

14. The reaction R of the body to a dose of drug D can be represented by the function

$$R(D) = D^2\left(\frac{C}{2} - \frac{D}{3}\right)$$

where C is the maximum amount that can be administered. R is the strength of the reaction, measured in millimeters mercury if blood pressure is being tested, or perhaps degrees, if change in body temperature is being measured. Find the dose that has the maximum sensitivity, i.e., where the rate of increase of R is the greatest. In other words, find where $R'(D)$ has a maximum. Graphically what can be said of R at this point?

4–3 IMPLICIT DIFFERENTIATION

Often a formula is given which does not define a function. For example, consider the formula

$$x^2 + y^2 = 1$$

whose graph, given in Fig. 4-20, is a circle of radius 1 with center at the origin.

The formula $x^2 + y^2 = 1$ does not define a function because for every value of x in the interval $(-1,1)$, there are two corresponding values of y which satisfy the equation. Geometrically, one sees from Fig. 4-20 that every vertical line $x = a$ for $-1 < a < 1$ intersects the circle in two points.

The derivative has been defined only for functions, so it may appear that we cannot talk about the derivative of the relation $x^2 + y^2 = 1$. However, the tangent line to the circle at a particular point is a familiar geometric quantity. It is that line which intersects the circle in one and only one point. For instance, the line $y = 1$ is the tangent line to the circle at the point $(0,1)$. Since the derivative measures the slope of the tangent line, one may suspect that the concept of the derivative can somehow be associated with the relation $x^2 + y^2 = 1$.

To do this, we will form functions from the relation $x^2 + y^2 = 1$ by suitably restricting the values of y. For example, if we restrict y to the values

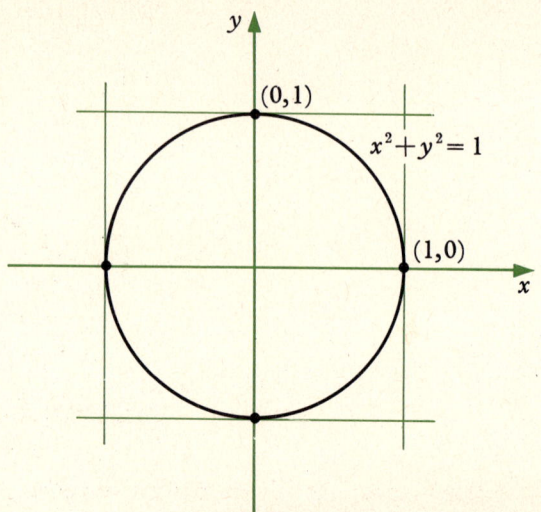

FIGURE 4-20

[0,1], we can consider all x in $[-1,1]$ and the graph that we arrive at is the upper half-circle. See Fig. 4-21.

The formula corresponding to $y \in [0,1]$ and $x \in [-1,1]$ is found by solving $x^2 + y^2 = 1$ for y.

$$x^2 + y^2 = 1$$
$$y^2 = 1 - x^2$$
$$y = \pm\sqrt{1-x^2}$$

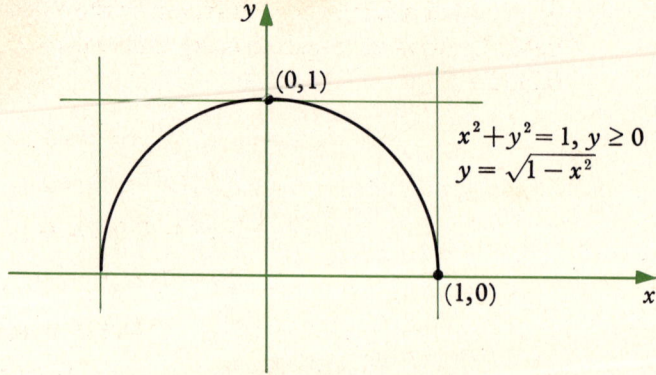

FIGURE 4-21

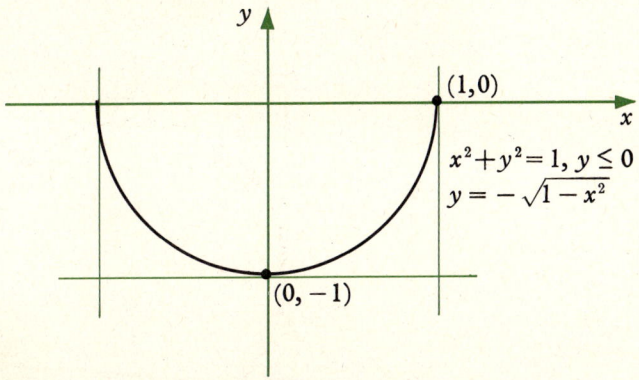

FIGURE 4-22

Since we are restricting y to positive values, we take

$$y = \sqrt{1-x^2}$$

Note that the above equation defines y as a function of x. Similarly, the lower half-circle (see Fig. 4-22) is given by $y = -\sqrt{1-x^2}$.

Other functions can be formed from $x^2 + y^2 = 1$ by suitably restricting the domain and range of $y = \sqrt{1-x^2}$ and $y = -\sqrt{1-x^2}$. Each of these functions is said to be defined *implicitly* by the relation $x^2 + y^2 = 1$.

Of course, when we write $f(x) = \sqrt{1-x^2}$ we have defined this function explicitly, but it behoves us to bypass the process of explicitly finding the functions which can be formed from a relation such as $x^2 + y^2 = 1$, and simply say that y is an *implicit* function of x.

Since the relation $x^2 + y^2 = 1$ is the union of two functions, it is possible to find the derivative of y with respect to x. Instead of solving for y, i.e., solving for the explicit functions above, we assume that y is a differentiable function of x and formally differentiate the equation $x^2 + y^2 = 1$ and solve for $D_x y$.

$$x^2 + y^2 = 1$$
$$D_x(x^2 + y^2) = D_x 1$$
$$D_x x^2 + D_x y^2 = 0$$
$$2x + 2y\, D_x y = 0$$
$$D_x y = -x/y$$

Observe that to calculate $D_x y^2$ we use the Chain Rule, in particular Rule D8 in Sec. 3-4 with $n = 2$. Thus $D_x y^2 = 2y\, D_x y$.

117

The significance of the fact that $D_x y = -x/y$ is that the slope of the tangent line to the circle $x^2 + y^2 = 1$ at any point (a, b) such that $b \neq 0$ is $-a/b$. (Why do we restrict b to be nonzero?)

EXAMPLE 4-18

The point $(\sqrt{2}/2, \sqrt{2}/2)$ lies on the circle $x^2 + y^2 = 1$. The slope of the tangent line to the circle at the point $(\sqrt{2}/2, \sqrt{2}/2)$ is $-(\sqrt{2}/2)/(\sqrt{2}/2) = -1$. Hence the tangent line is $y - (\sqrt{2}/2) = -[x - (\sqrt{2}/2)]$.

This technique of finding y' directly when a relation involving x and y is given is called *implicit differentiation*. Implicit differentiation is particularly useful when it is inconvenient or impossible to solve a given equation for y.

EXAMPLE 4-19

Consider the relation

$$xy + 2y + x - 1 = 0$$

To find y' one could solve this equation for y and then use the Quotient Rule. Let us first use implicit differentiation.

$$D_x(xy + 2y + x - 1) = D_x 0$$
$$D_x xy + D_x 2y + D_x x + D_x(-1) = 0$$
$$y + x D_x y + 2 D_x y + 1 = 0$$
$$D_x y (x + 2) = -1 - y$$
$$D_x y = \frac{-1 - y}{x + 2}$$

Note that to calculate $D_x xy$ one needs the Product Rule, Rule D6. Applying the Product Rule yields

$$D_x xy = y D_x x + x D_x y$$
$$= y + x D_x y$$

If one solves the above equation $xy + 2y + x - 1 = 0$ directly for y one gets

$$y = \frac{1 - x}{x + 2}$$

then, using the Quotient Rule,

$$y' = \frac{-(x+2) - (1-x)}{(x+2)^2} = \frac{-3}{(x+2)^2}$$

If the technique of implicit differentiation is valid, then the two expressions

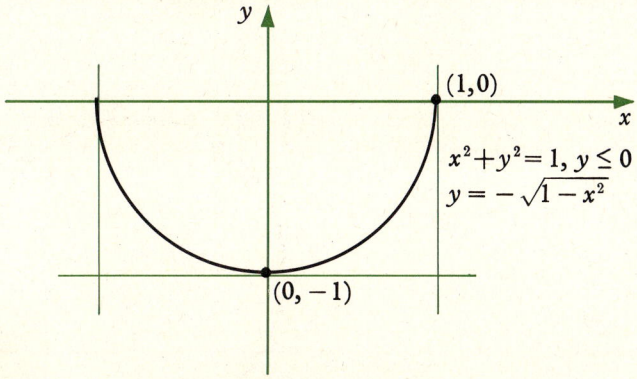

FIGURE 4-22

Since we are restricting y to positive values, we take

$$y = \sqrt{1-x^2}$$

Note that the above equation defines y as a function of x. Similarly, the lower half-circle (see Fig. 4-22) is given by $y = -\sqrt{1-x^2}$.

Other functions can be formed from $x^2+y^2=1$ by suitably restricting the domain and range of $y = \sqrt{1-x^2}$ and $y = -\sqrt{1-x^2}$. Each of these functions is said to be defined *implicitly* by the relation $x^2+y^2=1$.

Of course, when we write $f(x) = \sqrt{1-x^2}$ we have defined this function explicitly, but it behoves us to bypass the process of explicitly finding the functions which can be formed from a relation such as $x^2+y^2=1$, and simply say that y is an *implicit* function of x.

Since the relation $x^2+y^2=1$ is the union of two functions, it is possible to find the derivative of y with respect to x. Instead of solving for y, i.e., solving for the explicit functions above, we assume that y is a differentiable function of x and formally differentiate the equation $x^2+y^2=1$ and solve for $D_x y$.

$$x^2 + y^2 = 1$$
$$D_x(x^2+y^2) = D_x 1$$
$$D_x x^2 + D_x y^2 = 0$$
$$2x + 2y D_x y = 0$$
$$D_x y = -x/y$$

Observe that to calculate $D_x y^2$ we use the Chain Rule, in particular Rule D8 in Sec. 3-4 with $n = 2$. Thus $D_x y^2 = 2y D_x y$.

The significance of the fact that $D_x y = -x/y$ is that the slope of the tangent line to the circle $x^2 + y^2 = 1$ at any point (a, b) such that $b \neq 0$ is $-a/b$. (Why do we restrict b to be nonzero?)

EXAMPLE 4-18

The point $(\sqrt{2}/2, \sqrt{2}/2)$ lies on the circle $x^2 + y^2 = 1$. The slope of the tangent line to the circle at the point $(\sqrt{2}/2, \sqrt{2}/2)$ is $-(\sqrt{2}/2)/(\sqrt{2}/2) = -1$. Hence the tangent line is $y - (\sqrt{2}/2) = -[x - (\sqrt{2}/2)]$.

This technique of finding y' directly when a relation involving x and y is given is called *implicit differentiation*. Implicit differentiation is particularly useful when it is inconvenient or impossible to solve a given equation for y.

EXAMPLE 4-19

Consider the relation

$$xy + 2y + x - 1 = 0$$

To find y' one could solve this equation for y and then use the Quotient Rule. Let us first use implicit differentiation.

$$D_x(xy + 2y + x - 1) = D_x 0$$
$$D_x xy + D_x 2y + D_x x + D_x(-1) = 0$$
$$y + x D_x y + 2 D_x y + 1 = 0$$
$$D_x y (x + 2) = -1 - y$$
$$D_x y = \frac{-1 - y}{x + 2}$$

Note that to calculate $D_x xy$ one needs the Product Rule, Rule D6. Applying the Product Rule yields

$$D_x xy = y D_x x + x D_x y$$
$$= y + x D_x y$$

If one solves the above equation $xy + 2y + x - 1 = 0$ directly for y one gets

$$y = \frac{1 - x}{x + 2}$$

then, using the Quotient Rule,

$$y' = \frac{-(x+2) - (1-x)}{(x+2)^2} = \frac{-3}{(x+2)^2}$$

If the technique of implicit differentiation is valid, then the two expressions for y' must be equal. Substituting $y = (1-x)/(x+2)$ into $(-1-y)/(x+2)$ yields

$$\frac{-1-y}{x+2} = \frac{-1-[(1-x)/(x+2)]}{x+2}$$

$$= \frac{-(x+2)-(1-x)}{(x+2)^2}$$

$$= \frac{-3}{(x+2)^2}$$

as desired.

In the above example it was possible to solve for y to find y'. In the next few examples, implicit differentiation is the reasonable method to find y'.

EXAMPLE 4-20

Consider $x^2 + y^3 + xy^2 - 3 = 0$. To find $D_x y$, we differentiate both sides and get

$$D_x x^2 + D_x y^3 + D_x xy^2 - D_x 3 = 0$$

$$2x + 3y^2 D_x y + y^2 + 2xy D_x y = 0$$

$$D_x y (3y^2 + 2xy) = -2x - y^2$$

$$D_x y = \frac{-2x - y^2}{3y^2 + 2xy}$$

The point $(1,1)$ is on the curve. The slope of the tangent line at $(1,1)$ is given by

$$D_x y \Big|_{\substack{x=1 \\ y=1}} = \frac{-2-1}{3+2} = -\frac{3}{5}$$

Hence the tangent line to the curve at the point $(1,1)$ is $y-1 = (-3/5)(x-1)$.

EXAMPLE 4-21

To find the equation of the tangent line to the curve

$$4x^2 + xy^3 - x^4 y = 10$$

at the point $(1,2)$, we first find y' by differentiating both sides of the equation.

$$D_x 4x^2 + D_x xy^3 - D_x x^4 y = 0$$

$$8x + y^3 + 3xy^2 D_x y - 4x^3 y - x^4 D_x y = 0$$

$$D_x y (3xy^2 - x^4) = 4x^3 y - 8x - y^3$$

$$D_x y = \frac{4x^3 y - 8x - y^3}{3xy^2 - x^4}$$

$$D_x y \bigg|_{\substack{x=1 \\ y=2}} = \frac{8-8-8}{12-1} = -\frac{8}{11}$$

Hence the tangent line is $y - 2 = -\frac{8}{11}(x-1)$.

EXAMPLE 4-22

The volume V of a sphere is given by $V = \frac{4}{3}\pi r^3$ where r is the radius. If the sphere is being inflated, then the volume and the radius are changing with respect to time t. If the radius is increasing at the rate of 2 ft./min., it is possible to find the rate that the volume is increasing by implicitly differentiating the above formula with respect to time.

$$\begin{aligned} D_t V &= D_t(\tfrac{4}{3}\pi r^3) \\ &= \tfrac{4}{3}\pi\, D_t r^3 \qquad\qquad (Rule\ D2) \\ &= \tfrac{4}{3}\pi 3r^2\, D_t r \\ &= 4\pi r^2\, D_t r \\ &= 4\pi r^2 (2) \\ &= 8\pi r^2 \end{aligned}$$

Hence if the radius is 2 ft., the rate of change of the volume is $D_t V|_{r=2} = 32\pi$.

EXERCISE 4-3

1. Find $D_x y$.

   (a) $y + y^2 + x = 1$
   (b) $3y + 4x + 2 = 0$
   (c) $x + x^2 + y^3 + y = 1$
   (d) $3x + 5y^2 + 6 = 0$
   (e) $4x + y^{1/2} + y^2 = 0$
   (f) $x + xy + y = 3$
   (g) $x + xy^2 + y^2 = 3$
   (h) $x^2 - 5x^2 y^2 + y = 1$
   (i) $x^{1/2} - xy^{1/2} + y = 0$
   (j) $x + y^{-1} + xy^{-2} = 1$
   (k) $xy + x^2 y^2 + x^3 y^{1/2} = 1$
   (l) $(x+y)^{1/2} + x = 0$
   (m) $x(x-y)^{1/2} + y = 1$
   (n) $xy(x-y)^{1/2} + x = 1$

2. Consider the point $(a, \sqrt{1-a^2})$ on the unit circle, that is, the circle with radius 1 and center at the origin. Determine the slope of the tangent line to the circle at the point $(a, \sqrt{1-a^2})$ by considering the slope of the radius of the circle through the point $(a, \sqrt{1-a^2})$ and noting that this radius is perpendicular to the tangent. Does your answer agree with the formula given in the text?

3. Restrict the domain of $x^2 + y^2 = 4$ to define a function.

4. Express explicitly two functions which are defined implicitly by $x^2 + x + y^2 = 1$.

5. Find the equation of the tangent line to the curve at the given point.

(a) $x + xy - y^2 = 1$, $(1,1)$
(b) $2x - y^2 + y = 6$, $(3,1)$
(c) $4x + y^3 + xy^2 = 16$, $(1,2)$
(d) $x - 4y^{1/2} + x^2 y^2 = 9$, $(1,4)$
(e) $x^{1/2} - xy^{1/2} + y = 3$, $(1,4)$
(f) $x + 3y^{-1/2} + x^2 y^{1/2} = 5$, $(1,1)$
(g) $(x+y)^{1/2} + x = 5$, $(3,1)$
(h) $xy(x+y)^{1/2} + x = 7$, $(1,3)$

4-4 RELATED RATES

In Ex. 4-22 in the previous section one quantity, the volume V of a sphere, was expressed as a function of another quantity, the radius r. Both quantities were changing with respect to time t. The rate of change of V was related to the rate of change of r by a formula derived by implicitly differentiating the original formula with respect to t. Such a problem is referred to as a *related rates* problem.

EXAMPLE 4-23

A rock is thrown into a still pond and circular ripples move out, the radius of the disturbed region increasing at the rate of 3 ft./sec. To find the rate at which the surface area of the disturbed region is increasing, we note that the area A is a function of the radius r given by the formula $A = \pi r^2$. We are given that $D_t r = 3$ and the problem is to find $D_t A$. Implicitly differentiating the formula $A = \pi r^2$ with respect to time yields

$$D_t A = 2\pi r \, D_t r$$
$$= 2\pi r (3)$$
$$= 6\pi r$$

Thus when the farthest ripple is 20 ft. from the place where the rock struck the pond, the rate of change of the surface area is

$$D_t A \big|_{r=20} = 120\pi$$

where the left-hand side of the equation is read "the derivative of A with respect to t when $r = 20$".

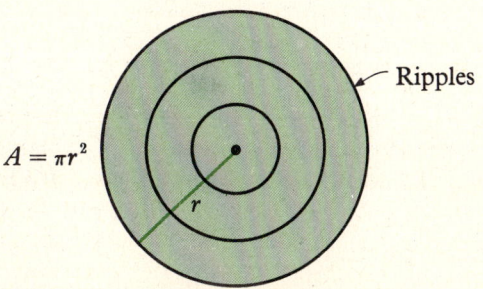

FIGURE 4-23

EXAMPLE 4-24

A company has determined that output y is related to input x, both measured in hundreds of units, by the formula

$$y = 2x^{1/2}$$

The firm determines that demand is increasing at the rate of 10 units per month, so that output is also increased at the rate of 10 units per month, i.e., $D_t y = 0.1$ where t is measured in months. If the current output is 800 units, i.e., $y = 8$ so that $x = 16$, to determine the rate of increase of x, we differentiate the equation with respect to t and get

$$D_t y = D_t 2x^{1/2}$$
$$= 2 \cdot \tfrac{1}{2} x^{-1/2} D_t x$$
$$= x^{-1/2} D_t x$$

Substituting the values $D_t y = 0.1$ and $x = 16$ into this equation yields

$$0.1 = \tfrac{1}{4} D_t x$$
$$D_t x = 0.4$$

Hence x must be increased by 40 units per month to meet the increase in demand.

EXAMPLE 4-25

A filter in the shape of an inverted cone is 10 cm high and has a radius of 20 cm. A solution is poured through the filter and residue gathers in the filter at the rate of 2 cm³/min. The volume of a cone is given by $V = \tfrac{1}{3}\pi r^2 h$ where r is the radius and h is the height. Since we are given $D_t V = 2$, we can find the rate at which the height of the residue is increasing if we can express V as a function of h alone, i.e., express r in terms of h, and then differentiate the formula with respect to time.

To express r in terms of h, compare the similar triangles shown in Fig. 4-24. Since the

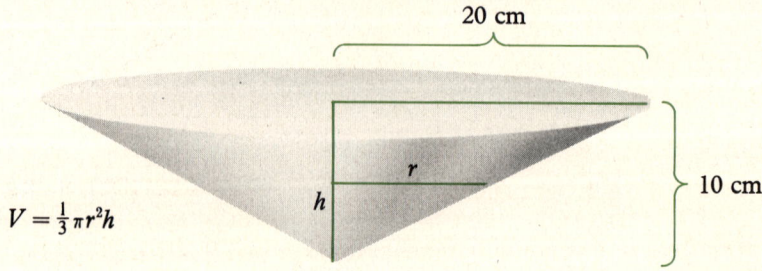

FIGURE 4-24

triangles are similar, we have $r/h = 20/10 = 2$ so that $r = 2h$. Substituting this relationship into the formula for V yields

$$V = \tfrac{1}{3}\pi(2h)^2 h$$
$$= \tfrac{4}{3}\pi h^3$$

Differentiating with respect to time yields

$$D_t V = \tfrac{4}{3}\pi 3 h^2 D_t h$$
$$2 = 4\pi h^2 D_t h$$
$$D_t h = \frac{2}{4\pi h^2} = \frac{1}{2\pi h^2}$$

If $h = 2$ then $D_t h|_{h=2} = 1/8\pi$ or about 0.04 cm/minute.

EXAMPLE 4-26

A tropical storm is 50 miles offshore and its path is perpendicular to the shore line. It is approaching the shore at the rate of 4 mph. Meteorologists studying the behavior of the storm from a truck on the shore wish to stay precisely 50 miles from the storm and remain on the shoreline. To determine the speed that the truck must maintain to remain 50 miles from the storm, we find a relationship between the distance y from the storm to the shore and the distance x that the truck has travelled. From Fig. 4-25 we have $x^2 + y^2 = 50^2$.

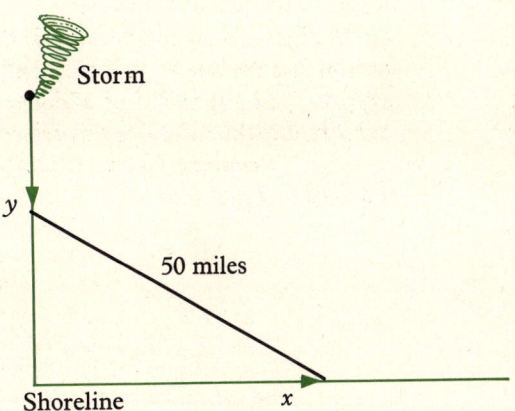

FIGURE 4-25

We are given that $D_t y = 4$ and the problem is to find $D_t x$. By differentiating $x^2 + y^2 = 2500$ we get

$$2x\, D_t x + 2y\, D_t y = 0$$

$$D_t x = \frac{-y D_t y}{x}$$

$$= \frac{-4y}{x}$$

Thus when the storm is 40 miles from shore, so that $y = 40$ and hence $x = 30$, the speed of the truck is

$$D_t x \bigg|_{\substack{x=30 \\ y=40}} = \frac{-4 \cdot 40}{30} = -\frac{16}{3}$$

or $5\frac{1}{3}$ mph. When the storm is 30 miles from shore, so that $y = 30$ and hence $x = 40$, the speed of the truck is

$$D_t x \bigg|_{\substack{x=40 \\ y=30}} = \frac{-4 \cdot 30}{40} = -3$$

or 3 mph.

EXAMPLE 4-27

Let us again consider the predator-prey model from ecology first introduced in Ex. 4-10. Let F be the size of the predator population and H the size of the prey population in a given territory. The relationship between F and H is of utmost importance in the balance of the environment. A mathematical model depicting this relationship should allow for certain fluctuations in F and H.

For example, as the predator population increases, more prey are killed for food and soon the prey population is so depleted that it cannot sustain the present predator population. Hence more predators die of lack of food and the predator population decreases. One such model uses the formula

$$0.2 F^{1/2} - 0.04 F + 2 H^{1/2} - 0.01 H = 20$$

If one of the rates $D_t F$ or $D_t H$ is known, then the other can be found by differentiating the above formula with respect to time. Suppose $D_t F$ is known and we desire to know $D_t H$. Differentiating, we get

$$0.1 F^{-1/2} D_t F - 0.04 D_t F + H^{-1/2} D_t H - 0.01 D_t H = 0$$

$$(H^{-1/2} - 0.01) D_t H = (0.04 - 0.1 F^{-1/2}) D_t F$$

$$D_t H = \frac{(0.04 - 0.1 F^{-1/2}) D_t F}{H^{-1/2} - 0.01}$$

For instance, if it is known that $D_t F = 0.02$ when $H = 1600$ and $F = 400$, so that $H^{-1/2} = 1/40 = 0.025$ and $F^{-1/2} = 1/20 = 0.05$, then

$$D_t H = \frac{(0.04 - 0.005)\,0.02}{0.025 - 0.01}$$

$$= \frac{.0007}{.015}$$

$$\cong .047$$

EXERCISE 4-4

1. A rock is thrown into a still pond and circular ripples move out, the radius of the disturbed region increasing at the rate of 2 ft./sec. Find the rate at which the surface area of the disturbed region is increasing when the farthest ripple is 10 ft. from the place where the rock struck the pond.

2. A firm determines that output y is related to input x, both measured in hundreds of units, by the formula $y = 3x^{1/3}$. If the demand is increasing at the rate of 20 units per month, at what rate must the input increase to meet this increase in demand if the current input is 2,700 units?

3. A filter in the shape of an inverted cone is 30 cm high and has a radius of 20 cm. A solution is poured through the filter and residue gathers in the filter at the rate of 5 cm^3/min. Find the rate at which the height of the residue is increasing when the height is 5 cm.

4. A tropical storm is 100 miles offshore and its path is perpendicular to the shore line. It is approaching the shore line at the rate of 5 mph. Meteorologists studying the behavior of the storm wish to stay precisely 100 miles from the storm and remain on the shoreline. Determine how fast they must travel along the shore when the storm is 60 miles from shore.

5. The relationship between the population size F of the predator and the population size H of the prey in a certain territory is determined to be

$$F^{1/2} - 0.3F + H^{1/2} - 0.1H = 30$$

Determine the rate of increase of the predator population if $H = 400$, $F = 100$ and $H' = 0.1$.

6. Water is being pumped into a cylindrical tank of radius 15 ft. at the rate of 25 ft.^{3}/min. How fast is the water level rising when the height of the water is 10 ft.?

7. Water is being pumped into a triangular trough at the rate of 10 ft.^{3}/min. If the depth of the trough is 12 ft., the width of the top is 12 ft., and the length of the top is 20 ft., at what rate is the height of the water rising when the height is 6 ft.?

8. A water tank is 30 ft. wide, 60 ft. long, and 10 ft. deep at one end and 15 ft. deep at the other end, with the depth decreasing linearly. Water is pumped in at the rate of 5 ft.^{3}/min. How fast is the water level rising when the water level is 3 ft. and 12 ft.?

9. A rubber balloon is being inflated with gas at a constant rate of 15 cm^3/sec. What is the rate of change of the radius when the volume is 45 cm^3?

10. A ladder 50 ft. long is standing against a building. The base of the ladder is moved away from the building at the rate of 3 ft./sec. How fast is the top of the ladder moving down the wall when the base of the ladder is 40 ft., 30 ft., and 20 ft. from the foot of the building?

11. Water is condensing on the surface of a spherical drop of water so that the surface remains a sphere and the volume is increasing at the rate of 5 in.^{3}/min. At what rate is the surface area and the radius increasing when the radius is 10 in.?

12. A 6 ft. tall man is walking away from a streetlight at the constant rate 3 ft./sec. If the streetlight is 20 ft. high, how fast is the length of the man's shadow increasing when he is 30 ft. from the streetlight?

13. A pulley, mounted on a dock, stands 10 ft. above the water and a boat is being towed in towards the dock. The rope is attached to the boat at a point 2 ft. above the water. If 1/2 ft. of rope is being drawn in each second, at what rate is the boat moving when it is 6 ft. from the dock?

14. The force of repulsion F between two electrical charges of like sign is inversely proportional to the distance d between them, where F is measured in newtons. It is observed that F is 10 when $d = 5$. If one charge approaches the other at the rate of 2 cm/min., at what rate is F increasing when $d = 5$?

15. The illuminance I of a light source varies inversely as the square of the distance d from the source, and $I = 20$ when $d = 5$. If d is increasing at the rate of 3 cm/min., at what rate is I decreasing when $d = 50$ cm?

4-5 THE MEAN VALUE THEOREM

We have thus far presented applications of the derivative for diverse fields outside of mathematics. In this section we give some theorems that will provide you with a more thorough understanding of the derivative.

One of the most useful theorems in mathematics is the *Mean Value Theorem*. In geometric terms it states that if a function is differentiable in the open interval (a, b) and continuous at a and b, then there exists a point $x_0 \in (a, b)$ such that the slope of the tangent line at x_0, $f'(x_0)$, is equal to the slope of the line containing $(a, f(a))$ and $(b, f(b))$, i.e., the latter and the tangent line at x_0 are parallel. (See Fig. 4-26.)

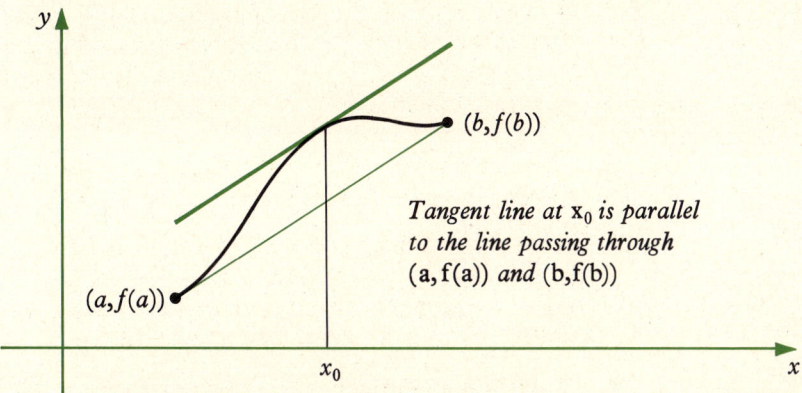

FIGURE 4-26

Theorem 4-2   *Mean Value Theorem*
If f is a function that is differentiable in the open interval (a, b) and continuous at a and b (and hence in $[a, b]$), then there exists at least one number $x_0 \in (a, b)$ such that

$$f'(x_0) = \frac{f(b) - f(a)}{b - a}$$

EXAMPLE 4-28
Consider the function $f(x) = x^2 + 2x + 1$. Now $f(x)$ is a polynomial function and is therefore differentiable on the whole real line, and in particular, is differentiable in $(0, 2)$, and certainly continuous at zero and 2. In this case $a = 0$, $b = 2$, so that

$$\frac{f(b) - f(a)}{b - a} = \frac{f(2) - f(0)}{2 - 0} = \frac{9 - 1}{2} = 4$$

while $f'(x) = 2x + 2$. If we set $f'(x) = 4$, and solve for x we have $2x + 2 = 4$, which implies $x = 1$. Hence

$$f'(1) = 4 = \frac{f(2) - f(0)}{2}$$

If we let $a = 1$, $b = 2$, and then solve

$$f'(x) = \frac{f(2) - f(1)}{2 - 1} = \frac{9 - 4}{1} = 5$$

we get $x = \frac{3}{2}$, so that $f'(\frac{3}{2}) = 5$. Hence $x_0 = \frac{3}{2}$ is the number that satisfies the conclusion of the Mean Value Theorem for $f(x) = x^2 + 2x + 1$ in $(1, 2)$.

We omit the proof of the Mean Value Theorem and now present some interesting theorems whose proofs are made accessible by the Theorem of the Mean. You will recognize the following theorem as Rule M-1.

Theorem 4-3 *If f is a function that is differentiable in the open interval (a, b) and continuous at a and b, and if $f'(x) > 0$ for all $x \in (a, b)$, then f is increasing in (a, b), i.e., if $a \leqslant x_1 < x_2 \leqslant b$, then $f(x_1) < f(x_2)$.*

PROOF Suppose $a \leqslant x_1 < x_2 \leqslant b$. We apply the Mean Value Theorem to f and (x_1, x_2). Hence there exists a number $x_0 \in (x_1, x_2)$ such that

$$f'(x_0) = \frac{f(x_2) - f(x_1)}{x_2 - x_1}$$

Note that $f'(x_0) > 0$ by hypothesis and since $x_2 - x_1 > 0$, we must have $f(x_2) - f(x_1) > 0$, which implies that $f(x_1) < f(x_2)$ as desired.

We know that if $f(x) = c$ is a constant function, then $f'(x) = 0$ for all x. The Mean Value Theorem can be used to show that the converse is also true, i.e., if $f'(x) = 0$ for all x, then $f(x) = c$ for some real number c.

Theorem 4-4 *If f is a function that is differentiable in the open interval (a, b) and continuous at a and b, and if $f'(x) = 0$ for all $x \in (a, b)$, then $f(x) = c$ for some real number c.*

PROOF Let x be any fixed number such that $x \in (a, b)$. We can apply the Mean Value Theorem to f and (a, x). Hence there exists a number $x_0 \in (a, x)$ such that

$$f'(x_0) = \frac{f(x) - f(a)}{x - a}$$

Now $f'(x_0) = 0$ by hypothesis and thus we are led to conclude that $f(x) = f(a)$. If we let $f(a) = c$, we have $f(x) = c$ for all $x \in (a, b)$ as desired.

We know that if $f(x) = g(x)$, then $f'(x) = g'(x)$. What about the converse? If $f'(x) = g'(x)$, it is not necessarily true that $f(x) = g(x)$. Note that

$$D_x x^2 = 2x = D_x(x^2 + 1)$$

The following theorem does give us some information on this idea, which will prove to be vital a little further on.

Theorem 4-5 *If f and g are functions that are differentiable in the open interval (a, b) and continuous at a and b, and if $f'(x) = g'(x)$ for all $x \in (a, b)$, then $f(x) = g(x) + c$ for some real number c and for every $x \in (a, b)$.*

PROOF Let $h(x) = f(x) - g(x)$. We want to show that $h(x)$ is a constant function on (a, b). We can apply Theorem 4-4 to $h(x)$ and (a, b) so that $h(x) = c$ for some real number c for all $x \in (a, b)$. Hence $f(x) - g(x) = c$ and so $f(x) = g(x) + c$.

EXERCISE 4-5

1. Verify that the functions satisfy the hypothesis of the Mean Value Theorem in the given open interval, and then find the number x_0 that satisfies the conclusion.

   (a) $f(x) = 3x + 1$, $(-3, 8)$

   (b) $f(x) = x^2 + 1$, $(-1, 1)$

   (c) $f(x) = x^2 + 1$, $(-1, 5)$

   (d) $f(x) = x^3 - 2x^2$, $(0, 1)$

   (e) $f(x) = x^3 - x^2 + 2$, $(-1, 1)$

   (f) $f(x) = \dfrac{1}{x}$, $(1, 2)$

   (g) $f(x) = x + \dfrac{1}{x}$, $(1, 2)$

2. State why the functions and the given intervals fail to satisfy the hypotheses of the Mean Value Theorem.

   (a) $f(x) = \dfrac{1}{x}$, $(-1, 1)$

   (b) $f(x) = \dfrac{1}{x}$, $(0, 1)$

   (c) $f(x) = [\![x]\!]$, $(0, \tfrac{1}{2})$

   (d) $f(x) = |x|$, $(-1, 1)$

3. Use the Mean Value Theorem to prove: If f is a function that is differentiable in $[a, b]$ and $f'(x) < 0$ for all $x \in (a, b)$, then f is decreasing in (a, b).

4. Use the Mean Value Theorem to prove Rolle's Theorem: If f is a function that is differentiable in (a, b) and continuous at a and b and if $f(a) = f(b) = 0$, then there exists a number $x_0 \in (a, b)$ such that $f'(x_0) = 0$.

5. Find an example of a function $f(x)$ and an interval (a, b) which satisfies the hypothesis of the Mean Value Theorem and for which there exists more than one number x_0 satisfying the conclusion.

6. Suppose f and g are two functions such that $f'(x) = g'(x)$ for all x and $g(1) = f(1)$. Show that $f(x) = g(x)$ by using Theorem 4-5.

7. Find two functions f and g such that $f'(x) = g'(x)$ for all x but that $f(x) \neq g(x)$.

4-6 L'HÔPITAL'S RULE

L'Hôpital's Rule is a very useful technique for evaluating certain limits. Most limits can be evaluated by simply plugging in the appropriate number. That is, if $f(x)$ and $g(x)$ are continuous at a, then

$$\lim_{x \to a} \frac{f(x)}{g(x)} = \frac{f(a)}{g(a)}$$

providing of course that $g(a) \neq 0$. For example,

$$\lim_{x \to 3} \frac{x^2 - 4}{x - 2} = \frac{3^2 - 4}{3 - 2} = \frac{5}{1} = 5$$

But if $g(a) = 0$ above, this method doesn't work. For example,

$$\lim_{x \to 2} \frac{x^2 - 4}{x - 2}$$

cannot be evaluated by plugging 2 into the numerator and denominator, since the result is $0/0$ which is meaningless. However, in Chap. 2 we discovered that this limit does exist and we evaluated it by factoring the numerator and cancelling the common factor in the denominator. That is,

$$\lim_{x \to 2} \frac{x^2 - 4}{x - 2} = \lim_{x \to 2} \frac{(x+2)(x-2)}{x - 2}$$
$$= \lim_{x \to 2} (x + 2)$$
$$= 4$$

It is often the case that $f(a) = g(a) = 0$ and $f(x)$ and $g(x)$ have no common factor, or the common factor is not apparent. In this case L'Hôpital's Rule can be applied.

Theorem 4-6 (*L'Hôpital's Rule*). *If $f(x)$ and $g(x)$ are differentiable functions at $x = a$ such that $f(a) = g(a) = 0$ but $g(x) \neq 0$ for all x close to a, then*

$$\lim_{x \to a} \frac{f(x)}{g(x)} = \lim_{x \to a} \frac{f'(x)}{g'(x)}$$

provided that $g'(a) \neq 0$.

More intuitively if $f(a)/g(a) = 0/0$, then consider $f'(a)/g'(a)$. If the latter exists it is the same number as $\lim_{x \to a} f(x)/g(x)$.

EXAMPLE 4-29

Consider

$$\lim_{x \to 2} \frac{x^2 - 4}{x - 2}$$

The function $(x^2 - 4)/(x - 2)$ satisfies the hypothesis of L'Hôpital's Rule with $f(x) = x^2 - 4$, $g(x) = x - 2$ and $a = 2$. Note that $f(2) = g(2) = 0$. Applying L'Hôpital's Rule with $f'(x) = 2x$ and $g'(x) = 1$ we get

$$\lim_{x \to 2} \frac{x^2 - 4}{x - 2} = \lim_{x \to 2} \frac{2x}{1} = 4$$

EXAMPLE 4-30

Consider

$$\lim_{x \to 1} \frac{x^3 - 2x + 1}{x - 1}$$

If $f(x) = x^3 - 2x + 1$ and $g(x) = x - 1$, then $f(1) = g(1) = 0$. The factorization of $f(x)$ is not apparent (it can be found by dividing $g(x)$ into $f(x)$) so that the limit is best evaluated by L'Hôpital's Rule. Accordingly,

$$\lim_{x \to 1} \frac{x^3 - 2x + 1}{x - 1} = \lim_{x \to 1} \frac{3x^2 - 2}{1} = 1$$

EXAMPLE 4-31

By L'Hôpital's Rule

$$\lim_{x \to 2} \frac{x^2 - 2x}{x^4 - 4x - 8} = \lim_{x \to 2} \frac{2x - 2}{4x^3 - 4} = \frac{2}{28} = \frac{1}{14}$$

It may be the case that L'Hôpital's Rule must be applied more than once, as in the next example.

EXAMPLE 4-32

By L'Hôpital's Rule

$$\lim_{x \to 0} \frac{x^3 - 2x^2}{x^4 + x^2} = \lim_{x \to 0} \frac{3x^2 - 4x}{4x^3 + 2x}$$

To evaluate the limit on the right we again apply L'Hôpital's Rule and get

$$\lim_{x \to 0} \frac{3x^2 - 4x}{4x^3 + 2x} = \lim_{x \to 0} \frac{6x - 4}{12x + 2} = \frac{-4}{2} = -2$$

Hence

$$\lim_{x \to 0} \frac{x^3 - 2x^2}{x^4 + x^2} = -2$$

Be advised that one can only apply L'Hôpital's Rule if $f(a) = g(a) = 0$. In the above example if we tried to apply the rule again, that is, to

$$\lim_{x \to 0} \frac{6x - 4}{12x + 2}$$

we would have this limit equal to

$$\lim_{x \to 0} \frac{6}{12} = \frac{1}{2}$$

which is the wrong answer.

EXAMPLE 4-33

Consider

$$\lim_{x \to 0} \frac{x^3 + x^2}{e^x - x - 1} = \lim_{x \to 0} \frac{3x^2 + 2x}{e^x - 1}$$

The numerator and denominator are both 0 when evaluated at 0, so we again apply L'Hôpital's Rule and get

$$\lim_{x \to 0} \frac{3x^2 + 2x}{e^x - 1} = \lim_{x \to 0} \frac{6x + 2}{e^x} = \frac{2}{1} = 2$$

EXERCISE 4-6

1. Evaluate the limits.

   (a) $\lim_{x \to 3} \dfrac{x^2 - 9}{x - 3}$

   (b) $\lim_{x \to 1} \dfrac{x^3 - 2x + 1}{x^2 - 1}$

   (c) $\lim_{x \to 1} \dfrac{x^4 - 2x^2 + 1}{x - 1}$

   (d) $\lim_{x \to 1} \dfrac{x^2 - 1}{x^4 - 2x^2 + 1}$

   (e) $\lim_{x \to 1} \dfrac{x^4 - 2x^2 + 1}{x^3 - 3x + 2}$

   (f) $\lim_{x \to 1} \dfrac{\ln x}{x^2 - 1}$

   (g) $\lim_{x \to 1} \dfrac{x \ln x}{x^3 - e^{1-x}}$

   (h) $\lim_{x \to 0} \dfrac{e^x - x - 1}{x^2}$

2. It may be the case that the numerator and denominator are both unbounded at a, i.e., $\lim_{x \to a} f(x) = \lim_{x \to a} g(x) = \pm \infty$. If so, L'Hôpital's Rule implies that $\lim_{x \to a} f'(x)/g'(x)$ is equal to the original limit, if it exists. For example, if $f(x) = \ln x$ and $g(x) = x^{-1}$, then $\lim_{x \to 0} \ln x = -\infty$ and $\lim_{x \to 0} x^{-1} = \infty$. Hence, applying L'Hôpital's Rule,

$$\lim_{x \to 0} \frac{\ln x}{x^{-1}} = \lim_{x \to 0} \frac{(1/x)}{-x^{-2}}$$

$$= \lim_{x \to 0} \frac{-x^2}{x}$$

$$= \lim_{x \to 0} -x = 0$$

Use this version of L'Hôpital's Rule to evaluate the limits.

(a) $\lim\limits_{x \to 0} \dfrac{\ln x}{x^{-2}}$

(b) $\lim\limits_{x \to 0} \dfrac{\ln x + x}{x^{-1}}$

(c) $\lim\limits_{x \to 0} x^x$

(Hint: $x^x = e^{\ln x^x} = e^{x \ln x}$)

4–7 ANTIDERIVATIVES

Thus far in our development of the calculus we have been concerned with deriving information about a function $f(x)$ by examining its derivative $f'(x)$. Often in practical situations we are given the derivative $f'(x)$ and we need to find $f(x)$ from it. If $f(x) = x^2$, then $f'(x) = 2x$. However, suppose $g'(x) = x^2$, can we, then, find $g(x)$? In this case it is easy to see that $g(x) = \frac{1}{3}x^3$ is one solution to our problem. We thus call $g(x) = \frac{1}{3}x^3$ an *antiderivative* of $f(x) = x^2$.

Definition $g(x)$ is an *antiderivative* of $f(x)$ if $g'(x) = f(x)$, that is, if $f(x)$ is the derivative of $g(x)$.

Note that $D_x(\frac{1}{3}x^3 + 7) = x^2$, so $g(x) = \frac{1}{3}x^3 + 7$ is also an antiderivative of $f(x) = x^2$, and in general, if $g(x)$ is an antiderivative of $f(x)$, then $g(x) + C$ for any constant C is also an antiderivative of $f(x)$, because $D_x C = 0$.

Suppose $g(x)$ is an antiderivative of $f(x)$, so $g'(x) = f(x)$. Then $g(x) + C$ for any constant C is an antiderivative; but could there exist another function, say, $h(x)$, which is an antiderivative of $f(x)$ and not of the form $g(x) + C$? If $h(x)$ is an antiderivative of $f(x)$, then $h'(x) = f(x) = g'(x)$. Theorem 4-5 (Sec. 4-5) allows us to conclude that $h(x) = g(x) + C$ for some constant C. Therefore every antiderivative of $f(x)$ is of the form $g(x) + C$.

The symbol $\int f(x)\,dx$ is used to indicate the antiderivative of $f(x)$. Thus $\int 2x\,dx = x^2 + C$. The dx is used to identify the variable. It plays the same role as the x in the notation for the derivative, D_x. Hence

$$\int x t^2\,dx = \frac{x^2 t^2}{2} + C$$

where t is not a function of x, whereas

$$\int x t^2\,dt = \frac{x t^3}{3} + C$$

where x is not a function of t.

Now suppose $D_x f(x) = g(x)$. We can then conclude that

$$\int g(x)\, dx = f(x) + C$$

by definition of the antiderivative. Therefore any formula involving a derivative can be restated involving antiderivatives.

Recall Rule D1: $D_x x^r = rx^{r-1}$. Equivalently, we can write

$$D_x \frac{x^{r+1}}{r+1} = x^r, \qquad r \neq -1$$

which states that the antiderivative of x^r is

$$\frac{x^{r+1}}{r+1}$$

hence

$$\int x^r\, dx = \frac{x^{r+1}}{r+1} + C$$

We record this as:

Rule I1 $\quad \int x^r\, dx = \dfrac{x^{r+1}}{r+1} + C$ *for all* $r \neq -1$

We can similarly rewrite Rules D2 and D3.

Rule I2 $\quad \int a f(x)\, dx = a \int f(x)\, dx$

Rule I3 $\quad \int (f(x) + g(x))\, dx = \int f(x)\, dx + \int g(x)\, dx$

EXAMPLE 4-34

Let $f(x) = x^2$ and $g(x) = x^3$. Then by Rules I2 and I1

$$\int 5x^2\, dx = 5 \int x^2\, dx = 5\frac{x^3}{3} + C$$

From Rules I3 and I1 we have

$$\int (x^2 + x^3)\, dx = \int x^2\, dx + \int x^3\, dx = \frac{x^3}{3} + \frac{x^4}{4} + C$$

We discuss rules D4 through D7 in a later chapter.

We can list the antiderivatives of familiar functions in a table for easy reference. Each entry in the table can be verified by finding the derivative of the function on the right and noting that it is equal to the function on the left.

TABLE 4-1

$f(x)$	$\int f(x)\,dx$
$x^r,\ r \neq -1$	$\dfrac{x^{r+1}}{r+1} + C$
$a_n x^n + a_{n-1} x^{n-1} + \cdots + a_0$	$\dfrac{a_n}{n+1} x^{n+1} + \dfrac{a_{n-1}}{n} x^n + \cdots + a_0 x + C$
$\dfrac{1}{x}$	$\ln|x| + C$
e^{ax}	$\dfrac{1}{a} e^{ax} + C$

Note that the first entry above is simply a restatement of Rule I1, while the second entry can be derived from Rules I1, I2, and I3. Since $1/x$ is defined for negative numbers and $\ln x$ is not, it is necessary to insert the absolute value of x, i.e., $\ln|x|$, in the third formula.

EXAMPLE 4-35

Let us calculate a few antiderivatives using Rules I1, I2, I3 and Table 4-1.

1. $\int (x+1)\,dx = \int x\,dx + \int 1\,dx = \dfrac{x^2}{2} + x + C$

2. $\int \left(x^3 + \dfrac{1}{x} \right) dx = \int x^3\,dx + \int \dfrac{1}{x}\,dx = \dfrac{x^4}{4} + \ln|x| + C$

3. $\int (e^x + x^{1/2})\,dx = \int e^x\,dx + \int x^{1/2}\,dx = e^x + \tfrac{2}{3} x^{3/2} + C$

4. $\int (e^{3x} + 5x^{-4/3})\,dx = \int e^{3x}\,dx + 5\int x^{-4/3}\,dx = \tfrac{1}{3} e^{3x} + \dfrac{5x^{-1/3}}{-1/3} + C =$
$\tfrac{1}{3} e^{3x} - 15 x^{-1/3} + C$

5. $\int \dfrac{6}{e^{5x}}\,dx = 6\int e^{-5x}\,dx = \dfrac{6 e^{-5x}}{-5} + C$

EXAMPLE 4-36

A company determines that the marginal cost MC function for a certain product is given by $MC = x^2 + 2x$ where the fixed cost is 15. By definition, $C'(x) = MC$ where $C(x)$ is the cost function. Hence

$$C(x) = \int MC$$
$$= \int (x^2 + 2x)\, dx$$
$$= \frac{x^3}{3} + x^2 + k$$

for some constant k. Since the fixed cost is 15, we have that $C(0) = 15 = k$. Hence

$$C(x) = \frac{x^3}{3} + x^2 + 15$$

EXAMPLE 4-37

We know that the acceleration due to gravity is $a(t) = -32$ ft/sec^2. We also know that the velocity $v(t)$ of a free falling object is $v(t) = \int a(t)\, dt$. Thus $v(t) = -32t + C$. To calculate C we note that the initial velocity v_0 is that value of $v(t)$ when $t = 0$, i.e., $v_0 = v(0)$. Thus $C = v_0$ and $v(t) = -32t + v_0$. The position $s(t)$ of the object is

$$s(t) = \int (-32t + v_0)\, dt$$
$$= -16t^2 + v_0 t + C_1$$

If the initial position of the object is s_0, then

$$s(t) = -16t^2 + v_0 t + s_0$$

which is the general equation of the position of a free falling object whose initial position and velocity are s_0 and v_0.

Many interesting applications of calculus lead to equations involving derivatives which are called *differential equations*. In fact, whenever we find an antiderivative, we are solving a differential equation of the form $f'(x) = g(x)$ where $g(x)$ is given and $f(x)$ is the unknown.

EXAMPLE 4-38

Solve the differential equation $f'(x) = 3x^2 + 4x - 5$. We have

$$f(x) = \int f'(x)\, dx$$
$$= \int (3x^2 + 4x - 5)\, dx$$
$$= x^3 + 2x^2 - 5x + C$$

We thus have a whole family of solutions. If one point of the solution is also given, which is called a *boundary value*, say $f(0) = 1$, then we get one particular function $f(x) = x^3 + 2x^2 - 5x + 1$.

One very common and powerful type of differential equation is the one of the form

$$f'(t) = kf(t)$$

for some constant k, which some mathematicians refer to as a *growth equation*. The next two examples illustrate one application in archaeology where the growth equation plays a fundamental role.

EXAMPLE 4-39

The rate at which radioactive material changes to lead is proportional to the amount of the material present, i.e., if $f(t)$ is the amount present, then $f'(t) = kf(t)$ for some constant k. Hence we seek a function $f(t)$ whose derivative is a constant times itself. We know of only one such function, namely $f(t) = Ce^{kt}$. Note that with this choice of $f(t)$, we have $f'(t) = Cke^{kt} = kf(t)$ as desired. In more advanced texts it is shown that this is the only solution.

EXAMPLE 4-40

Living plant and animal tissues contain radioactive carbon (C-14), which has a *half-life* of 5,568 years (at the end of that period of time one-half of the original amount remains). When the organism dies, no new carbon is received. We can develop a formula for the amount of C-14 remaining after t years. From Ex. 4-22, $f(t) = Ce^{kt}$. Now $f(0) = Ce^0 = C$, so C is the amount present when $t = 0$. When $t = 5568$, $f(0)/2$ is present, and thus

$$f(5568) = \frac{f(0)}{2} = f(0) e^{5568k}$$

Hence $\frac{1}{2} = e^{5568k}$, and $\ln \frac{1}{2} = \ln e^{5568k} = 5568k$ which gives us $k = \frac{1}{5568} \ln \frac{1}{2}$.

Archaeologists use this function to date uncovered ruins. Suppose that analysis of ashes found in an unearthed Egyptian settlement reveals that one-fifth of the C-14 present in the original ashes has decomposed. Then $4f(0)/5 = f(0)e^{kt}$ implies that $\ln \frac{4}{5} = \ln e^{kt} = kt$. Thus

$$t = \frac{1}{k} \ln \frac{4}{5} = 5568 \left(\frac{\ln \frac{4}{5}}{\ln \frac{1}{2}} \right) = 5568 \left(\frac{\ln 4 - \ln 5}{\ln 1 - \ln 2} \right)$$

$$\cong 5568 \left(\frac{1.386 - 1.609}{0 - 0.693} \right) \cong 1{,}730 \text{ years}$$

Hence the settlement existed approximately 1,730 years ago.

EXERCISE 4-7

1. Find the antiderivatives.

   (a) $\int (5x+7)\, dx$

   (b) $\int (8x^2 - 6x + 2)\, dx$

   (c) $\int (9x^3 - x^{1/2})\, dx$

   (d) $\int (x^2 - 1)^2\, dx$

   (e) $\int x(x^2 - 1)^2\, dx$

   (f) $\int (x^{3/5} - 6x^{1/2})\, dx$

   (g) $\int (x^{-1/2} - 7x^{1/2})\, dx$

   (h) $\int (x^2 - 6x^{-9/2})\, dx$

   (i) $\int e^{6x}\, dx$

   (j) $\int \left(\frac{1}{8x} + \frac{8}{x}\right) dx$

   (k) $\int \left(e^{2x} + \frac{1}{x}\right) dx$

   (l) $\int (e^{-8x} + \pi)\, dx$

2. Suppose $f(x)$ and $g(x)$ are both antiderivatives of the same function $h(x)$. If you know the shape of the graph of $f(x)$, what can you say about the graph of $g(x)$?

3. Rule I3 states that the antiderivative of a sum is the sum of the antiderivatives. You may be tempted to believe that the antiderivative of a product is the product of the antiderivatives. However, this is in general false. Let $f(x) = x$ and $g(x) = x^2$. Show that

   $$\int fg\, dx \neq \int f\, dx \cdot \int g\, dx$$

   In fact, it is a fairly difficult problem to find functions f and g such that

   $$\int fg\, dx = \int f\, dx \cdot \int g\, dx$$

   It would be instructive for you to try.

4. A company determines that the marginal cost MC function for a certain product is given by $MC = x^2 + 2x$. Find the cost function if the fixed cost is 10.

5. A company determines that the marginal revenue MR function for a certain product is given by $MR = x^3 - x + 10$. Find the revenue function.

6. The rate of change $R'(D)$ of the reaction of the body $R(D)$ to a dose of blood D is given by $R'(D) = DC - D^2$ where C is a constant. Find $R(D)$.

7. Suppose the rate of growth $f'(t)$ of the weight of the limb of a body $f(t)$ with respect to age t is given by $f'(t) = kt^n$ for some constant k. Find $f(t)$.

8. In thermodynamics it is assumed that the rate of increase of work W that an ideal gas makes when its volume V expands is given by $dW/dV = nRT/V$ where n and R are constants depending upon the particular gas and the temperature T is held constant. Find W in terms of V.

9. Infusion is the biological process of admitting a substance into the veins at a steady rate. Let A be the amount of glucose admitted and let V be the volume of the liquids in the body. If c is the concentration of glucose, then the rate of change in the concentration of the glucose with respect to time dc/dt is often assumed to be approximated by $dc/dt = k(A/V)$ where k and V are constants and A depends upon t. Suppose $A = t^2 + 3t$. Find $c(t)$ where the infusion of the concentration of glucose at the beginning ($t = 0$) was c_0.

10. Oxygen is transferred from the lungs to the rest of the body by a substance in the blood called hemoglobin. The rate of combination of hemoglobin and oxygen is proportional to the concentration of oxygen in the air. If we let C be the rate of combination of hemoglobin and oxygen molecules and t the concentration of oxygen in the air, we have $C(t) = kt$ for some constant k depending upon the pressure of oxygen among other factors. Find the function relating the amount of hemoglobin and oxygen molecules combined with the concentration of oxygen in the air.

11. A simplified mathematical model of a muscle divides operation of the muscle into two parts: one part contracts with force and the other stretches according to the laws of elasticity. This second force is a force of friction which is proportional to the rate of contraction. Let l be the length of the extension and F be force required of the muscle. Then we have $F = k\, dl/dt$. If $F = t^2 - t$, find l.

12. Suppose archaeologists determine that a certain campsite of American Indians contains ashes whose C-14 content is five-sixths of the original amount of ashes. Find the date of the campsite.

13. Suppose an object is thrown downward and it leaves the hand with a speed of 10 ft per sec. Find the position function $s(t)$, velocity function $v(t)$, and the acceleration function $a(t)$ for the object.

14. Suppose an object is thrown upward and it leaves the hand with a speed of 64 ft per sec. Find the position function $s(t)$, the velocity function $v(t)$, and the acceleration function $a(t)$ for the object. How far up does the object travel? When does it hit the ground?

5

The Integral

5-1 AREA AND THE DEFINITE INTEGRAL

The study of calculus can be divided into two main parts, differential and integral calculus. In differential calculus one uses the derivative to study problems dealing with motion or rate of change. Integral calculus deals with problems concerning length, area, and volume by investigating the concept called the *definite integral* of a function $f(x)$, written $\int_a^b f(x) \, d(x)$. The close connection between these two theories is given in the remarkably beautiful and important Fundamental Theorem of Calculus.

The rigorous definition of the definite integral is quite difficult and we put off the task of defining it until the end of this section. First we center our attention on one very important application of the definite integral, that of finding the area in the plane bounded by a curve. The role of this application parallels that of the slope of the tangent line which was used to define the derivative. That is, the geometrical significance of area not only provides insight to the definition but also gives us a firm base on which we can develop many of the properties of the definite integral.

You know how to find the area of certain regions, for example, a square, a rectangle, a triangle. The early Greeks even knew how to find areas of some regions that were defined by specific continuous curves. But the general question, that of finding the area underneath an arbitrary continuous curve, was not solved until about 1700 with the advent of the calculus.

The procedure used to calculate this type of area entails the use of approximations to the area. Let $f(x)$ be a function which is continuous and nonnegative in an interval $[a, b]$. We will develop a process for finding the area bounded by $f(x)$, the x-axis, and the lines $x = a$ and $x = b$ by first fitting appropriate rectangles into this area and approximating the area by summing the areas of these rectangles (see Fig. 5-1). Then we construct better and better approximations which will lead us to a limit process.

We divide the interval $[a, b]$ into n subintervals of equal width, $(b-a)/n$. These subintervals are

$$\left[a,\ a + \frac{b-a}{n}\right],\ \left[a + \frac{b-a}{n},\ a + 2 \cdot \frac{b-a}{n}\right],\ \ldots,\ \left[a + (n-1)\frac{b-a}{n},\ b\right]$$

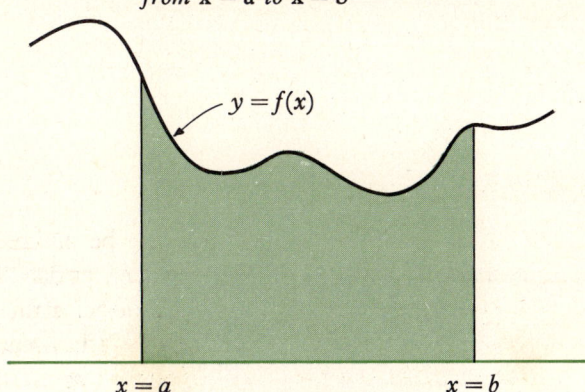

An area under the curve $y = f(x)$ from $x = a$ to $x = b$

FIGURE 5-1

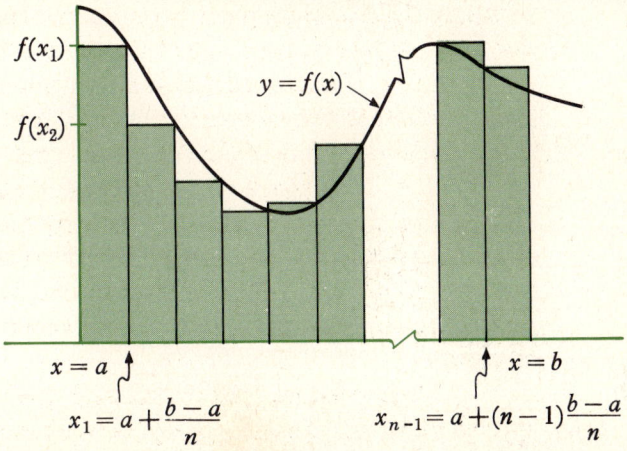

An approximation to the area under $y = f(x)$ from $x = a$ to $x = b$ by using rectangles

FIGURE 5-2

Let x_i be the right-hand endpoint of the ith subinterval which is
$$a + i \cdot \frac{b-a}{n}$$
We now construct n rectangles, each of width $(b-a)/n$ and length $f(x_i)$ for $i = 1, \ldots, n$. The area of each rectangle is
$$\frac{b-a}{n} f(x_i)$$
If we sum all of these areas we get a number that is an approximation to the area under $f(x)$ from $x = a$ to $x = b$, as in Fig. 5-2.

EXAMPLE 5-1

Let $f(x) = x^2$ and $a = 0$ and $b = 2$. If we choose $n = 4$, each rectangle has width $(2-0)/4 = 1/2$. Since
$$x_1 = \tfrac{1}{2} \qquad x_2 = 1 \qquad x_3 = \tfrac{3}{2} \qquad \text{and} \qquad x_4 = 2$$
we have
$$f(x_1) = f(\tfrac{1}{2}) = \tfrac{1}{4}$$
$$f(x_2) = f(1) = 1$$
$$f(x_3) = f(\tfrac{3}{2}) = \tfrac{9}{4}$$

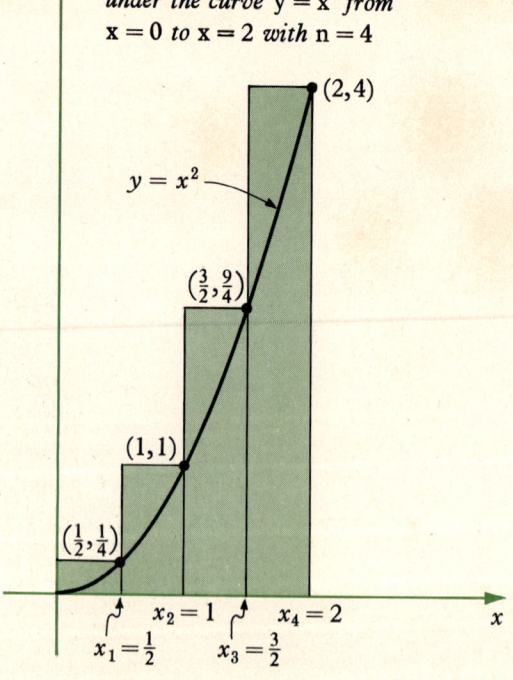

FIGURE 5-3

and
$$f(x_4) = f(2) = 4$$
Our approximation is

$$\tfrac{1}{2}f(x_1) + \tfrac{1}{2}f(x_2) + \tfrac{1}{2}f(x_3) + \tfrac{1}{2}f(x_4) = \tfrac{1}{2}(\tfrac{1}{4}+1+\tfrac{9}{4}+4) = \tfrac{1}{2}(\tfrac{30}{4}) = \tfrac{15}{4}$$

From Fig. 5-3 one can readily see that the number $\tfrac{15}{4}$ is larger than the area under the curve. For a closer approximation, or in other words, to reduce the amount of excess area included in our rectangles which is not in the area under $f(x) = x^2$, note that we may simply choose a larger

EXAMPLE 5-2

If we again let $f(x) = x^2$, $a = 0$, and $b = 2$ but now choose $n = 8$ in Fig. 5-4, then each rectangle has width $(2-0)/8 = 1/4$. Our approximation is

$$\tfrac{1}{4}f(x_1) + \tfrac{1}{4}f(x_2) + \tfrac{1}{4}f(x_3) + \tfrac{1}{4}f(x_4) + \cdots + \tfrac{1}{4}f(x_8)$$
$$= \tfrac{1}{4}(\tfrac{1}{16}+\tfrac{1}{4}+\tfrac{9}{16}+1+\tfrac{25}{16}+\tfrac{9}{4}+\tfrac{49}{16}+4) = \tfrac{51}{16}$$

To get a still closer approximation, we can again choose a larger n.

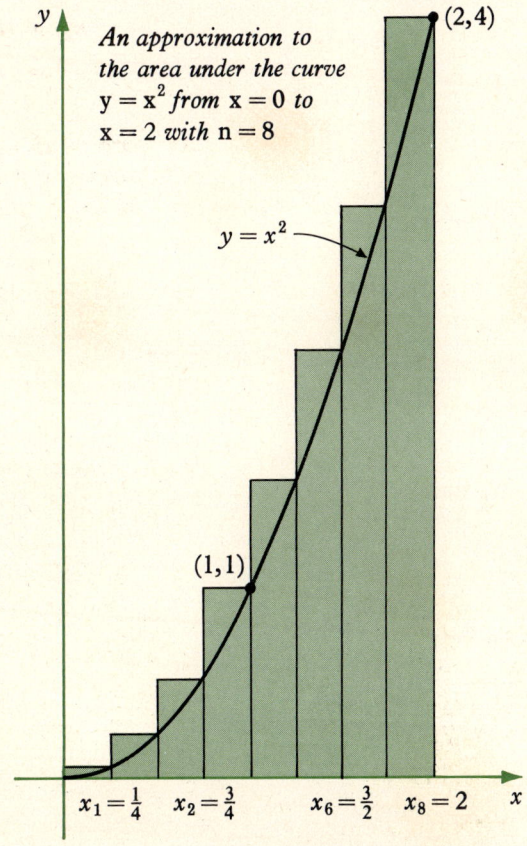

An approximation to the area under the curve $y = x^2$ from $x = 0$ to $x = 2$ with $n = 8$

FIGURE 5-4

EXAMPLE 5-3

A cardiologist measures his patients' physical endurance by a series of tests, the first of which is a ten-minute run on a treadmill. Initially the speed of the treadmill is set at 5 mph. The speed is gradually increased but only as the physical capacity of the patient allows. The cardiovascular system of the patient is monitored by a number of sensors which allow the speed of the treadmill to increase at a faster rate if the patient is in good physical condition. An assistant records the speed of the treadmill at two-minute intervals. The data is given in the following table.

Time (min)	0	2	4	6	8	10
Speed (mph)	5	5.4	5.61	5.7	5.9	6

Thus the initial speed was 5 mph at $t = 0$. After two minutes the patient was running at the speed of 5.4 mph. How far did the patient run in ten minutes?

Consider first the distance travelled in the first two minutes. Since we don't know his actual speed throughout the interval, we can only approximate how far he ran. If we assume that he ran at the speed of 5.4 mph for the full two minutes, then the distance travelled would be given by $(5.4)(2)(1/60) = 0.18$ miles since he ran for $2(1/60)$ hours. Of course, this is the highest speed which he attained in the interval, so he didn't quite run 0.18 miles but it is a fairly good approximation.

Using this same approximating technique in the second two-minute interval, from $t = 2$ to $t = 4$, the approximate distance travelled is $(5.61)(2)(1/60) = 0.187$ miles. This same procedure can be applied to the three remaining intervals and the data is given in a table and a bar graph.

Time (min)		0	2	4	6	8	10
Speed (mph)		5	5.4	5.61	5.7	5.91	6
Approximate	in interval	0	0.18	0.187	0.19	0.197	2.0
distance	total		0.18	0.367	0.557	0.754	0.954

When one considers the bar graph, one sees that the approximate distance travelled in a two-minute interval is the area of the appropriate rectangle, times 1/60, since the width of the rectangle represents the time, two minutes, and the height is the speed of the treadmill.

To get a better approximation to the total distance travelled, the speed could be recorded at more frequent subintervals, say every minute, or every 30 seconds. The appropriate table would have more entries and the bar graph would have more rectangles. Better approximations are obtained by letting the number of rectangles, i.e., the number of recordings, approach infinity. This is precisely the language used to discuss the definite integral.

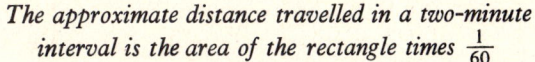

The approximate distance travelled in a two-minute interval is the area of the rectangle times $\frac{1}{60}$

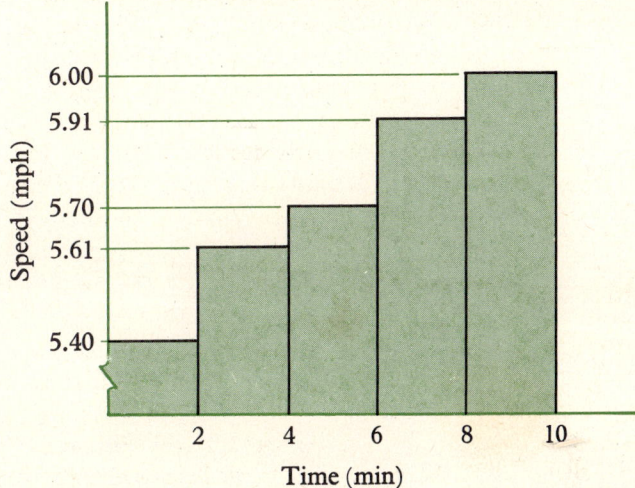

FIGURE 5-5

It is important to note the significance of the result of the limit process of summing an increasing number of rectangles. In Example 5-3, the heights of the rectangles are velocities. The result of the summing procedure is a distance, the distance that the patient ran in ten minutes. In Sec. 3-2, we discovered that the distance function is the antiderivative of the velocity function. In the next section, this close connection between the integral and the antiderivative will be explained via the Fundamental Theorem of Calculus.

It is important to note that for any continuous function $f(x)$, as n gets larger and larger, the numbers arrived at by our method of approximation get closer and closer to the area under the curve. This should be clear from Fig. 5-6. We are thus lead to consider the limit of this type of sum as n approaches infinity, or as $1/n$ approaches zero, and to treat the number

$$\lim_{1/n \to 0} \frac{b-a}{n}[f(x_1) + f(x_2) + \cdots + f(x_n)]$$

as the area of the region bounded by $f(x)$ the x-axis, $x = a$, and $x = b$. If $f(x)$ is continuous on $[a, b]$, then this limit always exists. Let's see how this limit behaves in an easy problem.

Closer approximations by choosing larger values of n

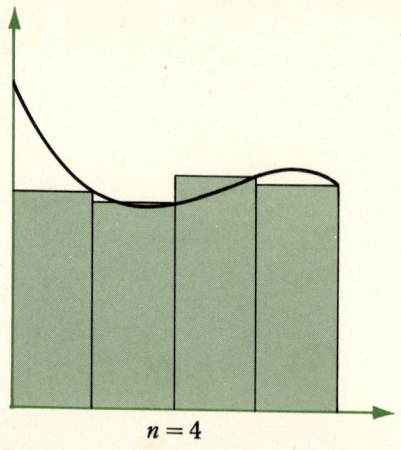

$n = 4$

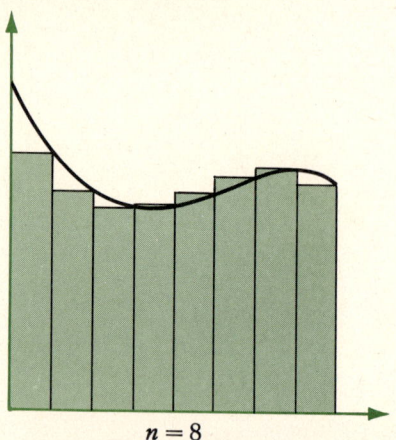

$n = 8$

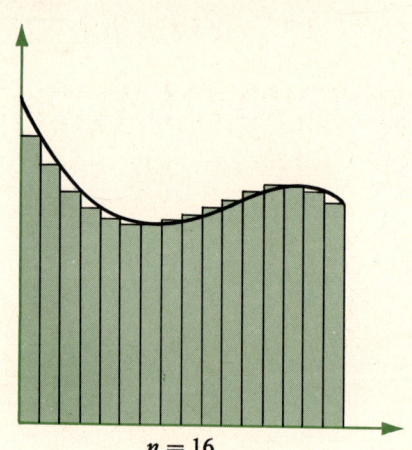

$n = 16$

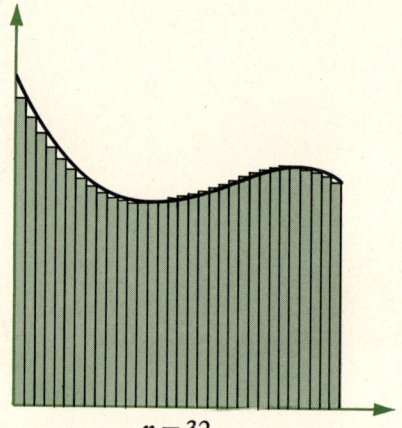

$n = 32$

FIGURE 5-6

EXAMPLE 5-4

Let $f(x) = 2x$, $a = 0$, and $b = 2$. From Fig. 5-7 one can see that the area bounded by $f(x)$, the x-axis, $x = 0$, and $x = 2$ is 4 since the region is a triangle. Now

$$\frac{b-a}{n} = \frac{2-0}{n} = \frac{2}{n}$$

while

$$f(x_1) = f\left(a + \frac{b-a}{n}\right) = f\left(\frac{2}{n}\right) = \frac{4}{n}$$

$$f(x_2) = f\left(\frac{4}{n}\right) = \frac{8}{n}$$

$$\cdots\cdots\cdots\cdots\cdots\cdots\cdots\cdots\cdots\cdots\cdots\cdots\cdots$$

$$f(x_n) = f(2) = 4$$

Hence

$$\frac{b-a}{n}[f(x_1) + \cdots + f(x_n)] = \frac{2}{n}\left(\frac{4}{n} + \frac{8}{n} + \frac{12}{n} + \cdots + 4\right)$$

$$= \left(\frac{2}{n}\right)\left(\frac{4}{n}\right)(1 + 2 + 3 + \cdots + n)$$

$$= \left(\frac{2}{n}\right)\left(\frac{4}{n}\right)\frac{n(n+1)}{2} = \frac{4(n+1)}{n} \ast$$

Therefore, the area is equal to

$$\lim_{1/n \to 0} \frac{b-a}{n}[f(x_1) + \cdots + f(x_n)] = \lim_{1/n \to 0} \frac{4(n+1)}{n} = 4$$

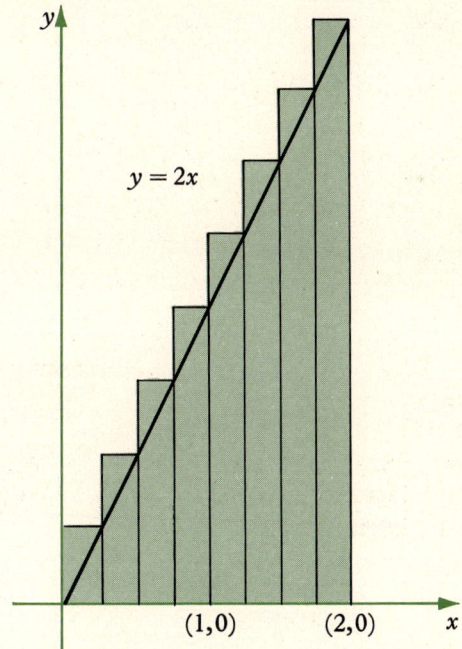

An approximation to the area under the curve y = 2x *from* x = 0 *to* x = 2

FIGURE 5-7

---

* See Appendix C for a proof that $1 + 2 + 3 + \cdots + n = n(n+1)/2$.

It is precisely this type of limit which is that concept called the *definite integral*.

 **Definition** Let $f(x)$ be a continuous function on the closed interval $[a, b]$. For each positive integer n, divide the interval $[a, b]$ into n equal sub-intervals of width $(b-a)/n$. Let x_i be the right-hand endpoint of the ith subinterval. Then we define the *definite* integral $\int_a^b f(x)\, dx$ by

$$\int_a^b f(x)\, dx = \lim_{1/n \to 0} \frac{b-a}{n}[f(x_1) + f(x_2) + \cdots + f(x_n)]$$

If $f(x)$ is a continuous function on $[a, b]$, then the integral $\int_a^b f(x)\, dx$ always exists. Sometimes the integral exists even if $f(x)$ is discontinuous on $[a, b]$. One word of caution is appropriate for those reading other texts on calculus. Ours is not the most common definition of the definite integral. Other texts initiate the discussion with the concept of a *partition* of the interval $[a, b]$, which divides $[a, b]$ into possibly unequal subintervals. Then where we chose the right-hand endpoint of each subinterval in order to calculate the width of our approximating intervals, others allow the selection of any arbitrary element from the subintervals. Even though on the surface our approach seems more restrictive, the two definitions are equivalent. It is our feeling that our definition is much easier to work with and to understand.

EXAMPLE 5-5

Let us compute $\int_0^1 x^2\, dx$ from the definition. We have

$$\frac{b-a}{n} = \frac{1-0}{n} = \frac{1}{n}$$

Then

$$\frac{b-a}{n}[f(x_1) + f(x_2) + \cdots + f(x_n)] = \frac{1}{n}\left[\left(\frac{1}{n}\right)^2 + \left(\frac{2}{n}\right)^2 + \left(\frac{3}{n}\right)^2 + \cdots + \left(\frac{n}{n}\right)^2\right]$$

$$= \frac{1}{n^3}(1^2 + 2^2 + 3^2 + \cdots + n^2)$$

$$= \frac{1}{n^3}\frac{n(n+1)(2n+1)}{6} \quad *$$

$$= \frac{1}{6}\frac{n+1}{n}\frac{2n+1}{n}$$

---

* See Appendix C for a proof that $1^2 + 2^2 + \cdots + n^2 = n(n+1)(2n+1)/6$.

Therefore,
$$\int_0^1 x^2\,dx = \lim_{1/n \to 0} \frac{1}{6}\frac{n+1}{n}\frac{2n+1}{n} = \frac{1}{6}\cdot 1 \cdot 2 = \frac{1}{3}$$

EXERCISE 5-1

1. Graph the functions and for choices of $n = 2, 4, 6, 8$ sketch the appropriate rectangles for a typical approximation. Which of these approximations do you think is greater than (less than, equal to) the area under the curve?
   (a) $f(x) = 2x^2$, $a = 0$, $b = 2$
   (b) $f(x) = x^2 + 1$, $a = 1$, $b = 3$
   (c) $f(x) = 2x^2 - 5$, $a = -1$, $b = 4$
   (d) $f(x) = 3x^2 - 2x - 1$, $a = 1$, $b = 2$
   (e) $f(x) = x^3 + 3$, $a = 0$, $b = 4$

2. Compute
$$\frac{b-a}{n}[f(x_1) + f(x_2) + \cdots + f(x_n)]$$
   (a) $f(x) = x^2 + 1$, $a = 0$, $b = 2$, $n = 4$
   (b) $f(x) = 3x - 1$, $a = 1$, $b = 5$, $n = 8$
   (c) $f(x) = x^2 + 1$, $a = 0$, $b = 4$, $n = 4$
   (d) $f(x) = x^3 + 1$, $a = -1$, $b = 1$, $n = 4$
   (e) $f(x) = 2x^2 + 3x$, $a = 0$, $b = 4$, $n = 8$

3. Use the definition to compute $\int_a^b f(x)\,dx$.
   (a) $\int_0^2 3x\,dx$ (d) $\int_0^2 x^2\,dx$
   (b) $\int_{-1}^5 2\,dx$ (e) $\int_0^1 (x^2 + 1)\,dx$
   (c) $\int_0^3 (3x - 1)\,dx$ (f) $\int_0^1 (2x^2 + 1)\,dx$

4. For arbitrary real numbers a and b use the definition of the definite integral to compute $\int_a^b f(x)\,dx$ using each function.
   (a) $f(x) = 2$ (c) $f(x) = x$
   (b) $f(x) = 3$ (d) $f(x) = x^2$

5-2 THE FUNDAMENTAL THEOREM OF CALCULUS

You can probably appreciate that if we had to use the definition to calculate every definite integral, our task would be tedious if not impossible. As we have mentioned before, the close connection between integrals and derivatives will ease our plight. In order to introduce the reasoning behind the Fundamental Theorem of Calculus, we now present a different approach to our area problem.

Consider the problem of finding the area bounded by the curve $f(x) = x^2$, the x-axis, $x = 0$, and $x = 2$ given in Fig. 5-3.

We define the "area function" $A(x)$ for all positive real numbers by assigning to a the number $A(a)$, which is the area bounded by $y = x^2$, the x-axis, $x = 0$, and $x = a$. From Sec. 5-1, one sees that $A(x) = \int_0^x t^2 \, dt$. Our desired area is thus $A(2) = \int_0^2 t^2 \, dt$. Instead of finding this number from the definition of the definite integral as in the last section, we will find an explicit formula for $A(x)$ by first finding a formula for the derivative $A'(x)$ of $A(x)$ and then using the methods of Sec. 4-7 concerning antiderivatives.

Let x be a fixed element in $(0, \infty)$ and let h be a small positive number. Then $A(x)$ and $A(x+h)$ are defined, and if we let $P = P(h) = A(x+h) - A(x)$ we see that P is a positive number, since it is the difference of the smaller area $A(x)$ from the larger area $A(x+h)$. Now regard P as the area of a rectangle whose base is h and whose height is a number g, which depends upon h. Note that g is between x^2 and $(x+h)^2$, which we can see from Fig. 5-8. If we now also *allow* h to take on negative values close to zero, in the same manner, we can see that $g(h) = P(h)/h$ is a positive number between x^2 and $(x+h)^2$. Hence $g(h)$ is defined for all numbers close but not equal to zero.

Recalling our definition of P we get

$$g(h) = \frac{P}{h} = \frac{A(x+h) - A(x)}{h}$$

Now as h gets smaller and smaller, note that $g(h)$ gets closer and closer to x^2. In other words,

$$\lim_{h \to 0} g(h) = x^2$$

By definition of the derivative, we have

$$A'(x) = \lim_{h \to 0} \frac{A(x+h) - A(x)}{h}$$

We therefore can conclude from the last two equations that

$$A'(x) = x^2$$

From Sec. 4-7 on antiderivatives, we have

$$A(x) = \frac{x^3}{3} + C$$

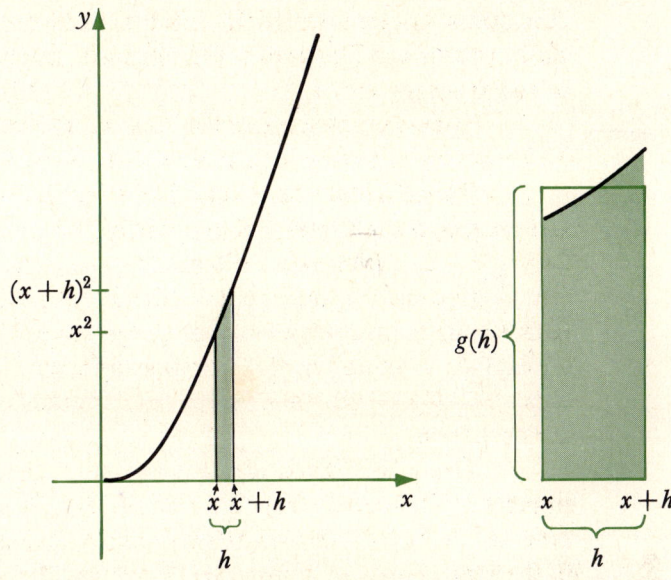

FIGURE 5-8

for some constant C. Since $A(0) = 0$, we have $C = 0$ and hence

$$A(x) = \frac{x^3}{3}$$

and so $A(2) = \frac{8}{3}$.

Notice that the number 2 was not significant in the previous discussion. We could just as well have started out to find the area bounded by $y = x^2$, $y = 0$, $x = 0$, and $x = a$ for any real number a, $a \neq 0$. The same procedure would show this area to be $A(a) = a^3/3$.

EXAMPLE 5-6

The area bounded by $y = x^2$, the x-axis, $x = 0$, and $x = 3$ is $A(3) = 3^3/3 = 27/3 = 9$. The area bounded by $y = x^2$, the x-axis, $x = 0$, and $x = 4$ is $A(4) = 4^3/3 = 64/3$. Thus the area bounded by $y = x^2$, the x-axis, $x = 3$, and $x = 4$ is

$$A(4) - A(3) = \frac{64}{3} - 9 = \frac{37}{3}$$

In general, the area bounded by $y = x^2$, the x-axis, $x = a$, and $x = b$ is $A(b) - A(a)$ for $b > a$.

Thus we see that there is a close connection between the two seemingly unrelated problems: finding antiderivatives and finding the area under a curve. The remarkable fact is that not only are they related but the solution of one depends heavily on the solution of the other.

Theorem 5-1 *The Fundamental Theorem of Calculus*
Suppose $f(x)$ is a continuous function on the closed interval $[a, b]$ such that $F(x)$ is an antiderivative of $f(x)$. Then

$$\int_a^b f(x)\, dx = F(b) - F(a)$$

We will adopt the notation $F(x)|_a^b = F(b) - F(a)$.

EXAMPLE 5-7

Let us compute some definite integrals using Theorem 5-1.

$$\int_0^1 2x\, dx = x^2 \Big|_0^1 = 1^2 - 0 = 1$$

$$\int_0^2 x^2\, dx = \frac{x^3}{3}\Big|_0^2 = \frac{2^3}{3} - 0 = \frac{8}{3}$$

$$\int_1^2 (x+x^2)\, dx = \frac{x^2}{2} + \frac{x^3}{3}\Big|_1^2 = \frac{2^2}{2} + \frac{2^3}{3} - \left(\frac{1^2}{2} + \frac{1^2}{3}\right) = \frac{28}{6} - \frac{5}{6} = \frac{23}{6}$$

$$\int_{-1}^2 x^5\, dx = \frac{x^6}{6}\Big|_{-1}^2 = \frac{64}{6} - \frac{1}{6} = \frac{63}{6} = \frac{21}{2}$$

$$\int_0^1 (1+x+x^2+x^3)\, dx = x + \frac{x^2}{2} + \frac{x^3}{3} + \frac{x^4}{4}\Big|_0^1 = 1 + \frac{1}{2} + \frac{1}{3} + \frac{1}{4} = \frac{25}{12}$$

$$\int_0^1 e^{2x}\, dx = \frac{1}{2}e^{2x}\Big|_0^1 = \frac{1}{2}e^{2(1)} - \frac{1}{2} = \frac{1}{2}e^2 - \frac{1}{2} = \frac{1}{2}(e^2 - 1)$$

Even though we have discussed the concept of area throughout this chapter, careful reading will show you that we have never actually defined what area is—it has remained an intuitive concept. Notice that in the definition of the integral, we never mentioned area. Now we are able to present the natural definition of area in terms of the integral.

Definition If $f(x)$ is a continuous, nonnegative function on $[a, b]$, then the area A bounded by $f(x)$, the x-axis, $x = a$, and $x = b$ is

$$A = \int_a^b f(x)\, dx$$

EXAMPLE 5-8

The area underneath $y = x^3$ from $x = 0$ to $x = 1$ is

$$\int_0^1 x^3\, dx = \left.\frac{x^4}{4}\right|_0^1 = \frac{1}{4}$$

Note that the area of the region bounded by $y = x^3$, the x-axis, and the lines $x = -1$ and $x = 0$ is the same as the area of the region bounded by $y = x^3$, the x-axis and the lines $x = 0$ and $x = 1$, which is $\frac{1}{4}$. Thus the area from $x = -1$ to $x = 1$ is $2 \cdot \frac{1}{4} = \frac{1}{2}$. Now suppose we tried to use Theorem 5-1 to find this area. We'd have

$$\int_{-1}^1 x^3\, dx = \left.\frac{x^4}{4}\right|_{-1}^1 = \frac{1}{4} - \frac{(-1)^4}{4} = \frac{1}{4} - \frac{1}{4} = 0$$

Why didn't we get the right answer? Because $f(x) = x^3$ is not nonnegative, i.e., $f(x) \leq 0$, in $[-1, 0]$.

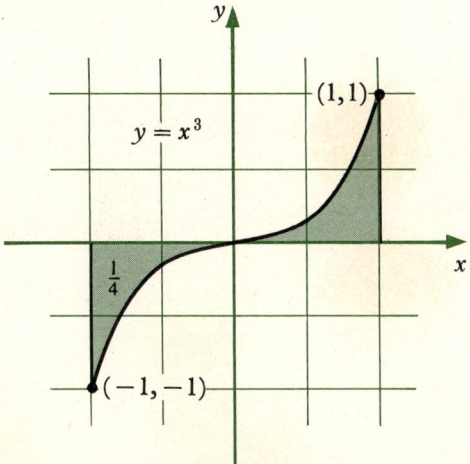

FIGURE 5-9

EXAMPLE 5-9

To find the area of the region bounded by $f(x) = x^2 - 4$ and the x-axis, one must note that $f(x) \leq 0$ in the interval $[-2, 2]$. Hence the area is given by

$$-\int_{-2}^2 (x^2 - 4)\, dx = -\left.\left(\frac{x^3}{3} - 4x\right)\right|_{-2}^2$$

$$= -\frac{2^3}{3} + 4 \cdot 2 - \left[-\frac{(-2)^3}{3} + 4(-2)\right]$$

$$= -\frac{8}{3} + 8 - \frac{8}{3} + 8$$

$$= \frac{32}{3}$$

We now wish to extend our methods to handle functions f such that $f(x) \leq 0$ on $[a, b]$. In Fig. 5-10 we can see that the area of the region bounded by $y = f(x)$, the x-axis, and the lines $x = a$ and $x = b$ where $f(x) \leq 0$ in $[a, b]$ is equal to the area bounded by $y = -f(x)$, the x-axis, and the lines $x = a$ and $x = b$.

Therefore, if $f(x) \leq 0$ on $[a, b]$, then the area bounded by $y = f(x)$, the x-axis, and the lines $x = a$ and $x = b$ is $\int_a^b -f(x)\,dx = -\int_a^b f(x)\,dx$.

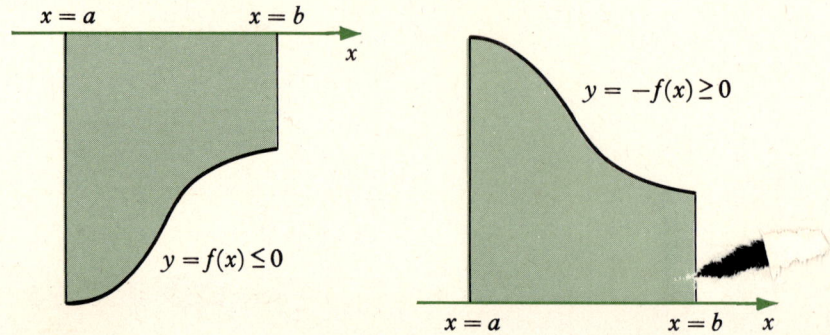

FIGURE 5-10

We are now able to compute areas of regions bounded by two curves.

EXAMPLE 5-10

We compute the area bounded by $f(x) = x^2$ and $g(x) = x+2$ sketched in Fig. 5-11. To find the points of intersection P and Q, we solve the equations simultaneously:

$$x^2 = x + 2$$
$$x^2 - x - 2 = 0$$
$$(x+1)(x-2) = 0$$
$$x = -1, 2$$

Hence the points of intersection are $(-1, 1)$ and $(2, 4)$. The area A under $y = x + 2$ is given by

$$A = \int_{-1}^{2} (x+2)\,dx = \left. \frac{x^2}{2} + 2x \right|_{-1}^{2} = \frac{15}{2}$$

The area B under $y = x^2$ is

$$B = \int_{-1}^{2} x^2\,dx = \left. \frac{x^3}{3} \right|_{-1}^{2} = 3$$

Thus the desired area is $A - B = \frac{15}{2} - 3 = \frac{9}{2}$. Note that we could have arrived at this same answer by considering $\int_{-1}^{2} (x+2-x^2)\,dx$.

The area bounded by $y = x + 2$ *and* $y = x^2$

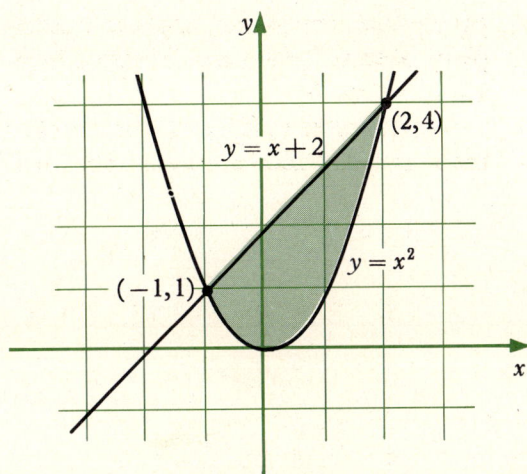

FIGURE 5-11

In general, if $f(x) \geq g(x)$ on $[a, b]$, then the area bounded by $f(x)$, $g(x)$, $x = a$, and $x = b$ is

$$\int_a^b [f(x) - g(x)]\, dx = \int_a^b f(x)\, dx - \int_a^b g(x)\, dx$$

The average value of a set of numbers $n_1, \ldots, n_m$ is given by

$$\frac{n_1 + n_2 + \cdots + n_m}{m}$$

In the same way we can define the "average" value of a function f in an interval $[a, b]$, called the *mean value* μ of f in $[a, b]$.

EXAMPLE 5-11

If we think of $\int_a^b f(x)\, dx$ as the area under $f(x)$ from a to b, and if we also think of μ to be a height, then the rectangle with height μ and width $b - a$ has the same area as that under $f(x)$ from a to b. In this way we get a geometric view of the mean μ. (See Fig. 5-12.)

EXAMPLE 5-12

A psychologist conducting a learning experiment determines that the expected achievement $E(t)$ of a subject who completes the experiment in time t is given by

$$E(t) = 3t^2 - \tfrac{1}{5}t^3$$

The mean value μ for the function y = f(x)
on the interval [a, b]

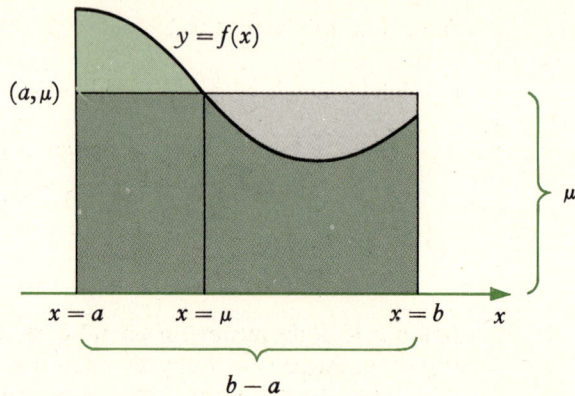

FIGURE 5-12

for $t \in [0, 10]$ where $E(t)$ is measured on a scale from zero to 100 and t is measured in minutes. One measure of central tendency of the experiment is the average value. The average value μ is equal to

$$\frac{1}{10} \int_0^{10} (3t^2 - \tfrac{1}{5}t^3)\, dt = \tfrac{1}{10}(t^3 - \tfrac{1}{20}t^4) \Big|_0^{10} = 50$$

Thus a score of 50 is average.

Often in applications one must sum a finite set of numbers, whether to find an average, a total distance, or perhaps a total income. Yet it is also common to have to "sum" over an infinite set of numbers or objects. The next few examples illustrate this phenomenon.

EXAMPLE 5-13

A company has determined that the approximate cost of the xth item of production is given by

$$f(x) = 40 - 4t + \tfrac{1}{5}t^2$$

where $f(x)$ is measured in dollars and x is measured in hundreds of units. The cost of the first 15 (hundred units) would then be given by

$$\int_0^{15} (40 - 4t + \tfrac{1}{5}t^2)\, dt = 40t - 2t^2 + \tfrac{1}{15}t^3 \Big|_0^{15}$$

$$= 40(15) - 2(15)^2 + \tfrac{1}{15}(15)^2$$

$$= 600 - 450 + 225$$
$$= 375$$

The cost of producing the 15th through the 30th would be given by

$$\int_{15}^{30} (40 - 4t + \tfrac{1}{5}t^2)\, dt = 40t - 2t^2 + \tfrac{1}{15}t^3 \Big|_{15}^{30}$$

$$= 40(30) - 2(30)^2 + \tfrac{1}{15}(30)^3 - 375$$

$$= 1200 - 1800 + 1800 - 375$$

$$= 825$$

EXAMPLE 5-14

In economics, the amount of capital stock $A(t)$ at time t is related to the rate of net investment $B(t)$ by the equation $D_t A(t) = B(t)$. Suppose that $B(t) = 0.9t^{1/2} + 1.1$ where $B(t)$ is measured in thousands of dollars per year. Then the total sum increase in capital stock in the eighth year is given by

$$\int_8^9 (0.9t^{1/2} + 1.1)\, dt = (0.9)(\tfrac{2}{3})t^{3/2} + 1.1t \Big|_8^9$$

$$= (0.6)9^{3/2} + (1.1)(9) - (0.6)8^{3/2} - (1.1)(8) \approx 4.1$$

EXAMPLE 5-15

Biologists assume that the flow of blood F across a cross-section of an arteriole of area A is given by $F = vA$ where v is the velocity of the blood. Since the flow in the arteriole is laminar, that is, due to the viscosity of the blood the layer closest to the wall will cling to the wall and its velocity will be zero while the velocity will increase with the distance from the wall and be greatest at the center. One can picture the flow as being layer upon layer of concentric tubes sliding together with the outermost tube remaining still and the innermost tube having the greatest velocity. The velocity of the blood in a given layer, or lamina, is given by

$$v(r) = k(R^2 - r^2)$$

where R is the radius of the arteriole, r is the distance between the layer and the center of the arteriole, both measured in centimeters, while k is a constant depending upon pressure differences, length of the arteriole, and the viscosity of blood. Since the area of the cross-section of the lamina r units from the center can be approximated to be

$$A(r) = 2\pi r\, \Delta r$$

where Δr is the radius of the lamina, we can take the flow in this lamina to be

$$F(r) = 2\pi k r (R^2 - r^2)\, \Delta r$$

Thus the total flow $\mathscr{F}$ in the arteriole, i.e., the volume per unit time, can be obtained by integrating $F(r)$ from $r=0$ to $r=R$, and hence continuously summing up the individual flows in the respective lamina. Therefore

$$\mathscr{F} = \int_0^R 2\pi kr(R^2-r^2)\,dr = \frac{\pi kR^4}{2}$$

Thus integration enables us to "sum" all the individual flows for every choice of r from zero to R.

EXAMPLE 5-16

The vast amount of recent research being conducted in the field of psychology dictates that mathematical methods be utilized for the organization of concepts. One such concept is the prediction of group behavior. It is hypothesized that the time $t(x)$ it takes a group to perform a task the xth time decreases as the number of times, x, the task is accomplished increases. The group tends to perfect its ability to perform the task. Suppose that for a certain group type and a specific task, empirical evidence shows that $t(x) = 0.3x^2 - 18x + 270$ where x is measured in seconds. Then the total time it takes the group to perform the task 10 times is predicted to be

$$\int_0^{10} (0.3x^2 - 18x + 270)\,dx = 0.1x^3 - 9x^2 + 270x \Big|_0^{10}$$

$$= 100 - 900 + 2700 = 1900 \text{ sec.}$$

The total time to perform the task 10 more times is predicted to be

$$\int_{10}^{20} (0.3x^2 - 18x + 270)\,dx = 0.1x^3 - 9x^2 + 270x \Big|_{10}^{20}$$

$$= 800 - 3600 + 5400 - 1900 = 700 \text{ sec.}$$

We now list some of the main properties of the definite integral. The proofs for continuous functions follow directly from Theorem 5-1.

Property 1 $\quad \int_a^a f(x)\,dx = 0$

Property 2 $\quad \int_a^b f(x)\,dx = -\int_b^a f(x)\,dx$

Property 3 $\quad \int_a^b f(x)\,dx + \int_b^c f(x)\,dx = \int_a^c f(x)\,dx$

Property 4 $\quad \int_a^b kf(x)\,dx = k\int_a^b f(x)\,dx$

Property 5 $\quad \int_a^b (f(x)+g(x))\,dx = \int_a^b f(x)\,dx + \int_a^b g(x)\,dx$

EXERCISE 5-2

1. Compute the definite integrals.

   (a) $\int_0^2 2\,dx$

   (b) $\int_0^2 3x\,dx$

   (c) $\int_{-1}^2 (3x+1)\,dx$

   (d) $\int_0^1 3x^2\,dx$

   (e) $\int_0^3 (x^2-2x)\,dx$

   (f) $\int_{-2}^0 (x^2-x)\,dx$

   (g) $\int_0^{-2} (x^2-x)\,dx$

   (h) $\int_{-1}^4 x^3\,dx$

   (i) $\int_0^3 (x^2+x^3)\,dx$

   (j) $\int_1^2 x^6\,dx$

   (k) $\int_0^1 (x+x^3+x^5)\,dx$

   (l) $\int_0^2 e^x\,dx$

   (m) $\int_0^2 2e^x\,dx$

   (n) $\int_0^1 (x+e^x)\,dx$

2. Find the area bounded by $y=x^2$, the x-axis, $x=3$, and $x=5$.
3. Find the area bounded by $y=x^2$, the x-axis, $x=2$, and $x=6$.
4. Find the area bounded by $y=x^2$, the x-axis, $x=0$, and $x=-2$.
5. Find the area bounded by $y=x^2$, the x-axis, $x=-1$, and $x=-3$.
6. Find the area bounded by $y=x^2+1$, the x-axis, $x=0$, and $x=2$.
7. Find the area bounded by $y=x^2+1$, the x-axis, $x=2$, and $x=4$.
8. Find the area bounded by $y=x^2+3$, the x-axis, $x=-1$, and $x=1$.
9. Find the area bounded by $y=x^3$, the x-axis, $x=0$, and $x=2$.
10. Find the area of the region bounded by $y=x^2+2x$, the x-axis, $x=3$, and $x=4$.
11. Find the area of the region bounded by $y=x^2+2x$ and $y=3x$.
12. Find the area of the region bounded by $y=x^3$ and $y=x$.
13. Find the area of the region bounded by $y=x^2-2$ and $y=-x^2+3x-3$.
14. Find the mean value μ of $f(x)$ in the interval $[a,b]$. Sketch $f(x)$ from $x=a$ to $x=b$ and in the same graph draw a rectangle of height μ and width $b-a$ for the following:

    (a) $f(x) = x^2$, $a = 0$, $b = 1$
    (b) $f(x) = x^2 + 1$, $a = 0$, $b = 1$
    (c) $f(x) = x + x^2$, $a = 0$, $b = 1$
    (d) $f(x) = x^2 + 2x + 1$, $a = -2$, $b = 0$

15. Suppose a company determines that the cost of producing the xth item of production is $f(x) = 0.03x^2 - 10x + 100$. What is the cost of producing the

first 10 items? 20 items? 50 items? If 100 items have been produced, what is the cost of producing the next 10 items? 20 items? 50 items?

16. Suppose a psychologist hypothesizes that it will take a specified group-type $t(x)$ seconds to complete a task the xth time where $t(x) = 0.06x^2 - 4x + 100$. Find the total time it should take the group to perform the task 10 times. If the group has performed the task 10 times, how much time should it take to perform the task another 10 times?

17. Suppose that the rate of net investment $B(t)$ of capital stock at time t is given by $B(t) = 0.76t^{1/3} + 1.2$ where $B(t)$ is measured in thousands of dollars. Find the total sum increase in capital stock in the ninth year.

18. A company purchases a new machine which will improve efficiency. It is predicted that $S(t)$ thousands of dollars will be saved in the tth month since the machine was purchased. If the monthly operating expenses are $E(t)$, then the net dollar gain in purchasing the machine is $G(t) = S(t) - E(t)$. If $S(t) = 50 - 0.1t - 0.01t^2$ and $E(t) = 0.01t^2$, find the total net gain in the first 10 months, and then in the first 20 months.

19. Suppose biologists determine that the blood flow $F(r)$ in a cross-section of a lamina (layer) that is r cm from the center of the arteriole is given by

$$F(r) = 2\pi k r(R^2 - r^2) + Cr$$

for constants k and C and where R is the radius of the lamina. Find the total flow in the cross-section of the arteriole.

20. Transit authorities assumed that average repair costs in dollars per mile for passenger cars that have been operated over m hundred thousand miles is given by $R(m) = km^{3/2} + d$ where k and d are constants. Find the average total repair cost for a passenger car that has been driven 300,000 miles.

21. Let $n(x)$ be the expected number of organisms in a bacteria culture with x percent probability of regenerating within the hour where $n(x) = x^2 + 2x$ is measured in millions per cubic foot. Explain why $\int_{10}^{20} n(x)\,dx$ approximates the number of organisms with a 10 to 20 percent chance to regenerate within the hour. Give an appropriate integral that estimates the number of organisms which have a less than 50 percent chance to regenerate within the hour.

22. A rocket has an acceleration of 12 ft per sec.^{2} Find the rocket's increase in velocity from $t = 0$ to $t = 5$. Compute the total distance that the rocket travels in the first 10 sec.

23. In psychology one studies deviation of individual or group accomplishment from a goal, where the deviation $D(t)$ depends upon time to and is measured numerically (perhaps number of words memorized, number of hits on a target, and so on) and the goal G is a constant. If $A(t)$ is a measurement of the accomplishment of the task, then $E(t) = A(t) - G$. Suppose for a given experiment $A(t) = 2t^3 - \frac{9}{2}t^2 + 3t$ and $G = 10$. Find the total deviation from the goal after $t = 3$ where t is measured in minutes.

5-3 TECHNIQUES OF INTEGRATION. I: Substitution

Now that we see that in order to utilize this powerful tool, the definite integral, it is necessary to calculate antiderivatives, we must develop techniques for finding antiderivatives. The first method we deal with, the method of substitution, is derived from the integral analogue of the Chain Rule, Rule D7, in Chap. 3. If the function $h(x)$ is the composition of two functions,
$$h(x) = f(g(x))$$
then the Chain Rule states that
$$h'(x) = f'(g(x)) \cdot g'(x)$$

Integration by substitution is the technique of recognizing when an integrand, i.e., a function to be integrated, is of the form
$$f(g(x)) \cdot g'(x)$$
such that if we set $u = g(x)$, then $\int f(u)\, du$ is easy to calculate. Let us see how this helps solve the problem of integrating $f(g(x)) \cdot g'(x)$. Suppose
$$F(u) = \int f(u)\, du$$
and so $F'(u) = f(u)$. If we set $h(x) = F(g(x))$, then by the Chain Rule we have
$$h'(x) = F'(g(x)) \cdot g'(x) = f(g(x)) \cdot g'(x)$$
Therefore
$$\int f(g(x)) \cdot g'(x)\, dx = h(x) = F(g(x))$$

EXAMPLE 5-17

Consider $\int (x^2+1)^2 (2x)\, dx$. If we set
$$u = g(x) = x^2 + 1 \quad \text{and} \quad f(x) = x^2$$
then the integrand is of the form $f(g(x)) \cdot g'(x)$. Also
$$\int f(u)\, du = \int u^2\, du = \frac{u^3}{3}$$
Thus
$$F(u) = \frac{u^3}{3}$$
Therefore,
$$\int (x^2+1)^2 (2x)\, dx = \frac{(x^2+1)^3}{3} + C$$

CHAP. 5   The Integral

In order to simplify our discussions, we employ a common tool of mathematics, a substitution of variable. If we set $u = g(x)$, we make a formal substitution of

$$du \quad \text{for} \quad g'(x)\,dx$$

Recall that we mentioned that the dx in expression of the integral is nothing more than a symbol. Similarly du above is nothing more than a symbol. Actually what we are doing is playing a game of symbol manipulation. We play the game for expediency, because it works.

The method of substitution proceeds as follows: to find $\int f(g(x)) \cdot g'(x)\,dx$:
1. Let $u = g(x)$
2. Let $du = g'(x)\,dx$
3. Find $\int f(u)\,du = F(u)$
4. Set $\int f(g(x)) \cdot g'(x)\,dx = F(g(x))$

The key step of course is the selection of $u = g(x)$ so that $\int f(u)\,du$ is easy to calculate. One important point to always keep in mind is that any technique of integration is a "trial and error" method. Be prepared to make selections for $g(x)$ which do not work, and if a selection doesn't work, make another. You should always check your answer by differentiating.

EXAMPLE 5-18
Consider $\int \sqrt{x+2}\,dx$. We proceed as follows:
1. Let $u = g(x) = x+2$
2. $du = dx$ so that $\sqrt{x+2}\,dx = u^{1/2}\,du$
3. $\int u^{1/2}\,du = \tfrac{2}{3}u^{3/2} + C$
4. Thus $\int \sqrt{x+2}\,dx = \tfrac{2}{3}(x+2)^{3/2} + C$

EXAMPLE 5-19
Consider $\int x(1-x^2)^{3/2}\,dx$. We proceed as follows:
1. Let $u = g(x) = 1-x^2$
2. Then $du = -2x\,dx$ and so $x(1-x^2)^{3/2}\,dx = u^{3/2}(-\tfrac{1}{2}\,du)$
3. $\int -\tfrac{1}{2}u^{3/2}\,du = (-\tfrac{1}{2})(\tfrac{2}{5})u^{5/2} = -\tfrac{1}{5}u^{5/2} + C$
4. Therefore $\int x(1-x^2)^{3/2}\,dx = -\tfrac{1}{5}(1-x^2)^{5/2} + C$

EXAMPLE 5-20
Consider $\int \dfrac{e^x\,dx}{2+e^x}$. We proceed as follows:
1. Let $u = 2 + e^x$
2. $du = e^x\,dx$ so that $\dfrac{e^x\,dx}{2+e^x} = \dfrac{du}{u}$

3. $\int \dfrac{du}{u} = \ln u + C$

4. Therefore $\int \dfrac{e^x\, dx}{2+e^x} = \ln(2+e^x) + C$

EXAMPLE 5-21

Consider $\int_0^1 (2x+1)(x^2+x)^{1/2}\, dx$. We first use the technique of substitution to find the antiderivative of $(2x+1)(x^2+x)^{1/2}$.

1. Let $u = x^2 + x$
2. $du = (2x+1)\, dx$ so that $(2x+1)(x^2+x)^{1/2}\, dx = u^{1/2}\, du$
3. $\int u^{1/2}\, du = \tfrac{2}{3} u^{3/2}$
4. Therefore $\int_0^1 (2x+1)(x^2+x)^{1/2}\, dx = \tfrac{2}{3}(x^2+x)^{3/2}\Big|_0^1 = \tfrac{1}{3}\, 2^{5/2}$

EXAMPLE 5-22

Consider $\int_1^e \dfrac{\ln x\, dx}{x}$. We proceed as follows:

1. Let $u = \ln x$
2. $du = \dfrac{1}{x}\, dx$ so that $\int \dfrac{\ln x\, dx}{x} = \int u\, du$
3. $\int u\, du = \dfrac{u^2}{2}$
4. Therefore $\int_1^e \dfrac{\ln x\, dx}{x} = \dfrac{(\ln x)^2}{2}\Big|_1^e = \dfrac{1}{2}$

EXERCISE 5-3

1. Evaluate the integrals first by expanding the integrand and then integrating, and secondly by substitution.

   (a) $\int x(x^2+1)\, dx$

   (b) $\int x^2(x^3+1)^2\, dx$

   (c) $\int (4x^3+1)(x^4+x)^2\, dx$

   (d) $\int (x^2+x)(2x^3+3x^2)^2\, dx$

2. Evaluate each of the integrals.

   (a) $\int x^2(1+x^3)^{1/2}\, dx$

   (b) $\int x^2(x^3+5)^{-1/3}\, dx$

   (c) $\int x(x^2+1)^{2/3}\, dx$

   (d) $\int (x^2+x)^{1/3}(2x+1)\, dx$

   (e) $\int (4t^3+t)^{-1/8}(12t^2+1)\, dt$

   (f) $\int (x^4+2x^2+4x)^5(x^3+x+1)\, dx$

(g) $\int (x^{4/3}+1)^6 x^{1/3}\, dx$

(k) $\int \dfrac{\log x^2}{x}\, dx$

(h) $\int xe^{x^2+1}\, dx$

(l) $\int \dfrac{(\log x)^2}{x}\, dx$

(i) $\int (e^{x^3+3x})(x^2+1)\, dx$

(m) $\int \dfrac{3x^3}{x^4+1}\, dx$

(j) $\int (e^x+1)^{1/2} e^x\, dx$

3. Calculate the definite integrals by first finding the antiderivative of the integrand using the method of substitution.

(a) $\displaystyle\int_1^2 x(x^2-1)^{1/2}\, dx$

(b) $\displaystyle\int_0^1 x^2(x^3+1)^{1/3}\, dx$

(c) $\displaystyle\int_1^2 (x+1)(x^2+2x)^{-1/2}\, dx$

(d) $\displaystyle\int_0^2 \dfrac{2x\, dx}{x^2+1}$

(e) $\displaystyle\int_0^1 e^{x^3+x}(x^2+\tfrac{1}{3})\, dx$

(f) $\displaystyle\int_1^2 \dfrac{(\log x)^3\, dx}{x}$

4. The rate of net investment $B(t)$ at time t is given by $B(t) = t\sqrt{t^2+1}$ where $B(t)$ is measured in thousands of dollars. Find the total sum increase in capital stock in the first year.

5. Let $n(x)$ be the expected number of organisms in a bacteria culture with x percent probability of regenerating within the hour. If $n(x) = \sqrt{3x+1}$ where $n(x)$ is measured in millions per cubic foot, find the number of organisms with a 0–5 percent chance to regenerate.

6. Under certain conditions a bacteria culture will grow at a rate given by $w'(t) = \sqrt{2t+1}$ where w is the weight in milligrams of the culture at time t, given in hours. Find the change in weight of the culture from $t = 0$ to $t = 4$.

5–4 TECHNIQUES OF INTEGRATION II: Integration by Parts

We remarked previously that the antiderivative of a product is not the product of the antiderivatives. However, we can utilize the product formula for derivatives in a modified fashion to handle rather complicated

integrands. Recall the product formula: If $h(x) = f(x)g(x)$, then
$$h'(x) = f(x)g'(x) + g(x)f'(x)$$
Now $h(x)$ is the antiderivative of $h'(x)$, and the antiderivative of a sum is the sum of the antiderivatives, so we have
$$h(x) = f(x)g(x) = \int f(x)g'(x)\,dx + \int g(x)f'(x)\,dx$$
If we let
$$u = f(x) \quad \text{and} \quad v = g(x)$$
so that
$$du = f'(x)\,dx \quad \text{and} \quad dv = g'(x)\,dx$$
by rearranging the above equation we have
$$\int u\,dv = uv - \int v\,du$$
This is the equation used in the technique of integration by parts. The following examples will illustrate the application of this equation.

EXAMPLE 5-23

Up till now we have not been able to calculate the seemingly easy integral $\int \ln x\,dx$. We can use integration by parts by letting
$$u = \ln x \quad \text{and} \quad v = x$$
Then
$$du = \frac{1}{x}\,dx \quad \text{and} \quad dv = dx$$
Thus
$$\int \ln x\,dx = x \ln x - \int x\left(\frac{1}{x}\right)dx$$
$$= x \ln x - \int dx$$
$$= x \ln x - x + C$$

EXAMPLE 5-24

Consider $\int xe^x\,dx$. We can set
$$u = x \quad \text{and} \quad dv = e^x\,dx$$
so that
$$du = dx \quad \text{and} \quad v = \int dv = \int e^x\,dx = e^x$$
Hence
$$\int xe^x\,dx = xe^x - \int e^x\,dx$$
$$= xe^x - e^x + C$$

It is important again to note that the technique of integration by parts is a "trial and error" method. For instance, in the last example, you may have been tempted to let

$$u = e^x \quad \text{and} \quad dv = x\, dx$$

Then

$$du = e^x\, dx \quad \text{and} \quad v = \int dv = \int x\, dx = \frac{x^2}{2}$$

We would then have

$$\int xe^x\, dx = \left(\frac{x^2}{2}\right)e^x - \int \left(\frac{x^2}{2}\right)e^x\, dx$$

Yet to solve the problem we now have to calculate $\int (x^2/2) e^x\, dx$, a much harder problem than the one we started with. The trick with integration by parts is to choose u and dv so that $\int v\, du$ is easily calculated.

Sometimes one has to apply the technique more than once to solve the problem as the next example illustrates.

EXAMPLE 5-25

Consider $\int x^2 e^x\, dx$. The idea here is to try and rid ourselves of the x^2. To this end we let

$$u = x^2 \quad \text{and} \quad dv = e^x\, dx$$

Then

$$du = 2x\, dx \quad \text{and} \quad v = \int dv = \int e^x\, dx = e^x$$

Hence

$$\int x^2 e^x\, dx = x^2 e^x - \int 2xe^x\, dx$$

To solve the integral on the right we must use integration by parts as we did in Ex. 5-24. Therefore

$$\int x^2 e^x\, dx = x^2 e^x - \int 2xe^x\, dx = x^2 e^x - 2xe^x + 2e^x + C$$

Sometimes it is necessary to utilize both substitution and integration by parts in the same problem.

EXAMPLE 5-26

Consider $\int_3^8 (x\, dx)/\sqrt{x+1}$. We first find the antiderivative of $x/\sqrt{x+1}$. Using integration by parts, we let

$$u = x \quad \text{and} \quad dv = \frac{dx}{\sqrt{x+1}}$$

Then $du = dx$, but to find v we must integrate, since

$$v = \int dv = \int \frac{dx}{\sqrt{x+1}}$$

To evaluate this integral we use substitution and let $w = x+1$, so that $dw = dx$ and

$$\int \frac{dx}{\sqrt{x+1}} = \int \frac{dw}{\sqrt{w}} = 2w^{1/2} = 2\sqrt{x+1}$$

From above, we now have

$$\int \frac{x\,dx}{\sqrt{x+1}} = x(2\sqrt{x+1}) - 2\int \sqrt{x+1}\,dx$$

To evaluate the integral on the right we again use substitution, letting $r = x+1$ so that $dr = dx$. Thus

$$\int 2\sqrt{x+1}\,dx = 2\int \sqrt{x+1}\,dx = 2\int \sqrt{r}\,dr = 2(\tfrac{2}{3})r^{3/2} = \tfrac{4}{3}(x+1)^{3/2}$$

Hence

$$\int \frac{x\,dx}{\sqrt{x+1}} = 2x\sqrt{x+1} - \frac{4}{3}(x+1)^{3/2}$$

Now we can insert the limits of integration.

$$\int_3^8 \frac{x\,dx}{\sqrt{x+1}} = 2x\sqrt{x+1} - \frac{4}{3}(x+1)^{3/2} \Big|_3^8$$

$$= 2 \cdot 8 \cdot 3 - \frac{4}{3} \cdot 27 - 2 \cdot 3 \cdot 2 + \frac{4}{3} \cdot 8$$

$$= \frac{32}{3}$$

EXERCISE 5-4

1. Find the antiderivatives by using integration by parts.

    (a) $\int xe^{2x}\,dx$

    (b) $\int x \ln x\,dx$

    (c) $\int x\sqrt{x+1}\,dx$

    (d) $\int x^2 \sqrt{x+1}\,dx$

    (e) $\int x^3 e^x\,dx$

    (f) $\int x^2 \ln x\,dx$

(g) $\displaystyle\int xe^{-x}\,dx$

(h) $\displaystyle\int x^{1/2}\ln x\,dx$

(i) $\displaystyle\int \frac{x\,dx}{\sqrt{2-x}}$

(j) $\displaystyle\int \frac{xe^x\,dx}{e^x+1}$

2. Evaluate the definite integrals by first using integration by parts to find the antiderivative of the integrand.

(a) $\displaystyle\int_0^1 xe^x\,dx$

(b) $\displaystyle\int_0^3 x\sqrt{x+1}\,dx$

(c) $\displaystyle\int_0^2 xe^{-x}\,dx$

(d) $\displaystyle\int_1^2 x^3\ln x\,dx$

(e) $\displaystyle\int_0^3 x^3\sqrt{x^2+1}\,dx$

3. A table of integrals has been provided in the back of the book. Often it is not possible to find the integrand you are dealing with in a table and so the techniques mentioned in this chapter are necessary. Use the table of integrals to calculate the following antiderivatives.

(a) $\displaystyle\int \frac{dx}{\sqrt{x^2+9}}$

(b) $\displaystyle\int \frac{dx}{\sqrt{x^2-16}}$

(c) $\displaystyle\int \frac{dx}{x^2-8}$

(d) $\displaystyle\int \frac{x\,dx}{3x+5}$

(e) $\displaystyle\int \frac{dx}{x(5x+7)}$

(f) $\displaystyle\int x^4\ln x\,dx$

5–5 TECHNIQUES OF INTEGRATION III: PARTIAL FRACTIONS*

We have devised means to integrate certain types of rational functions, for example

$$\frac{1}{2x+3} \qquad \frac{x}{2x^2+3} \qquad \frac{4x+3}{2x^2+3x}$$

Note that the degree of the numerator is less than the degree of the denominator in each of the above rational functions. If this is the case, the function is called a *proper* rational fraction. If the degree of the numerator is greater than or equal to the degree of the denominator, the function is called an *improper* rational function. If the numerator is not of lower degree than the denominator, one can always use long division to express the improper rational function as the sum of a polynomial and a proper rational function.

---

* This section should be omitted if Sec. 1-6 was omitted.

Hence we shall restrict our attention to the integration of proper rational functions.

EXAMPLE 5-27

The rational function

$$f(x) = \frac{3x^3 - 8x^2 + 3x + 1}{x^2 - 2x}$$

is an improper rational function. By long division we have

$$\begin{array}{r} 3x-2 \\ x^2-2x \overline{\smash{\big)}\,3x^3-8x^2+3x+1} \\ \underline{3x^3-6x^2 } \\ -2x^2+3x \\ \underline{-2x^2+4x } \\ -x+1 \end{array}$$

and hence we can write

$$f(x) = \frac{3x^3 - 8x^2 + 3x}{x^2 - 2x} = 3x - 2 + \frac{-x+1}{x^2 - 2x}$$

where $f(x)$ is the sum of the polynomial function $3x-2$ and the proper rational function $(-x+1)/(x^2-2x)$.

We first recall the method of adding two fractions.

EXAMPLE 5-28

The sum of the two fractions

$$\frac{1}{2x+1} + \frac{1}{x-1}$$

is accomplished by first finding the least common denominator, namely $(2x+1)(x-1)$ and then multiplying the numerator and denominator of the first fraction by $x-1$ and the second by $2x+1$. Thus

$$\frac{1}{2x+1} + \frac{1}{x-1} = \frac{x-1}{(x-1)(2x+1)} + \frac{2x+1}{(2x+1)(x-1)}$$

$$= \frac{3x}{(2x+1)(x-1)}$$

Integration by partial fractions requires the student to reverse this procedure. That is, we start with a proper rational function, such as $3x/(2x^2-x-1)$ and express it as the sum of two or more fractions. Thus the integral of the original rational function will be equal to the sum of the integrals of the fractions.

EXAMPLE 5-29

Consider the integral

$$\int \frac{3x\,dx}{2x^2 - x - 1}$$

Since $2x^2 - x - 1 = (2x+1)(x-1)$, we write

$$\int \frac{3x\,dx}{2x^2 - x - 1} = \int \frac{3x\,dx}{(2x+1)(x-1)}$$

From Ex. 5-2 we have

$$\int \frac{3x\,dx}{(2x+1)(x-1)} = \int \frac{dx}{2x+1} + \int \frac{dx}{x-1}$$

$$= \tfrac{1}{2}\ln|2x+1| + \ln|x-1| + C$$

The process of expressing a fraction as the sum of simpler fractions, as was done in Ex. 5-29, requires a knowledge of the process of summing fractions. We demonstrate the technique with the fraction which we have previously encountered.

EXAMPLE 5-30

To express the fraction

$$\frac{3x}{(2x+1)(x-1)}$$

as the sum of two (or more) fractions, we look for the possible denominators which, when combined to make the above fraction, could produce the denominator given, namely $(2x+1)(x-1)$. The only two such denominators are $2x+1$ and $x-1$. Hence, we assume that the fraction can be expressed as

$$\frac{3x}{(2x+1)(x-1)} = \frac{A}{2x+1} + \frac{B}{x-1}$$

where A and B are to be determined. Of course, we know from Ex. 5-28 that A and B both must be 1, but we will demonstrate the technique by showing how we would have determined A and B if we started with the above fraction. We now take the sum of the two fractions containing A and B and get

$$\frac{A(x-1)}{(2x+1)(x-1)} + \frac{B(2x+1)}{(x-1)(2x+1)} = \frac{A(x-1) + B(2x+1)}{(2x+1)(x-1)}$$

Hence

$$\frac{3x}{(2x+1)(x-1)} = \frac{Ax + 2Bx - A + B}{(2x+1)(x-1)}$$

Since the denominators are equal (note that we *chose* the denominators above so that this would be the case) the numerators must be equal, and so

$$3x = (A+2B)x - A + B$$

These are two polynomials, hence the coefficients of each power of x must be equal. Thus

$$3 = A + 2B$$
$$0 = -A + B$$

Here we express $3x$ as $3x+0$. The second equation yields $A = B$. Substituting $A = B$ into the first equation yields $3 = 3A$ implying that $A = 1$. Hence $B = 1$ as we predicted.

We can apply this procedure for rational functions of the form

$$\frac{ax + b}{cx^2 + dx + e}$$

where a or b may be zero. In Ex. 5-30, $a = 3$, $b = 0$, $c = 2$, $d = e = -1$. Let us consider some examples.

EXAMPLE 5-31

Consider the integral

$$\int \frac{(x-3)\,dx}{3x^2 + 2x - 1}$$

We first factor $3x^2 + 2x - 1 = (3x-1)(x+1)$ and we write

$$\frac{x-3}{3x^2 + 2x - 1} = \frac{x-3}{(3x-1)(x+1)}$$

$$= \frac{A}{3x-1} + \frac{B}{x+1}$$

$$= \frac{A(x+1) + B(3x-1)}{(3x-1)(x+1)}$$

Equating numerators yields

$$x - 3 = A(x+1) + B(3x-1)$$
$$= (A+3B)x + A - B$$

Equating like coefficients gives us

$$1 = A + 3B \qquad \text{(coefficients of } x\text{)}$$
$$-3 = A - B \qquad \text{(constant terms)}$$

Subtracting the second equation from the first yields $4 = 4B$, $B = 1$, and hence $A = -2$. Thus

$$\frac{x-3}{3x^2+2x-1} = \frac{-2}{3x-1} + \frac{1}{x+1}$$

$$\int \frac{(x-3)\,dx}{3x^2+2x-1} = \int \frac{-2\,dx}{3x-1} + \int \frac{dx}{x+1}$$

$$= -\tfrac{2}{3}\ln|3x-1| + \ln|x+1| + C$$

EXAMPLE 5-32

Consider the integral

$$\int \frac{(x^3-9x+1)\,dx}{x^2-9}$$

The integrand is an improper rational function so we use long division to express

$$\frac{x^3-9x+1}{x^2-9} = x + \frac{1}{x^2-9}$$

The problem now entails integrating the proper rational function $1/(x^2-9)$. To this end we use the technique of partial fractions and write

$$\frac{1}{x^2-9} = \frac{1}{(x+3)(x-3)}$$

$$= \frac{A}{x+3} + \frac{B}{x-3}$$

$$= \frac{A(x-3) + B(x+3)}{(x+3)(x-3)}$$

$$= \frac{(A+B)x - 3A + 3B}{(x+3)(x-3)}$$

$$0 = A + B$$

$$1 = -3A + 3B$$

Hence $A = -\tfrac{1}{6}$ and $B = \tfrac{1}{6}$. Therefore

$$\int \frac{dx}{x^2-9} = \int \frac{-\tfrac{1}{6}\,dx}{x+3} + \int \frac{\tfrac{1}{6}\,dx}{x-3}$$

$$= -\tfrac{1}{6}\ln(x+3) + \tfrac{1}{6}\ln(x-3) + C$$

Returning to the original problem we have

$$\int \frac{(x^3-9x+1)\,dx}{x^2-9} = \int x\,dx + \int \frac{dx}{x^2-9}$$

$$= \tfrac{1}{2}x^2 - \tfrac{1}{6}\ln|x+3| + \tfrac{1}{6}\ln|x+3| + C$$

If the denominator is not readily factored, the quadratic formula can be used to find the factors. If they are complex numbers, then a knowledge of trigonometry is required. All the functions presented in this text can be easily factored.

The technique of partial fractions can be applied to rational functions whose denominator is of degree greater than two. In general, the partial fractions are chosen so that every possible factor of the denominator of the original fraction is made a denominator of one of the fractions.

EXAMPLE 5-33

Consider the integral

$$\int \frac{(x^2 + 2x + 4)\, dx}{(x+2)^2 x}$$

The integrand is a proper rational function because the denominator is of degree three whereas the degree of the numerator is two. The factors of the denominator are $(x+2)^2$, $(x+2)$, and x. Hence we write

$$\frac{x^2 + 2x + 4}{(x+2)^2 x} = \frac{A}{(x+2)^2} + \frac{B}{x+2} + \frac{C}{x}$$

$$= \frac{Ax + Bx(x+2) + C(x+2)^2}{(x+2)^2 x}$$

$$= \frac{(B+C)x^2 + (A+2B+4C)x + 4C}{(x+2)^2 x}$$

Equating the numerators and then equating coefficients of like powers of x yields

$$1 = B + C$$
$$2 = A + 2B + 4C$$
$$4 = 4C$$

Hence $C = 1$, $B = 0$, and $A = -2$. Thus

$$\frac{x^2 + 2x + 4}{(x+2)^2 x} = \frac{-2}{(x+2)^2} + \frac{1}{x}$$

and so

$$\int \frac{(x^2 + 2x + 4)\, dx}{(x+2)^2 x} = \int \frac{-2\, dx}{(x+2)^2} + \int \frac{dx}{x}$$

To evaluate the first integral on the right, one uses the substitution $u = x + 2$, $du = dx$ and so

$$\int \frac{-2\,dx}{(x+2)^2} = \int \frac{-2\,du}{u^2}$$
$$= 2u^{-1}$$
$$= 2(x+2)^{-1}$$

Therefore

$$\int \frac{(x^2+2x+4)\,dx}{(x+2)^2 x} = 2(x+2)^{-1} + \ln x + C$$

EXAMPLE 5-34

Consider the integral

$$\int \frac{2x^2 - x - 4}{x(x-1)(x+2)}\,dx$$

The factors of the denominator are x, $x-1$, and $x+2$ and thus we write

$$\frac{2x^2 - x - 4}{x(x-1)(x+2)} = \frac{A}{x} + \frac{B}{x-1} + \frac{C}{x+2}$$

$$= \frac{A(x-1)(x+2) + Bx(x+2) + Cx(x-1)}{x(x-1)(x+2)}$$

$$= \frac{(A+B+C)x^2 + (A+2B-C)x - 2A}{x(x-1)(x+2)}$$

$$2 = A + B + C$$
$$-1 = A + 2B - C$$
$$-4 = -2A$$

Hence $A = 2$. Substituting this value into the first two equations and adding the equations yields $B = -1$ and $C = 1$. Hence

$$\frac{2x^2 - x - 4}{x(x-1)(x+2)} = \frac{2}{x} - \frac{1}{x-1} + \frac{1}{x+2}$$

Therefore

$$\int \frac{2x^2 - x - 4}{x(x-1)(x+2)}\,dx = \int \frac{2\,dx}{x} - \int \frac{dx}{x-1} + \int \frac{dx}{x+2}$$

$$= 2\ln|x| - \ln|x-1| + \ln|x+2|$$

EXERCISE 5-5

1. Evaluate the integrals.

(a) $\displaystyle\int \frac{dx}{2x^2 - x - 1}$

(b) $\displaystyle\int \frac{(x+1)\,dx}{2x^2 - x - 1}$

(c) $\displaystyle\int \frac{x\,dx}{x^2 + 2x + 1}$

(d) $\displaystyle\int \frac{(x-1)\,dx}{3x^2 - 4x}$

(e) $\displaystyle\int \frac{x\,dx}{(x-1)(x-2)}$

(f) $\displaystyle\int \frac{dx}{x(x-5)}$

(g) $\displaystyle\int \frac{(x^3+1)\,dx}{x^2 - x}$

(h) $\displaystyle\int \frac{x^4\,dx}{x^2 - 3x + 2}$

(i) $\displaystyle\int \frac{dx}{x(x-1)^2}$

(j) $\displaystyle\int \frac{(x+3)\,dx}{x(x+2)^2}$

(k) $\displaystyle\int \frac{x\,dx}{(x-1)(x-2)^2}$

(l) $\displaystyle\int \frac{x^2\,dx}{(x-1)^3}$

2. Evaluate the definite integrals.

(a) $\displaystyle\int_2^4 \frac{(x-1)\,dx}{x(x+1)}$

(b) $\displaystyle\int_4^5 \frac{(3-x)\,dx}{x(x+3)}$

(c) $\displaystyle\int_2^3 \frac{2\,dx}{x^2 - 1}$

(d) $\displaystyle\int_2^3 \frac{dx}{x(x-1)^2}$

(e) $\displaystyle\int_1^2 \frac{(3x^2 + 3x + 1)\,dx}{x^2(2x+1)}$

(f) $\displaystyle\int_2^3 \frac{(x^2 - 3x + 3)\,dx}{(x-2)^2(x-1)}$

5–6 TABLES OF INTEGRALS

Most of the integrands that we have encountered thus far have required a direct application of one particular technique of integration. It is often the case, however, that an integration problem requires lengthy and cumbersome manipulation. For such problems, a table of integrals is usually helpful.

The table in Appendix A provides a list of antiderivatives which is intended to acquaint the student with the use of such tables. The reader is encouraged to consult a more extensive table, such as the one found in *Mathematical Handbook of Formulas and Tables* which contains more than 500 antiderivatives.

The use of a table of integrals is usually quite straightforward, but there are a few pitfalls which we will now try to point out. It is important to notice, especially when using rather large tables, that the table is organized by the similarity of the integrands. The first few listings are generally the elementary formulas, presented in this text as Rules. These are usually followed by integrands involving $ax+b$ for constants a and b, then those containing ax^2+b, $(ax+b)^2$, and so on.

Because the table is so arranged, many integrands which are very similar to the one confronting the student will be bunched together. It is very easy to make a mistake in choosing the proper formula. Consider for example the integral

$$\int \frac{dx}{x(3x+2)}$$

Referring to the table, formulas 19, 20, and 22 are all similar and each could easily be mistaken for the correct formula, namely formula 21. Note that formula 19 is incorrect because there is no x appearing in the denominator of the integrand of the formula. Similarly, formulas 20 and 22 are incorrect because $(ax+b)^2$ appears in the integrand rather than $ax+b$.

Often some algebraic manipulation is necessary before one uses the table.

EXAMPLE 5-35

Consider the integral

$$\int \frac{dx}{\sqrt{1+4x^2}}$$

Note that the correct formula to use is formula 11, but it cannot be applied immediately because the coefficient of x^2 in the problem is 4, whereas the coefficient of x^2 in the formula is 1. It is necessary to factor the coefficient of x^2 out of the radical sign and then out of the integral so that the coefficient of x^2 in the problem agrees with that in the formula. We use the fact that $\sqrt{1+4x^2} = \sqrt{4(\frac{1}{4}+x^2)} = 2\sqrt{\frac{1}{4}+x^2}$. Hence

$$\int \frac{dx}{\sqrt{1+4x^2}} = \int \frac{dx}{2\sqrt{\frac{1}{4}+x^2}}$$

$$= \tfrac{1}{2} \int \frac{dx}{\sqrt{\frac{1}{4}+x^2}}$$

Now formula 11 can be applied with $a = \tfrac{1}{2}$, so that $a^2 = \tfrac{1}{4}$. Noting that the factor $\tfrac{1}{2}$ must be retained, we have

$$\int \frac{dx}{\sqrt{1+4x^2}} = \tfrac{1}{2} \ln \left| x + \sqrt{\tfrac{1}{4}+x^2} \right|$$

Most tables do not include integrands which can be simplified by an elementary substitution. Hence it is often the case that a simple substitution of a variable must precede the application of a formula in the table.

EXAMPLE 5-36

Consider the integral

$$\int \frac{x\,dx}{\sqrt{x^4-1}}$$

The reader should first convince himself that no formula in the table has an integrand of this form. In fact, no x^4 can be found in any integrand in the table. If we make the substitution $u = x^2$, then $u^2 = x^4$ and $du = 2x\,dx$, $\frac{1}{2}du = x\,dx$. Hence

$$\int \frac{x\,dx}{\sqrt{x^4-1}} = \frac{1}{2}\int \frac{du}{\sqrt{u^2-1}}$$

Formula 12 can be applied to the integrand on the right. Thus

$$\int \frac{du}{\sqrt{u^2-1}} = \ln|u + \sqrt{u^2 \pm 1}|$$

and therefore

$$\int \frac{x\,dx}{\sqrt{x^4-1}} = \frac{1}{2}\ln|u + \sqrt{u^2 \pm 1}|$$

$$= \frac{1}{2}\ln|x^2 + \sqrt{x^4 \pm 1}|$$

Formula 29 is an example of a recursive formula. It may have to be used more than once in a given problem.

EXAMPLE 5-37

Consider the integral

$$\int x^3 e^{5x}\,dx$$

We apply formula 29 with $n = 3$ and $a = 5$ and get

$$\int x^3 e^{5x}\,dx = \tfrac{1}{5}x^3 e^{5x} - \tfrac{3}{5}\int x^2 e^{5x}\,dx$$

To evaluate the integral on the right we again apply formula 29 with $n = 2$ and $a = 5$ and get

$$\int x^2 e^{5x}\,dx = \tfrac{1}{5}x^2 e^{5x} - \tfrac{2}{5}\int x e^{5x}\,dx$$

To evaluate the integral on the right, we again apply formula 29 with $n = 1$ and $a = 5$ and get

$$\int xe^{5x}\,dx = \tfrac{1}{5}xe^{5x} - \tfrac{1}{5}\int e^{5x}\,dx$$
$$= \tfrac{1}{5}xe^{5x} - \tfrac{1}{25}e^{5x}$$

Now we put the pieces of the puzzle back together. From the last three equations we have

$$\int x^3 e^{5x}\,dx = \tfrac{1}{5}x^3 e^{5x} - \tfrac{3}{5}\int x^2 e^{5x}\,dx$$
$$= \tfrac{1}{5}x^3 e^{5x} - \tfrac{3}{5}\left[\tfrac{1}{5}x^2 e^{5x} - \tfrac{2}{5}\int xe^{5x}\,dx\right]$$
$$= \tfrac{1}{5}x^3 e^{5x} - \tfrac{3}{25}x^2 e^{5x} + \tfrac{6}{25}\left[\tfrac{1}{5}xe^{5x} - \tfrac{1}{25}e^{5x}\right]$$
$$= \tfrac{1}{5}x^3 e^{5x} - \tfrac{3}{25}x^2 e^{5x} + \tfrac{6}{125}xe^{5x} - \tfrac{6}{625}e^{5x}$$

To evaluate a definite integral via the table of integrals one first finds the antiderivative of the integrand by the table and then evaluates the antiderivative at the limits of integration.

EXAMPLE 5-38

Consider the definite integral

$$\int_2^3 \frac{dx}{x(2x-3)}$$

To find the antiderivative of the integrand we apply formula 21 with $a = 2$ and $b = -3$. Hence

$$\int \frac{dx}{x(2x-3)} = \frac{1}{-3}\ln\left|\frac{x}{2x-3}\right|$$

Inserting the limits of integration yields

$$\int_2^3 \frac{dx}{x(2x-3)} = -\frac{1}{3}\ln\left|\frac{x}{2x-3}\right|\Big|_2^3$$
$$= -\frac{1}{3}\ln\left|\frac{3}{3}\right| + \frac{1}{3}\ln\left|\frac{2}{1}\right|$$
$$= \frac{1}{3}\ln 2$$

EXERCISE 5-6

1. Use the table of integrals to evaluate the antiderivatives.

   (a) $\displaystyle\int \frac{dx}{\sqrt{1+9x^2}}$

   (b) $\displaystyle\int \frac{dx}{\sqrt{9x^2-1}}$

(c) $\displaystyle\int \frac{dx}{x\sqrt{1-9x^2}}$

(d) $\displaystyle\int \frac{dx}{x\sqrt{1+9x^2}}$

(e) $\displaystyle\int \frac{dx}{x\sqrt{9+16x^2}}$

(f) $\displaystyle\int \frac{dx}{9-16x^2}$

(g) $\displaystyle\int \frac{dx}{4x^2-9}$

(h) $\displaystyle\int \frac{x\,dx}{2x+3}$

(i) $\displaystyle\int \frac{x\,dx}{(2x+3)^2}$

(j) $\displaystyle\int \frac{dx}{x(5x-1)^2}$

(k) $\displaystyle\int \frac{dx}{5x^2-x}$

(l) $\displaystyle\int \sqrt{16+3x^2}\,dx$

2. Use the table of integrals to evaluate the antiderivatives.

(a) $\displaystyle\int x^2 e^{7x}\,dx$

(b) $\displaystyle\int x^4 e^{5x}\,dx$

(c) $\displaystyle\int x^4 e^{3x}\,dx$

(d) $\displaystyle\int x^4 \ln x\,dx$

3. Evaluate the following.

(a) $\displaystyle\int_1^2 \frac{dx}{x\sqrt{1+4x^2}}$

(b) $\displaystyle\int_0^1 \frac{x\,dx}{2x+1}$

(c) $\displaystyle\int_1^2 \frac{dx}{x(5x-1)}$

(d) $\displaystyle\int_8^{10} \frac{x\,dx}{(x-7)^2}$

(e) $\displaystyle\int_{-1}^1 \frac{dx}{16-9x^2}$

(f) $\displaystyle\int_1^3 x^5 \ln x\,dx$

4. Use substitution and the table of integrals to evaluate the antiderivatives.

(a) $\displaystyle\int \frac{x\,dx}{\sqrt{x^4-1}}$

(b) $\displaystyle\int \frac{x\,dx}{1-4x^4}$

(c) $\displaystyle\int \frac{x^2\,dx}{\sqrt{1+x^6}}$

(d) $\displaystyle\int x\ln(2x+3)\,dx$

(e) $\displaystyle\int \frac{e^x\,dx}{e^{2x}-9}$

(f) $\displaystyle\int \frac{e^x\,dx}{3e^x+5}$

(g) $\displaystyle\int \frac{e^{2x}\,dx}{3e^x+5}$

(h) $\displaystyle\int \frac{x^3\,dx}{(3x^2+1)^2}$

6

Applications of the Integral

6-1 NUMERICAL INTEGRATION: TRAPEZOIDAL RULE

When we defined the definite integral, we used approximations to the area under the curve which consisted of sums of areas of certain rectangles. The definite integral was then defined using a limit process. Up to this point, however, we have calculated definite integrals by means of antiderivatives according to the Fundamental Theorem of Calculus. Often it is impractical or even impossible to find the correct antiderivative. For example, the definite integral $\int_0^2 (x^3+1)^{1/2}\,dx$ cannot be evaluated by the methods which we have previously employed.

In such cases, it is necessary to find an approximation to the definite integral. One method would be to use rectangles as we did in Sec. 5-1. In this section and the next, we present two additional methods for approximating integrals, each usually yielding a closer approximation than with rectangles.

Instead of approximating the area under the curve $y = f(x)$ with rectangles, we now use trapezoids. It is clear from Fig. 6-1 that the sum of the areas of the trapezoids can better approximate the integral than the sum of the areas of the rectangles.

Consider the problem of approximating the definite integral $\int_a^b f(x)\,dx$. We first divide the interval $[a, b]$ into n equal subintervals, each having length $(b-a)/n$. The elements in this subdivision are

$$a = x_0, x_1, x_2, \ldots, x_n = b$$

Observe that $x_k = a + k(b-a)/n$.

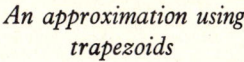

An approximation using rectangles

An approximation using trapezoids

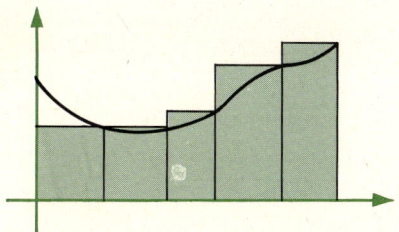
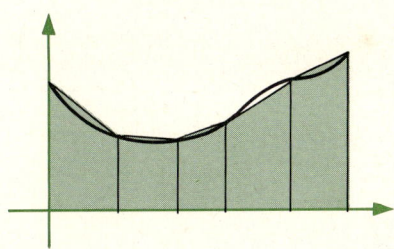

FIGURE 6-1

The area A_k of a trapezoid whose base is the interval $[x_{k-1}, x_k]$, and thus whose respective heights are $f(x_{k-1})$ and $f(x_k)$, is given by

$$A_k = (x_k - x_{k-1})f(x_{k-1}) + \tfrac{1}{2}(x_k - x_{k-1})[f(x_k) - f(x_{k-1})]$$

that is, the sum of the area of the rectangle of height $f(x_{k-1})$ and the area of the triangle of height $f(x_{k-1}) - f(x_k)$. Noting that $x_k - x_{k-1} = (b-a)/n$, we can simplify the above formula for A_k to

$$A_k = \left(\frac{b-a}{2n}\right)[f(x_{k-1}) + f(x_k)]$$

Our approximation entails summing all the trapezoids, and thus

$$\int_a^b f(x)\,dx \cong A_1 + A_2 + \cdots + A_n$$

$$= \frac{b-a}{2n}[f(x_0)+f(x_1)] + \frac{b-a}{2n}[f(x_1)+f(x_2)]$$

$$+ \cdots + \frac{b-a}{2n}[f(x_{n-1})+f(x_n)]$$

We can factor out $(b-a)/2n$ from each term. Noting that each $f(x_k)$ appears twice, except $f(x_0)$ and $f(x_n)$, i.e., $f(a)$ and $f(b)$, we get

$$\int_a^b f(x)\,dx \cong \frac{b-a}{2n}[f(a)+2f(x_1)+2f(x_2)+\cdots+2f(x_{n-1})+f(b)]$$

which is known as the *Trapezoidal Rule*.

EXAMPLE 6-1

To demonstrate the effectiveness of the Trapezoidal Rule we first apply the technique to the integral whose value we can easily calculate, and thus we can readily calculate the error. We will use the Trapezoidal Rule to

approximate
$$\int_0^1 x^2\,dx$$
using ten trapezoids. Thus $n = 10$, $a = 0$, $b = 1$, $x_k = k/10$ and hence $f(x_k) = k^2/100$. Substituting these values into the formula yields

$$\int_0^1 x^2\,dx$$

$$\cong \frac{1}{20}\left[0 + 2\left(\frac{1}{10}\right)^2 + 2\left(\frac{2}{10}\right)^2 + \cdots + 2\left(\frac{9}{10}\right)^2 + \left(\frac{10}{10}\right)^2\right]$$

$$= \frac{1}{20}[0.02 + 0.08 + 0.18 + 0.32 + 0.50 + 0.72 + 0.98 + 1.28 + 1.62 + 1.0]$$

$$= \frac{1}{20}[7.70]$$

$$= .335$$

By the Fundamental Theorem of Calculus

$$\int_0^1 x^2\,dx = \frac{x^3}{3}\bigg|_0^1 = \frac{1}{3} = 0.333\ldots$$

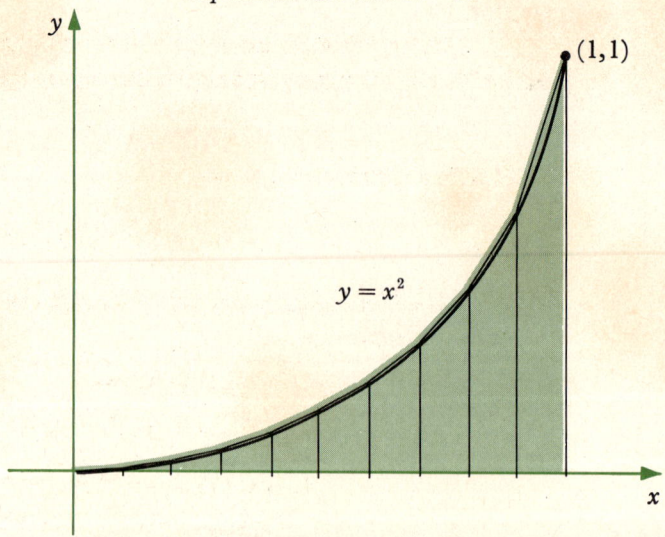

An approximation of $\int_0^1 x^2\,dx$ by the trapezoidal rule with $n = 10$

FIGURE 6-2

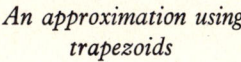

| An approximation using rectangles | An approximation using trapezoids |

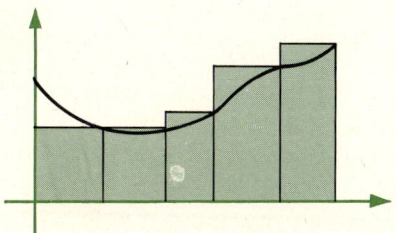 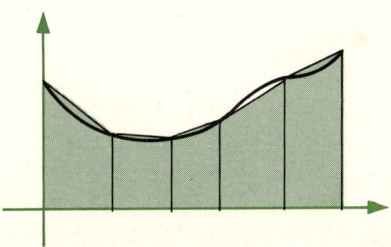

FIGURE 6-1

The area A_k of a trapezoid whose base is the interval $[x_{k-1}, x_k]$, and thus whose respective heights are $f(x_{k-1})$ and $f(x_k)$, is given by

$$A_k = (x_k - x_{k-1})f(x_{k-1}) + \tfrac{1}{2}(x_k - x_{k-1})[f(x_k) - f(x_{k-1})]$$

that is, the sum of the area of the rectangle of height $f(x_{k-1})$ and the area of the triangle of height $f(x_{k-1}) - f(x_k)$. Noting that $x_k - x_{k-1} = (b-a)/n$, we can simplify the above formula for A_k to

$$A_k = \left(\frac{b-a}{2n}\right)[f(x_{k-1}) + f(x_k)]$$

Our approximation entails summing all the trapezoids, and thus

$$\int_a^b f(x)\,dx \cong A_1 + A_2 + \cdots + A_n$$

$$= \frac{b-a}{2n}[f(x_0) + f(x_1)] + \frac{b-a}{2n}[f(x_1) + f(x_2)]$$

$$+ \cdots + \frac{b-a}{2n}[f(x_{n-1}) + f(x_n)]$$

We can factor out $(b-a)/2n$ from each term. Noting that each $f(x_k)$ appears twice, except $f(x_0)$ and $f(x_n)$, i.e., $f(a)$ and $f(b)$, we get

$$\int_a^b f(x)\,dx \cong \frac{b-a}{2n}[f(a) + 2f(x_1) + 2f(x_2) + \cdots + 2f(x_{n-1}) + f(b)]$$

which is known as the *Trapezoidal Rule*.

EXAMPLE 6-1

To demonstrate the effectiveness of the Trapezoidal Rule we first apply the technique to the integral whose value we can easily calculate, and thus we can readily calculate the error. We will use the Trapezoidal Rule to

approximate
$$\int_0^1 x^2\,dx$$
using ten trapezoids. Thus $n = 10$, $a = 0$, $b = 1$, $x_k = k/10$ and hence $f(x_k) = k^2/100$. Substituting these values into the formula yields

$$\int_0^1 x^2\,dx$$

$$\cong \frac{1}{20}\left[0 + 2\left(\frac{1}{10}\right)^2 + 2\left(\frac{2}{10}\right)^2 + \cdots + 2\left(\frac{9}{10}\right)^2 + \left(\frac{10}{10}\right)^2\right]$$

$$= \frac{1}{20}[0.02 + 0.08 + 0.18 + 0.32 + 0.50 + 0.72 + 0.98 + 1.28 + 1.62 + 1.0]$$

$$= \frac{1}{20}[7.70]$$

$$= .335$$

By the Fundamental Theorem of Calculus

$$\int_0^1 x^2\,dx = \left.\frac{x^3}{3}\right|_0^1 = \frac{1}{3} = 0.333\ldots$$

An approximation of $\int_0^1 x^2\,dx$ by the trapezoidal rule with $n = 10$

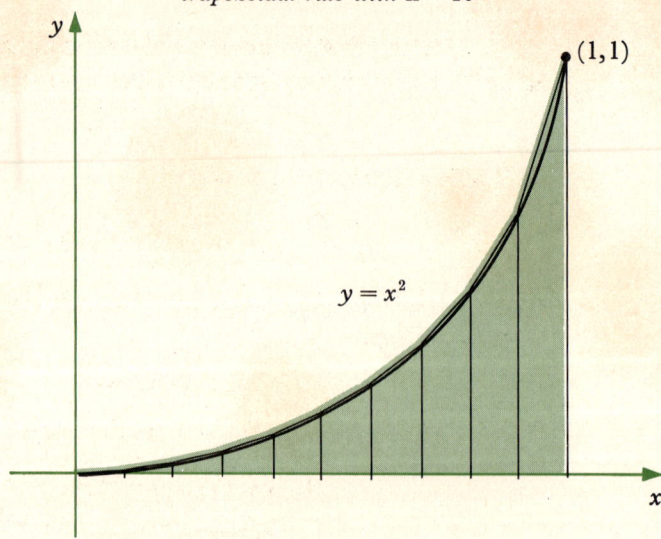

FIGURE 6-2

SEC. 6-1   Numerical Integration: Trapezoidal Rule

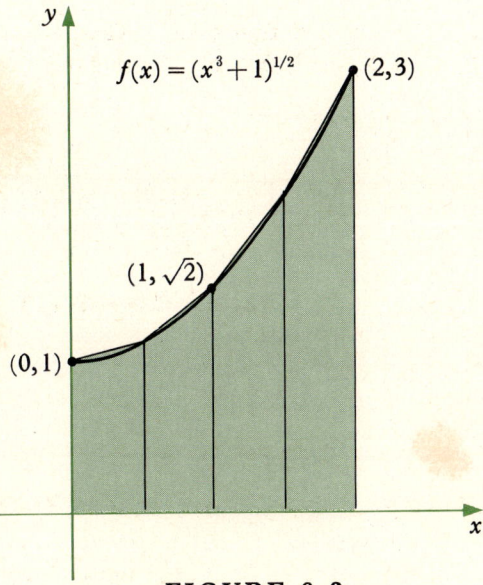

*An approximation of $\int_0^2 (x^3+1)^{1/2}\,dx$
by the trapezoidal rule with* n = 4

FIGURE 6-3

We see that the approximation by the Trapezoidal Rule is quite good. An even better approximation could be obtained by increasing the number of trapezoids.

We mentioned earlier that the integral $\int_0^2 (x^3+1)^{1/2}\,dx$ cannot be evaluated by the methods which we have previously introduced.

EXAMPLE 6-2

We approximate the integral

$$\int_0^2 (x^3+1)^{1/2}\,dx$$

by the Trapezoidal Rule with four trapezoids. Thus $n=4$, $a=0$, $b=2$, $x_k = k/2$ and thus

$$f(x_0) = f(0) = 1$$
$$f(x_1) = f(1/2) = \sqrt{9/8} \cong 1.07$$
$$f(x_2) = f(1) = \sqrt{2} \cong 1.41$$
$$f(x_3) = f(3/2) = \sqrt{35/8} \cong 2.09$$
$$f(x_4) = f(2) = \sqrt{9} = 3$$

183

*An approximation of $\int_0^1 e^{-x^2}\,dx$
by the trapezoidal rule with* n = 5

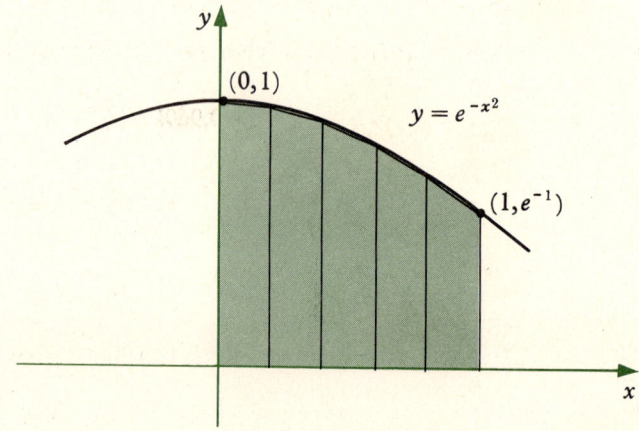

FIGURE 6-4

Substituting these values into the formula yields

$$\int_0^2 (x^3+1)^{1/2}\,dx \cong \frac{1}{4}[1+2(1.07)+2(1.41)+2(2.09)+3]$$

$$= \frac{1}{4}[1+2.14+2.82+4.18+3]$$

$$= \frac{1}{4}[13.14]$$

$$= 3.24$$

The function $y = e^{-x^2}$ has no antiderivative and thus to evaluate $\int_0^1 e^{-x^2}\,dx$ one must use an approximation such as the Trapezoidal Rule.

EXAMPLE 6-3

We will use the Trapezoidal Rule with five trapezoids to approximate

$$\int_0^1 e^{-x^2}\,dx$$

Thus $n = 5$, $a = 0$, $b = 1$, $x_k = k/5$ and thus

$$f(x_0) = f(0) = e^0 = 1$$
$$f(x_1) = f(1/5) = e^{-1/25} = e^{-0.04} \cong 0.960$$
$$f(x_2) = f(2/5) = e^{-4/25} = e^{-0.16} \cong 0.825$$

$$f(x_3) = f(3/5) = e^{-9/25} = e^{-0.36} \cong 0.698$$

$$f(x_4) = f(4/5) = e^{-16/25} = e^{-0.64} \cong 0.527$$

$$f(x_5) = f(5/5) = e^{-1} \cong 0.368$$

Substituting these values into the formula yields

$$\int_0^1 e^{-x^2} dx \cong \frac{1}{10}[1 + 2(0.960) + 2(0.825) + 2(0.698) + 2(0.527) + 0.368]$$

$$= \frac{1}{10}[1 + 1.920 + 1.650 + 1.396 + 1.054 + 0.368]$$

$$= \frac{1}{10}[7.388]$$

$$= 0.7388$$

EXERCISE 6-1

1. Use the Trapezoidal Rule to approximate $\int_a^b x^2 \, dx$ for the given values of a, b, and n.
   (a) $a = 0, b = 1, n = 12$
   (b) $a = 0, b = 1, n = 20$
   (c) $a = 0, b = 2, n = 10$
   (d) $a = 0, b = 3, n = 10$

2. Use the Trapezoidal Rule to approximate $\int_a^b x^3 \, dx$ for the given values of a, b, and n.
   (a) $a = 0, b = 1, n = 10$
   (b) $a = 0, b = 1, n = 12$
   (c) $a = 0, b = 2, n = 10$
   (d) $a = 0, b = 3, n = 10$

3. Use the Trapezoidal Rule to approximate $\int_a^b (1/x) \, dx$ for the given values of a, b, and n.
   (a) $a = 1, b = 2, n = 5$
   (b) $a = 1, b = 2, n = 4$
   (c) $a = 1, b = 3, n = 5$
   (d) $a = 1, b = 4, n = 5$

4. Use the Trapezoidal Rule to approximate $\int_a^b (x^3 + 1)^{1/2} \, dx$ for the given values of a, b, and n.
   (a) $a = 0, b = 2, n = 8$
   (b) $a = 0, b = 4, n = 4$
   (c) $a = 0, b = 4, n = 8$
   (d) $a = 0, b = 1, n = 4$

5. Use the Trapezoidal Rule to approximate $\int_a^b e^{-x^2}\, dx$ for the given values of a, b, and n.
   - (a) $a = 0, b = 1, n = 3$
   - (b) $a = 0, b = 1, n = 4$
   - (c) $a = 0, b = 1, n = 10$
   - (d) $a = 0, b = 2, n = 4$

6. Approximate the integrals by means of the Trapezoidal Rule using five trapezoids.
   - (a) $\int_0^2 e^{-0.5x^2}\, dx$
   - (b) $\int_0^1 \dfrac{1}{1+x^2}\, dx$
   - (c) $\int_0^1 (1-x^2)^{1/2}\, dx$

6-2 NUMERICAL INTEGRATION: SIMPSON'S RULE

When we defined the integral of $f(x)$ we approximated the area under $y = f(x)$ over each subinterval $[x_{k-1}, x_k]$ by the constant function $y = f(x_k)$. Hence the area considered was a rectangle. (See Fig. 5-2.) The Trapezoidal Rule utilizes a similar approach, but the constant function is replaced with a linear function $y = mx + b$ where m and b were chosen so that the line would pass through the two points $(x_{k-1}, f(x_{k-1}))$ and $(x_k, f(x_k))$. Hence the area considered was a trapezoid. (See Fig. 6-1.)

We now consider a technique of numerical integration which is based upon replacing the curve over the subintervals by a parabola, i.e., a polynomial function of degree two of the form $y = Ax^2 + Bx + C$. This method, while a bit more complicated than the Trapezoidal Rule, is superior to the

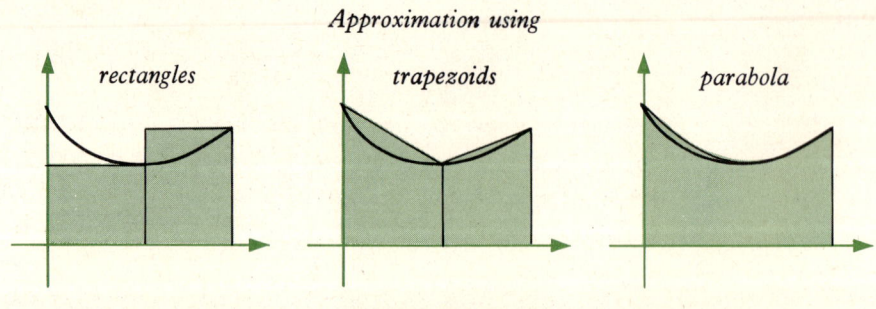

FIGURE 6-5

Trapezoidal Rule in that it usually yields a closer approximation and it is readily adapted to use on computers.

Consider the area A under the parabola $y = Ax^2 + Bx + C$ from $x = -a$ to $x = a$. If we let $y_0 = y(-a)$, $y_1 = y(0)$ and $y_2 = y(a)$, then it can be shown that

$$A = \int_{-a}^{a} (Ax^2 + Bx + C)\, dx$$

$$= \frac{a}{3}(y_0 + 4y_1 + y_2)$$

To approximate $\int_a^b f(x)\, dx$ we first subdivide the interval $[a, b]$ into n equal subintervals, each having length $(b-a)/n$. The elements of the subdivision are $a = x_0, x_1, x_2, \ldots, x_n = b$. Observe that $x_k = a + k(b-a)/n$.

If we let $y_k = f(x_k)$ for $k = 0, 1, 2, \ldots, n$, then the area A_k under the parabola $y = Ax^2 + Bx + C$ which passes through the points (x_{k-1}, y_{k-1}), (x_k, y_k) and (x_{k+1}, y_{k+1}) from x_{k-1} to x_{k+1} is given by

$$A_k = \int_{x_{k-1}}^{x_{k+1}} (Ax^2 + Bx + C)\, dx$$

$$= \frac{b-a}{3n} [y_{k-1} + 4y_k + y_{k+1}]$$

The number n must be an *even* number because we must apply the above formula to successive pairs of intervals. The approximation entails summing all these areas A_k. Note that the subscript of A_k must be an odd integer. This results in the following formula.

$$\int_a^b f(x)\, dx \cong A_1 + A_3 + \cdots + A_{n-1}$$

$$= \frac{b-a}{3n}[y_0 + 4y_1 + y_2] + \frac{b-a}{3n}[y_2 + 4y_3 + y_4]$$

$$+ \cdots + \frac{b-a}{3n}[y_{n-2} + 4y_{n-1} + y_n]$$

Observe that we can simplify this formula by factoring out the expression $(b-a)/3n$ from each term and by noting that y_0 and y_n appear only once, while all y_i's for odd numbers i have a coefficient of 4 and all y_i's for even numbers i appear twice in the formula. *Simpson's Rule* is thus given by

$$\int_a^b f(x)\, dx \cong \left(\frac{b-a}{3n}\right)(y_0 + 4y_1 + 2y_2 + 4y_3 + 2y_4 + 4y_5 + 2y_6 + \cdots + 4y_{n-1} + y_n)$$

CHAP. 6   Applications of the Integral

EXAMPLE 6-4
To approximate
$$\int_0^1 x^2 \, dx$$
by Simpson's Rule, we choose $n = 10$ and so $a = 0$, $b = 1$, $x_k = k/10$ and $y_k = (k/10)^2 = k^2/100$. Substituting these values into the formula yields

$$\int_0^1 x^2 \, dx$$

$$\cong \frac{1}{30}\left[0 + 4\left(\frac{1}{100}\right) + 2\left(\frac{4}{100}\right) + \cdots + \left(\frac{100}{100}\right)\right]$$

$$= \frac{1}{30}[0 + 0.04 + 0.08 + 0.36 + 0.32 + 1.0 + 0.72 + 1.96 + 1.28 + 3.24 + 1]$$

$$= \frac{1}{30}(10) = \frac{1}{3}$$

EXAMPLE 6-5
To approximate
$$\int_1^3 \frac{1}{x} \, dx$$
we choose $n = 8$ and so $a = 1$, $b = 3$, $x_k = 1 + (k/4) = (4+k)/4$ and $y_k = 4/(4+k)$. Substituting these values into the formula yields

$$\int_1^3 \frac{1}{x} \, dx$$

$$\cong \frac{2}{3 \cdot 8}\left[1 + 4\left(\frac{4}{5}\right) + 2\left(\frac{2}{3}\right) + 4\left(\frac{4}{7}\right) + 2\left(\frac{1}{2}\right) + 4\left(\frac{4}{9}\right) + 2\left(\frac{2}{5}\right) + 4\left(\frac{4}{11}\right) + \left(\frac{1}{3}\right)\right]$$

$$= \frac{1}{12}\left[1 + \left(\frac{16}{5}\right) + \left(\frac{4}{3}\right) + \left(\frac{16}{7}\right) + 1 + \left(\frac{16}{9}\right) + \left(\frac{4}{5}\right) + \left(\frac{16}{11}\right) + \left(\frac{1}{3}\right)\right]$$

$$= \frac{1}{12}\left[\frac{9137}{693}\right] \cong 1.1$$

EXERCISE 6-2

1. Use Simpson's Rule to evaluate $\int_a^b x^2 \, dx$ using the given values of n, a, and b.
   - (a) $n = 4$, $a = 0$, $b = 1$
   - (b) $n = 10$, $a = 0$, $b = 1$
   - (c) $n = 4$, $a = 0$, $b = 5$
   - (d) $n = 10$, $a = 0$, $b = 5$

2. Use Simpson's Rule to evaluate $\int_a^b (1/x)\,dx$ using the given values of n, a, and b.
   - (a) $n = 6, a = 1, b = 3$
   - (b) $n = 8, a = 1, b = 5$
   - (c) $n = 10, a = 1, b = 3$
   - (d) $n = 10, a = 1, b = 5$
3. Use Simpsons' Rule to evaluate the integrals.
   - (a) $\displaystyle\int_0^2 (1+x)^{-1}\,dx,\ n = 4$
   - (b) $\displaystyle\int_0^1 (1+x^2)^{-1}\,dx,\ n = 4$
   - (c) $\displaystyle\int_0^1 e^{-x^2}\,dx,\ n = 4$
   - (d) $\displaystyle\int_1^2 \ln x,\ n = 4$

6–3 INFINITE LIMITS AND IMPROPER INTEGRALS

When we consider the limit of $f(x)$ as x approaches a,

$$\lim_{x \to a} f(x)$$

we are concerned with the behavior of $f(x)$ when x is close to a. It is often the case that we are concerned with the behavior of $f(x)$ as x becomes very large, or as x increases without bound. For example, as x becomes very large, $1/x$ approaches zero and $x/(x+1)$ approaches 1.

For notational purposes, we utilize the symbol ∞, read "infinity", and we write "$x \to \infty$", read "x approaches infinity", to mean that x increases without bound. Similarly, by "$x \to -\infty$", read "x approaches minus infinity", we mean x decreases without bound.

We will be content to provide an intuitive definition of the concept of an *infinite limit*.

Definition By $\lim_{x \to \infty} f(x)$ is meant that number L such that as x approaches infinity, i.e., as x increases without bound, $f(x)$ approaches L. We similarly define $\lim_{x \to -\infty} f(x)$.

One way of calculating infinite limits is to notice that as x approaches ∞, $1/x$ approaches zero from the right. Hence

$$\lim_{x \to \infty} f(x) = \lim_{1/x \to 0^+} f(x)$$

and
$$\lim_{x \to -\infty} f(x) = \lim_{1/x \to 0^-} f(x)$$

EXAMPLE 6-6

As we mentioned before, we have

$$\lim_{x \to \infty} \frac{1}{x} = 0$$

$$\lim_{x \to \infty} \frac{x}{x+1} = \lim_{x \to \infty} \frac{1}{1 + 1/x} = 1 \qquad \text{\textit{divide numerator and denominator by }} x$$

We can also see that

$$\lim_{x \to \infty} e^{-x} = \lim_{x \to \infty} \frac{1}{e^x} = 0$$

$$\lim_{x \to \infty} \frac{2x+1}{x} = \lim_{x \to \infty} \left(2 + \frac{1}{x}\right) = 2$$

$$\lim_{x \to -\infty} \frac{1}{x} = 0$$

$$\lim_{x \to -\infty} \frac{x}{x+1} = 1$$

$$\lim_{x \to \infty} e^{1/x} = \lim_{1/x \to 0^+} e^{1/x} = 1$$

$$\lim_{x \to \infty} \frac{3x^2 + x}{2x^2 - 1} = \lim_{x \to \infty} \frac{3 + 1/x}{2 - 1/x^2} = \frac{3}{2}$$

The graphs of $y = x/(x+1)$ and $y = e^{-x}$ are given in Fig. 6-6.

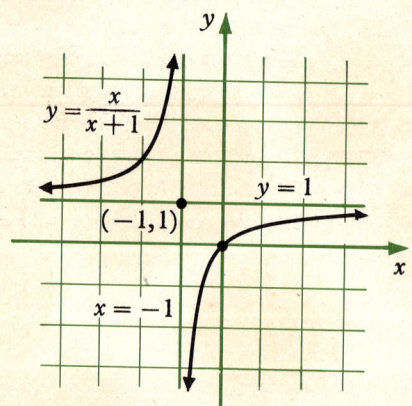
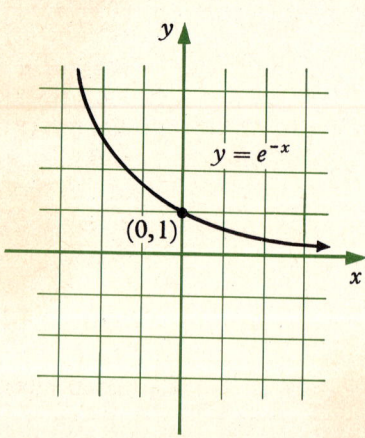

FIGURE 6-6

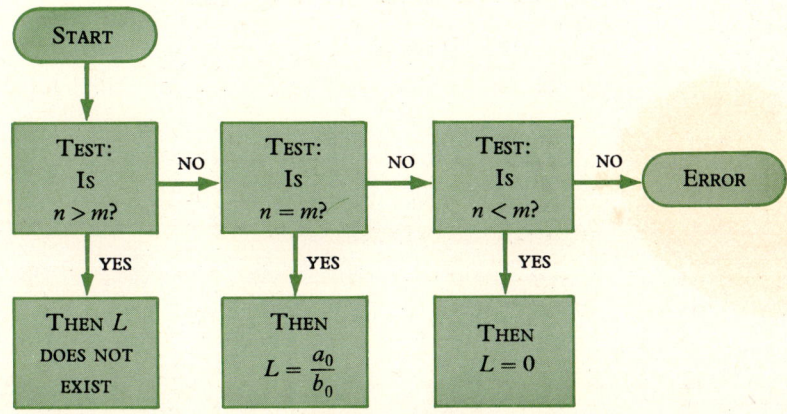

FIGURE 6-7

All the basic limit theorems mentioned in Chap. 2 hold for infinite limits and we leave their formulation to the reader. We provide the flowchart (Fig. 6-7) on the calculation of the infinite limit of a rational function.

Up to this point we have dealt only with the integral of a function that is continuous in the closed interval $[a, b]$. We can extend this concept to infinite intervals $[a, \infty)$, $(-\infty, b]$, and even $(-\infty, \infty)$ by the use of infinite limits.

EXAMPLE 6-7

Consider $\int_1^\infty 1/x^2 \, dx$. If we consider this expression as the number representing the area of the region bounded by $y = 1/x^2$, the x-axis, and to the right of $x = 1$, we might believe that no such number could be feasible since this region is in a sense "infinite", i.e., it has no bound on the right. Yet mathematicians have developed a plausible definition for such a number which has invaluable applications. To define this number we merely look at areas bounded by $y = 1/x^2$, the x-axis, $x = 1$, and $x = b$, and then let b increase without bound while we search for a limit, as illustrated in Fig. 6-8. Hence we are looking at

$$\lim_{b \to \infty} \int_1^b \frac{1}{x^2} \, dx = \lim_{b \to \infty} -x^{-1} \Big|_1^b = \lim_{b \to \infty} \left(-\frac{1}{b} + 1\right) = 1$$

Thus we say that

$$\int_1^\infty \frac{1}{x^2} \, dx = 1$$

CHAP. 6   Applications of the Integral

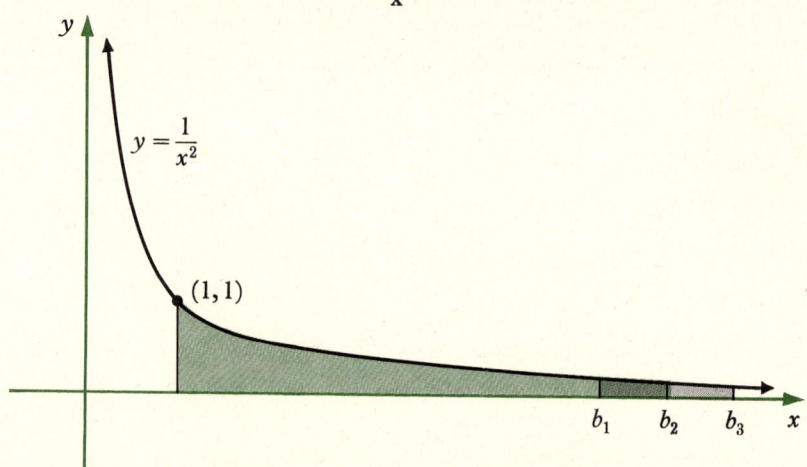

FIGURE 6-8

The previous example leads us to the following definition.

Definition   We define the following *improper integrals*:

$$\int_a^\infty f(x)\,dx = \lim_{b \to \infty} \int_a^b f(x)\,dx$$

$$\int_{-\infty}^a f(x)\,dx = \lim_{b \to -\infty} \int_b^a f(x)\,dx$$

Of course these integrals are certainly legal mathematical concepts—the term "improper" is meant only to differentiate this concept from the definite integral defined in Sec. 5-1.

As evidenced by the next example, sometimes an improper integral does not exist. When an improper integral exists it is called *convergent*, when it doesn't it is called *divergent*.

EXAMPLE 6-8

Consider the improper integral

$$\int_1^\infty \frac{1}{x}\,dx = \lim_{b \to \infty} \ln x \Big|_1^b = \lim_{b \to \infty} \ln b$$

which does not exist because $\ln x$ increases without bound as $x \to \infty$, which is sometimes expressed as

$$\lim_{b \to \infty} \ln b = \infty$$

Hence the integral $\int_1^\infty \frac{1}{x} dx$ does not exist.

EXAMPLE 6-9
Consider
$$\int_0^\infty e^{-x} dx = \lim_{b \to \infty} -e^{-x} \Big|_0^b = \lim_{b \to \infty} (-e^{-b} + e^0) = 1$$

EXERCISE 6-3

1. Find $\lim_{x \to \infty} f(x)$, if it exists.

   (a) $f(x) = \frac{1}{x^3}$

   (b) $f(x) = \frac{2}{x^4}$

   (c) $f(x) = 5x^{-5}$

   (d) $f(x) = 3x$

   (e) $f(x) = \frac{2x+1}{x}$

   (f) $f(x) = \frac{5x^2+1}{2x^2}$

   (g) $f(x) = \frac{1-x}{1+x}$

   (h) $f(x) = \frac{1-x}{1+x^2}$

   (i) $f(x) = \frac{x^3 - x^2}{x - 1}$

   (j) $f(x) = \frac{x^3 - x^2}{x^2 - 1}$

   (k) $f(x) = \frac{x^3 - x^2}{x^3 - 1}$

   (l) $f(x) = \frac{x^3 - x^2}{x^4 - 1}$

   (m) $f(x) = e^{-x^2}$

   (n) $f(x) = \frac{1}{\ln x}$

   (o) $f(x) = \frac{\sqrt{x^2+1}}{2x}$

   (p) $f(x) = \frac{7x^6 - x + 1}{-2x^6 + 1}$

2. Calculate the integrals, if they exist.

   (a) $\int_1^\infty 3x^{-2} dx$

   (b) $\int_1^\infty 2x^{-3} dx$

   (c) $\int_0^\infty e^{-2x} dx$

   (d) $\int_0^\infty e^x dx$

   (e) $\int_0^\infty xe^{-x^2} dx$

   (f) $\int_4^\infty (1+2x)^{-3/2} dx$

CHAP. 6   Applications of the Integral

3. One can also define the improper integral:

$$\int_{-\infty}^{\infty} f(x)\,dx = \int_{-\infty}^{0} f(x)\,dx + \int_{0}^{\infty} f(x)\,dx$$

Calculate the integrals, if they exist.

(a) $\displaystyle\int_{-\infty}^{\infty} x^{-3}\,dx$ (b) $\displaystyle\int_{-\infty}^{\infty} x(x^2+1)^{-3}\,dx$

4. Another type of improper integral can be defined when the integrand is discontinuous at a given point. Suppose $f(x)$ is continuous in $[a, c)$ but not continuous at c. (See Exer. 2-1, Problem 18, for the definition of right- and left-hand limits.) Then we define

$$\int_a^c f(x)\,dx = \lim_{t \to c^-} \int_a^t f(x)\,dx$$

Hence

$$\int_{1/2}^{1} [\![x]\!]\,dx = \lim_{t \to 1^-} \int_{1/2}^{t} [\![x]\!]\,dx = \lim_{t \to 1} \int_{1/2}^{t} 0\,dx = 0$$

Also

$$\int_0^1 x^{-1/2}\,dx = \lim_{t \to 0^+} \int_t^1 x^{-1/2}\,dx = \lim_{t \to 0^+} 2x^{1/2}\Big|_t^1 = \lim_{t \to 0^+}(2 - 2t^{1/2}) = 2$$

Calculate the integrals, if they exist.

(a) $\displaystyle\int_0^4 x^{1/2}\,dx$ (c) $\displaystyle\int_0^1 x^{-1}\,dx$

(b) $\displaystyle\int_{1.1}^{2} [\![x]\!]\,dx$ (d) $\displaystyle\int_1^2 x(1-x^2)^{-1/2}\,dx$

6-4 VOLUME AND WORK

2nd section

As we indicated previously, integrals can be used to calculate volumes. The process used to define the volume of a solid is analogous to that used for defining the area of a region.

Consider the volume of the solid in Fig. 6-9. We approximate this volume by sectioning it into small cylinder-shaped solids, whose volume we can calculate.

The volume V of a cylinder-shaped solid as pictured in Fig. 6-10 with cross-sectional area $A(x)$ and height h is $A(x)h$.

Consider the solid pictured in Fig. 6-11. It is bounded by the planes perpendicular to the x-axis and through the lines $x = a$ and $x = b$ and centered about the x-axis. If we divide the interval $[a, b]$ into n equal

A cylinder-shaped solid

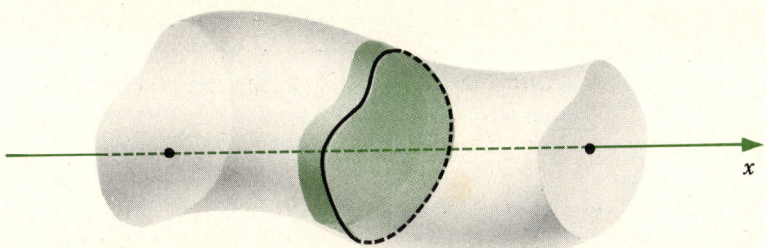

FIGURE 6-9

parts by the numbers $a = x_0, x_1, x_2, \ldots, x_n = b$ where

$$x_i = a + i\frac{b-a}{n}$$

then the volume of each cylindrical solid is

$$A(x_i)\left(\frac{b-a}{n}\right)$$

where $A(x_i)$ is the area of the cross-sectional region perpendicular to the x-axis and through the line $x = x_i$.

Our approximation of the volume of the entire figure is then arrived at by summing these volumes, and is given by

$$\frac{b-a}{n}[A(x_1) + A(x_2) + \cdots + A(x_n)]$$

The volume of a cylinder-shaped solid

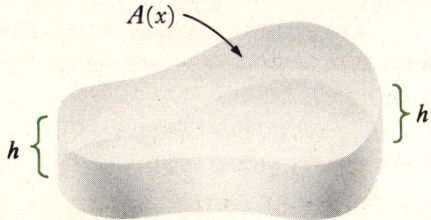

FIGURE 6-10

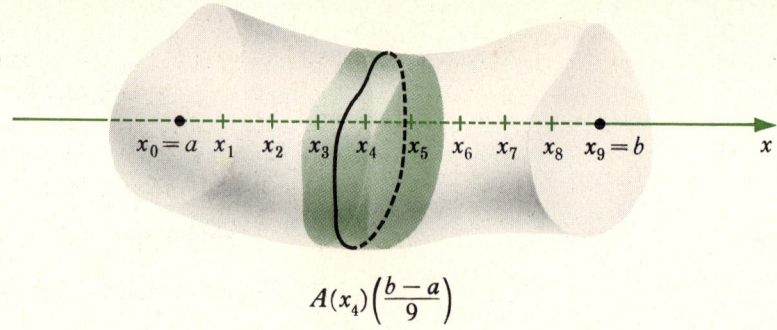

$$A(x_4)\left(\frac{b-a}{9}\right)$$

FIGURE 6-11

We can then define the volume V of the solid to be given by

$$V = \lim_{n\to\infty} \frac{b-a}{n}[A(x_1) + \cdots + A(x_n)] = \int_a^b A(x)\,dx$$

The following examples illustrate the idea.

EXAMPLE 6-10

Let us first calculate the volume of a cylinder of height a and radius r. If we set the coordinate axes as in Fig. 6-12, we see that for any x in the interval $[0, a]$, the cross-sectional area $A(x)$, being the area of a circle of radius r, is given by $A(x) = \pi r^2$. Hence the volume of the cylinder is

$$\int_0^a \pi r^2\,dx = \pi r^2 x\Big|_0^a = \pi a r^2$$

This is the usual formula for the volume of a cylinder.

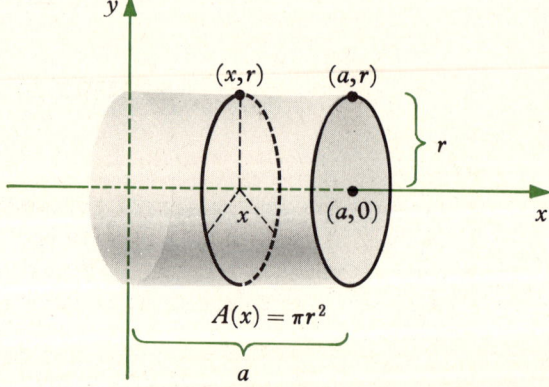

FIGURE 6-12

EXAMPLE 6-11

Let us now calculate the volume of a sphere of radius r by first setting the coordinate axes as in Fig. 6-13. For each number x in the interval $[-r, r]$ the cross-sectional area $A(x)$, being the area of the circle with radius y, is given by $A(x) = \pi y^2$. Now $x^2 + y^2 = r^2$ since every point (x,y) is on the circle in the plane of the x- and y-axes. Hence

$$A(x) = \pi(r^2 - x^2)$$

for every x in $[-r, r]$. Hence the volume V of the sphere is

$$V = \int_{-r}^{r} \pi(r^2 - x^2)\, dx = \pi\left(r^2 x - \frac{x^3}{3}\right)\bigg|_{-r}^{r} = \frac{4}{3}\pi r^3$$

This is the usual formula for the volume of a sphere of radius r.

The volume of a sphere

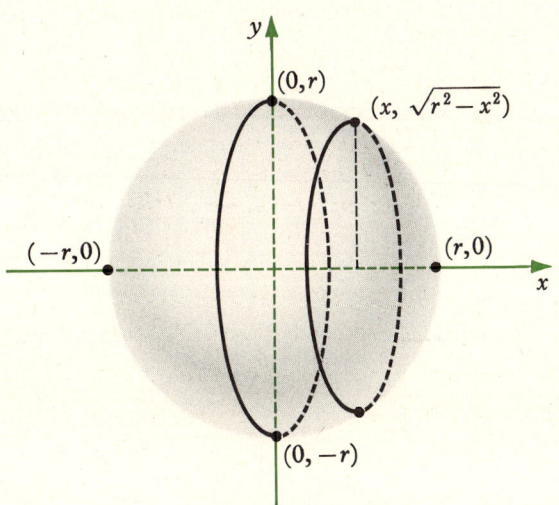

FIGURE 6-13

EXAMPLE 6-12

Consider the solid generated by rotating about the x-axis the region bounded by the x- and y-axes, $x = y$, and $x = a$ described in Fig. 6-14. The solid so generated is a cone. For each x in the interval $[0, a]$, we have $A(x) = \pi x^2$, since the area is a circle with radius $y = x$. Hence the volume V is

$$V = \int_{0}^{a} \pi x^2\, dx = \pi \frac{x^3}{3}\bigg|_{0}^{a} = \frac{\pi a^3}{3}$$

The volume of a cone

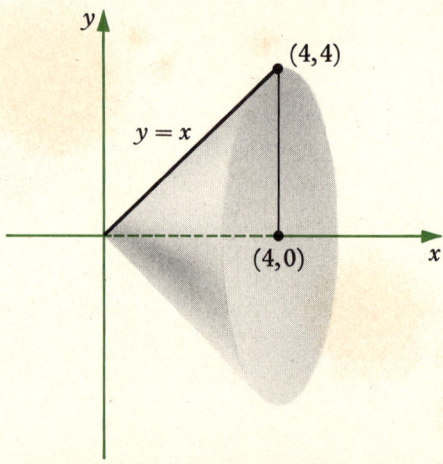

FIGURE 6-14

which is the usual formula for the volume of a cone where the radius of the base equals the height.

EXAMPLE 6-13

Consider the solid generated by rotating about the x-axis the region bounded by the x- and y-axes, $x = 4$, and the parabola $y^2 = x$. This solid is similar to the headlight of a car and is illustrated in Fig. 6-15. For each x in the interval $[0, 4]$, we have $A(x) = \pi y^2 = \pi x$, since the region is a circle with radius y. Hence the volume V is given by

$$V = \int_0^4 \pi x \, dx = \left. \frac{\pi x^2}{2} \right|_0^4 = 8\pi$$

The solids described in Examples 6-12 and 6-13 are usually referred to as "solids of revolution".

In physics one encounters the concept of *work*. If a constant force, F, measured in pounds is applied to an object and moves it d ft, then one defines the work W done on the object by

$$W = Fd$$

where W is measured in foot-pounds. If we have a variable force the calculus is needed to define the concept of work. Suppose the object under force moves from a to b while a force of $f(x)$ units if being exerted on the object when it is at the point with coordinate x. If we divide the interval

The volume of a headlight-shaped object

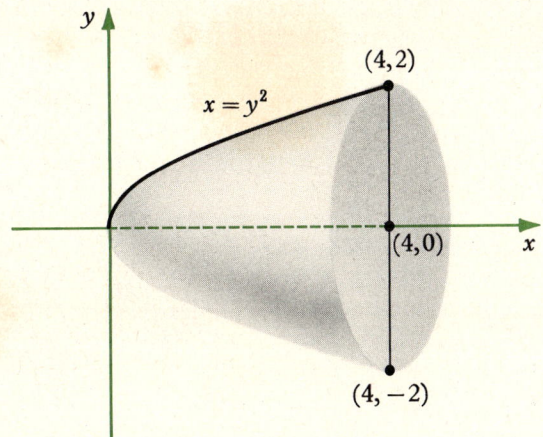

FIGURE 6-15

$[a, b]$ into n equal subintervals by the points

$$a = x_0, x_1, x_2, \ldots, x_n = b \quad \text{where } x_i = a + i\frac{b-a}{n}$$

we can approximate the amount of work done by moving the object from x_i to x_{i+1} by

$$W_i = f(x_i)\frac{b-a}{n}$$

That is, we are assuming the force is constant on the interval $[x_i, x_{i+1}]$. Our approximation to the total work is the sum of these W_i, that is,

$$\frac{b-a}{n}[f(x_1) + f(x_2) + \cdots + f(x_n)]$$

We can then define the amount of work, W, done in moving the object from a to b by a force $f(x)$ by

$$W = \lim_{n \to \infty} \frac{b-a}{n}[f(x_1) + \cdots + f(x_n)]$$

$$= \int_a^b f(x)\, dx$$

EXAMPLE 6-14

Hooke's law states that the force $F(x)$ required to hold a spring extended within its elastic limit a total of x units beyond its natural length is proportional to the distance x, that is, $F(x) = kx$ for some constant k. Suppose

that a force of 30 lb is needed to stretch a spring 10 ft. Then $30 = k \cdot 10$, so that $k = 3$ and thus
$$F(x) = 3x$$
for this particular spring. The amount of work W done in stretching the spring 10 ft beyond its natural length is
$$W = \int_0^{10} 3x \, dx = \left.\frac{3x^2}{2}\right|_0^{10} = 150 \text{ ft-lb}$$
If the spring has been extended 3 ft and we extend it to 6 ft, then the work done is
$$W = \int_3^6 3x \, dx = \left.\frac{3x^2}{2}\right|_3^6 = \frac{81}{2} \text{ ft-lb}$$

The following example shows how to calculate the *escape velocity* of a rocket, i.e., that velocity which a rocket must achieve before it shuts down its engines in order to escape the earth's gravitational field.

EXAMPLE 6-15

Newton's Second Law of Motion states that
$$F(x) = ma(x)$$
where F is the force exerted on a body, m is its mass, and a is its acceleration when the body is at the point with coordinate x. The work done in moving the body from x_0 to x_1 is
$$W = \int_{x_0}^{x_1} F(x) \, dx = \int_{x_0}^{x_1} ma(x) \, dx$$
To find the antiderivative of this function we must make two fairly difficult substitutions. Let us first assume that at time t the body is at position $s(t)$, that is, $x = s(t)$. Then
$$dx = s'(t) \, dt = v(t) \, dt$$
since by definition the velocity $v(t)$ is equal to $s'(t)$. Hence
$$\int ma(x) \, dx = m \int a(x) \, dx = m \int a(s(t)) v(t) \, dt$$
Now we can simply write $a(s(t)) = a(t)$, since we are simply writing the acceleration function as a function of t instead of x, so that
$$\int a(x) \, dx = \int a(t) v(t) \, dt$$
Now $a(t) = v'(t)$, so we have
$$\int a(x) \, dx = \int v'(t) v(t) \, dt$$

We now let $u = v(t)$, and so $du = v'(t)\,dt$, and hence

$$\int a(x)\,dx = \int u\,du = \frac{u^2}{2} = \frac{v(t)^2}{2}$$

Therefore

$$\int ma(x)\,dx = \frac{mv(t)^2}{2}$$

Now when $x = x_0$ we have $t = t_0$, and when $x = x_1$ we have $t = t_1$, hence

$$\int_{x_0}^{x_1} ma(x)\,dx = \frac{mv(t)^2}{2}\bigg|_{t_0}^{t_1} = \frac{mv(t_1)^2}{2} - \frac{mv(t_0)^2}{2}$$

If we let $v_0 = v(t_0)$ and $v_1 = u(t_1)$, then

$$W = \tfrac{1}{2}mv_1^2 - \tfrac{1}{2}mv_2^2$$

The expression $\tfrac{1}{2}mv^2$ is often called *kinetic energy*. The gravitational force of a rocket obeys an inverse square law called Newton's Universal Gravitation law, which has the form

$$F(x) = \frac{-mG}{x^2}$$

where G is a constant. The value of G is about 1.427×10^{16} ft^3/sec^2, or 1427 with 13 zeroes after it. Combining this with our previous expression for W, we have

$$W = \frac{1}{2}mv_1^2 - \frac{1}{2}mv_0^2 = \int_{x_0}^{x_1} -\frac{mG}{x^2}\,dx = mG\left(\frac{1}{x_1} - \frac{1}{x_0}\right)$$

Dividing by $-2m$ has the effect of rewriting the equation in the form

$$v_0^2 - v_1^2 = 2G\left(\frac{1}{x_0} - \frac{1}{x_1}\right)$$

Now if we desire that speed v_0 which will enable the rocket to escape the earth's gravitational pull, we require that any velocity attained after v_0, i.e., any velocity v_1, must never be zero. Hence we require

$$v_0^2 \geq 2G\left(\frac{1}{x_0} - \frac{1}{x_1}\right)$$

In addition note that as t increases without bound, the position x of the rocket also must increase without bound. Hence, as $t \to \infty$, $x \to \infty$, and thus $1/x \to 0$. That is, we require that

$$\lim_{t \to \infty} \frac{1}{x} = 0$$

Hence
$$\lim_{t \to \infty} v_0^2 = v_0^2 \geq \lim_{t \to \infty} 2G\left(\frac{1}{x_0} - \frac{1}{x_1}\right) = \frac{2G}{x_0}$$

Therefore to escape the earth's gravitational field any rocket must attain a velocity in excess of $2G/x_0$, which is called the escape velocity and is approximately 7 miles per second.

EXERCISE 6-4

1. Find the volumes of the solids generated by rotating the following areas about the x-axis, where each area is bounded by the given curves.
   (a) $y = 2x, y = 0, x = 2$
   (b) $y = x^2, y = 0, x = 1$
   (c) $y = 2x^2, y = 0, x = 1$
   (d) $y = \sqrt{x}, y = 0, x = 4$
   (e) $y = \sqrt{4+x}, y = 0, x = 0$
   (f) $y = 1/x, y = 0, x = 1, x = 2$
   (g) $y = x^2 - 1, y = 0$
   (h) $y = x^3, y = 0, x = 1$
   (i) $y = x^2, y = x$
   (j) $y = e^x, y = 0, x = 0, x = 1$

2. The method used to find volumes of solids of revolution can be generalized to find volumes of solids generated by revolving an area about a line not the x-axis. A clear mental picture of the solid obtained is helpful. Find the volume of the solid generated by revolving the areas bounded by the following curves about the indicated line.
   (a) $y = x, y = 0$, and $x = 2$ about the line $x = 3$
   (b) $y = x^2, y = 0$, and $x = 1$ about the line $x = 3$
   (c) $y = x, y = 0$, and $x = 4$ about the line $x = -1$

3. An object weighing 150 lb is lifted 25 ft. Find the amount of work done on the object.

4. Find the work done when a force of $f(x)$ lb is applied to an object to move it from x_0 to x_1 for the following choices of $f(x)$, x_0, and x_1.
   (a) $f(x) = 2x, x_0 = 0, x_1 = 4$
   (b) $f(x) = x^2, x_0 = 0, x_1 = 3$
   (c) $f(x) = x^2 - x, x_0 = 0, x_1 = 5$
   (d) $f(x) = \dfrac{1}{(x+1)^2}, x_0 = 1, x_2 = 2$
   (e) $f(x) = 5x^3 - x, x_0 = 0, x_1 = 1$

5. Find the work done in stretching a spring from its natural length of 10 in. to a length of 20 in. if a force of 6 lb is needed to extend it 12 ft.

6. A spring of natural length 6 in. requires a force of 3 lb to extend it to a length of 10 in. Find the work done in stretching the spring from a length of 8 in. to a length of 18 in.

6-5 PROBABILITY

One of the most interesting and powerful mathematical systems is the theory of probability. Its methods of prediction, or perhaps of organization and measurement of unpredictability, have been utilized in practically every branch of science, from the financial expert studying the stock market to the psychologist conducting a learning experiment to a biologist studying heredity.

The development of the techniques of the theory of probability relies on calculus. In this section we briefly outline the methods of the theory of probability and indicate how calculus fits into its structure.

If an experiment is conducted, whether it be the experiment of tossing a coin, conducting a poll, observing a rat run through a maze, or deciding which stock to buy, whether it is a real or conceptual experiment, the *probability* that a certain outcome will be observed can intuitively be thought of as that fraction of the total outcomes in which that particular outcome would occur if the experiment were conducted a great many times.

The task of assigning probabilities to outcomes is usually a fairly obvious matter, and so to explicitly define the procedure might seem confusing. A few examples should adequately illustrate the idea.

EXAMPLE 6-16

If the experiment consists of tossing a coin, the outcomes would be to observe a head and to observe a tails. If we have a fair coin, the probability of observing a head is $\frac{1}{2}$, and of observing a tails is also $\frac{1}{2}$. This is a conceptual experiment. No coin is "perfectly" fair. To see whether a coin is "fair", you would want to toss the coin many times and record the number of heads observed. The more times you toss the coin the better. If you toss the coin 1000 times and only 250 heads are observed, you may wish to assign a probability of $\frac{1}{4}$ to observing a head, and $\frac{3}{4}$ to observing a tails with that particular coin.

EXAMPLE 6-17

If the experiment is to roll a die, the probability assigned to the outcome of a particular face being observed face up is $\frac{1}{6}$.

EXAMPLE 6-18

If we select a card from an ordinary deck of 52 cards, the probability of drawing a club is $\frac{1}{4}$, a picture card is $\frac{12}{52} = \frac{3}{13}$, the ace of hearts is $\frac{1}{52}$.

EXAMPLE 6-19

A psychologist observes a rat running through a maze. At one point in the maze the rat is confronted with three doors. After observing a great many rats run through the maze, the psychologist notes that half the rats ran through the middle door, half of the rest ran through the left, and the others through the right door. He then assigns the probability of $\frac{1}{2}$ to the outcome that the rat will run through the middle door, and $\frac{1}{4}$ to running through each of the side doors.

EXAMPLE 6-20

Consider the experiment defined by selecting one card from an urn containing one ace, two 2's, three 3's, four 4's, five 5's, five 6's, four 7's, three 8's, two 9's and one 10. The probability of observing the ten possible outcomes, drawing an ace through ten, are given in the following table. There are a total of thirty cards in the urn. If we let x represent an outcome then $P(x)$ is the probability that x will be drawn.

x	$P(x)$
1	$\frac{1}{30}$
2	$\frac{2}{30} = \frac{1}{15}$
3	$\frac{3}{30} = \frac{1}{10}$
4	$\frac{4}{30} = \frac{2}{15}$
5	$\frac{5}{30} = \frac{1}{6}$
6	$\frac{5}{30} = \frac{1}{6}$
7	$\frac{2}{15}$
8	$\frac{1}{10}$
9	$\frac{1}{15}$
10	$\frac{1}{30}$

The function $P(x)$ is called a *probability density function*. An alternate method for recording these data is with a graph, as in Fig. 6-16.

This type of probability density function is called *discreet* because its domain is a finite set, in this case $\{1, 2, 3, 4, 5, 6, 7, 8, 9, 10\}$. It is often the case in applications that the domain of the probability density function $P(x)$ is an interval, or perhaps it is so large that the domain is assumed to be an interval, and then $P(x)$ is assumed to be a *continuous* function. The following examples illustrate the meaning of a continuous probability density function.

EXAMPLE 6-21

Psychologists often measure the time x it takes an individual or a group to learn a task. Then any positive real number x could be an outcome. Suppose a psychologist predicts that a certain group-type will have a probability

Discrete distribution function

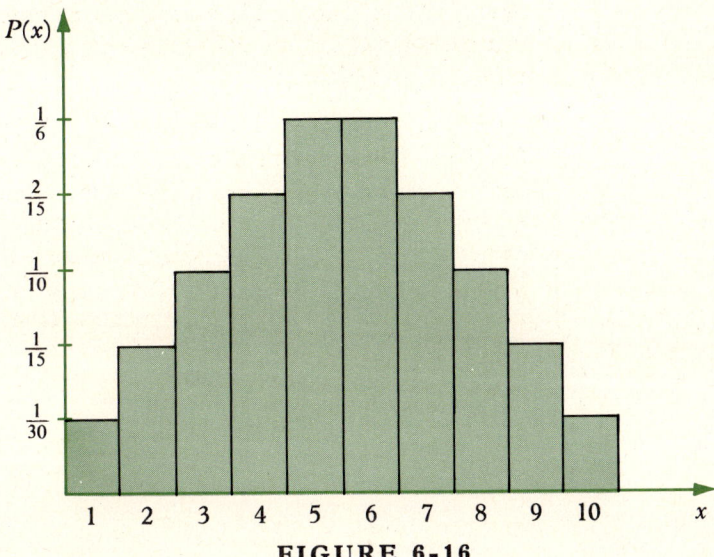

FIGURE 6-16

density function given by

$$P(x) = \tfrac{3}{500}(10x - x^2)$$

where x is measured in minutes. Note that $P(5) = \tfrac{3}{20}$ is a maximum. Often we want to know what the probability is that a group will complete the task in a certain time period, say, in the first 5 minutes. Then we must sum the individual probabilities for each time x for x in $[0, 5]$. Of course by this we mean

$$\int_0^5 P(x)\,dx = \int_0^5 \frac{3}{500}(10x - x^2)\,dx = \frac{3}{500}\left(5x^2 - \frac{x^3}{3}\right)\Big|_0^5 = \frac{1}{2}$$

Thus the probability that a group will finish in the first 5 minutes is $\tfrac{1}{2}$. The probability that a group will finish between $x = 2$ and $x = 5$ is

$$\int_2^5 P(x)\,dx = \int_2^5 \tfrac{3}{500}(10x - x^2)\,dx = \tfrac{99}{250} \cong \tfrac{2}{5}$$

From this example we can see that areas under the curve of the probability density function $P(x)$ represent probabilities. Hence the probability that an arbitrary outcome between the real numbers a and b will be observed is

$$\int_a^b P(x)\,dx$$

EXAMPLE 6-22

A farmer strives to optimize the productivity of his land while utilizing his time to a maximum. Analysis of his work load can help him decide how much time and effort should be spent on each crop in a specific time period. Suppose he determined from empirical evidence that the probability density function $P(x)$ for a certain crop to not return a profit if he spends only x minutes per acre per month is given by

$$P(x) = \frac{1}{(x+1)^2}$$

Then the probability, taking into consideration many factors such as supply, demand, weather, that he will not yield a profit if he spends between 3 and 4 minutes per acre per month is

$$\int_3^4 \frac{dx}{(x+1)^2} = -(x+1)^{-1} \Big|_3^4 = -\frac{1}{5} + \frac{1}{4} = \frac{1}{20}$$

while between 4 and 5 minutes the probability is

$$\int_4^5 \frac{dx}{(1+x)^2} = -(x+1)^{-1} \Big|_4^5 = -\frac{1}{6} + \frac{1}{5} = \frac{1}{30}$$

For $P(x)$ to qualify as a probability density function, it is required to enjoy a few simple properties:

Property 1   The domain of $P(x)$, or the set of outcomes x, must be a set of real numbers.

Property 2 $P(x) \geq 0$ for all x, since a negative probability makes no sense.

Property 3   If the domain of $P(x)$ is an interval $[a, b]$ (a and b could be $-\infty$ or ∞), then

$$\int_a^b P(x)\, dx = 1$$

Property 3 is necessary because if we think of $\int_a^b P(x)\, dx$ as the sum of the probabilities of *all* the possible outcomes, and if we assume that we must observe some outcome, then we are dealing with a certainty whose probability is 1.

EXAMPLE 6-23

A very common and useful probability density function found in countless applications is the function $f(x) = ae^{-ax}$ where a is some positive constant. The graph of $f(x) = ae^{-ax}$ is given in Fig. 6-17. To ensure that $f(x)$ is a probability density function, we restrict the domain to $[0, \infty)$. Let us check to make sure that this function is a legal probability density function for any $a > 0$. Since $e^x > 0$ for all x, we have $e^{-x} = 1/e^x > 0$ for all x, and since $a > 0$, $ae^{-ax} > 0$ for all x. Consider

$$\int_0^\infty ae^{-ax}\,dx = \lim_{b\to\infty} \int_0^b ae^{-ax}\,dx$$

$$= \lim_{b\to\infty} -e^{-ax}\Big|_0^b$$

$$= \lim_{b\to\infty} (-e^{-ab} + 1)$$

$$= 1$$

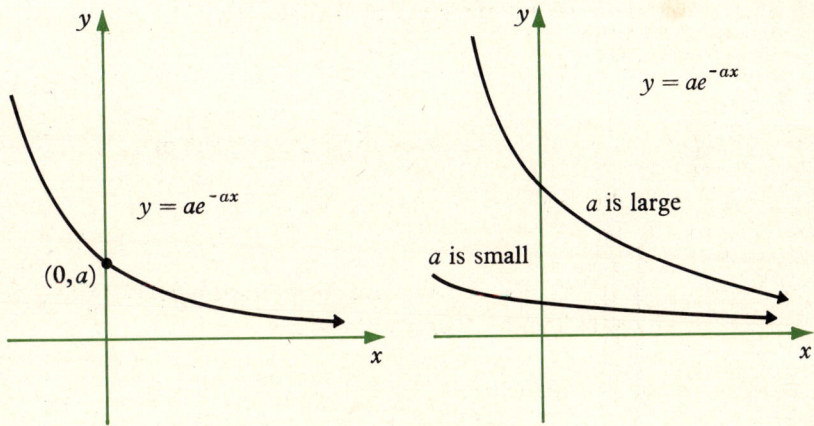

FIGURE 6-17

The following example is a simplified model designed to illustrate how the theory of probability is of paramount importance in economics.

EXAMPLE 6-24

Manufacturing companies often utilize probability models in decision-making processes. For example, suppose a machine will produce one defective item in 100 when it is performing normally. If an inspector checks the machine, he will test a stream of items and record the intervals in which defectives appear. Starting from the first defective one, the probability density function is assumed to be $P(x) = ae^{-ax}$. Now $P(0)$ is of course $\frac{1}{100}$ if the machine is performing normally. Hence, for normal operation, the probability density function is

$$P(x) = 0.01 e^{-0.01x}$$

The probability that the next defective item will be observed within the first

100 is

$$\int_0^{100} 0.01 e^{-0.01x}\, dx = -e^{0.01x}\Big|_0^{100} = 1 - e^{-1} \cong 0.632$$

Thus the probability that the next defective will not be within the next 100 is $1 - 0.632 = 0.368$.

There are a number of key decisions to be made here. The company must decide how many defectives in a given interval can be financially tolerated. The decision to call in a repairman is tricky. Two types of errors may be made. The company first designs a test to determine if the machine is malfunctioning, i.e., they select an N such that if the next defective is recorded within the next N items, the test conclusion is that the machine is malfunctioning. The determination of this N must be balanced by the probabilities of committing the following types of errors: (1) The machine is actually performing normally and the test conclusion is that it is malfunctioning, and (2) the machine is actually malfunctioning and the test conclusion is that it is performing normally. Generally the decisions as to which error to minimize and at what levels the probabilities of errors should be set are cost decisions, the cost of a repairman versus the cost of producing too many defectives.

EXERCISE 6-5

1. Given the following experiments, list the possible outcomes and the probability of each occurring.
   (a) Toss a die and record the number of dots on the upper face.
   (b) Toss two dice and record the sum of the number of dots on the two upper faces.
   (c) Select a card from a deck and record the number on the card, recording 10 for a face card.
   (d) Toss two coins and record the number of heads.
   (e) Toss three coins and record the number of heads.
   (f) Select one ball from an urn containing three red and four white balls and record the number of red balls selected.
   (g) Select two balls from an urn containing two red and three white balls and record the number of red balls selected.

2. Give the probability distribution function for each experiment in Problem 1 in a table and a graph. Which of these functions are discreet?

3. Show that the functions $P(x)$ are probability distribution functions in the given intervals.
   (a) $P(x) = \frac{3}{500}(10x - x^2)$, $[0, 10]$
   (b) $P(x) = \frac{6}{125}(5x - x^2)$, $[0, 5]$

(c) $P(x) = \dfrac{1}{(x+1)^2}$, $[0, \infty]$

(d) $P(x) = \dfrac{4}{(x+1)^2}$, $[3, \infty)$

(e) $P(x) = \dfrac{2x}{(x^2+1)^2}$, $[0, \infty)$

(f) $P(x) = xe^{-x}$, $[0, \infty)$

4. Show that for any function $f(x)$ that is continuous, nonnegative, and nonzero on $[a,b]$ (i.e., $f(x) \neq 0$ for all $x \in [a,b]$) there exists a constant C such that $P(x) = C f(x)$ is a probability distribution function. (*Hint*: Consider $\int_a^b f(x)\, dx$.)

5. Given the following functions $f(x)$ and intervals, find a constant C such that $P(x) = C f(x)$ is a probability distribution function in the given interval.

(a) $f(x) = \dfrac{1}{(x+1)^2}$, $[4, 6]$

(b) $f(x) = \dfrac{1}{(x+1)^2}$, $[4, \infty)$

(c) $f(x) = \dfrac{1}{(x+2)^2}$, $[0, \infty)$

(d) $f(x) = 10x - x^2$, $[0, 10]$

(e) $f(x) = 7x - x^2$, $[0, 2]$

6. From Ex. 6-19 and the table below, find the probability that the next defective will occur within the following intervals.
   (a) Within the next 200 observed.
   (b) Within the next 300 observed.
   (c) Not within the next 200 observed.
   (d) Not within the next 300 observed.

x	e^{-x}
1	0.368
2	0.135
3	0.050
0.1	0.905
0.2	0.815

7. From Problem 6, what N should the company in Ex. 6-14 choose in order to to be 95 percent sure that if the machine is operating normally, the next defective will be within the next N items observed.

8. The demand for an inventory item has a probability density function $P(x)$ given by
$$P(x) = 0.01e^{-0.01x}$$
for x items being requested over a month's period. What is the probability that less than 10 items will be requested? less than 100? more than 10? Use the table in Problem 6.

9. The number of accidents reported on a certain stretch of highway per year has a probability density function given by
$$P(x) = 0.02e^{-0.02x}$$
for x accidents reported in the following year. What is the probability that the number of accidents reported will be less than 10? less than 100? more than 10?

10. The probability density function $P(x)$ for a subject in a psychological testing program making a certain choice after x seconds is given by
$$P(x) = 0.3e^{-0.3x}$$
What is the probability that the subject will make the choice in less than 10 sec? more than 10 sec?

6-6 APPLICATIONS FROM ECONOMICS

In this section we present several applications from the field of economics. In each, the concepts defined rely heavily on an area beneath a curve, and thus depend on the calculus.

The output P, or productivity, of an economic unit such as labor, machinery, or land is a function of the size of the input x. If $P(x)$ is a measure of the output of x units of input, then the rate of change of $P(x)$ with respect to x, that is, the derivative $P'(x)$, is called the *marginal physical productivity*, or simply marginal product. Marginal productivity measures the effects of an increase or decrease of input and helps answer such questions as whether an increase in labor or land would be beneficial in increasing profit.

Consider the economic model of a landlord who hires labor to work his land. One man produces a large marginal product, which could be considered to be a wage rate determined by the amount of work done, in this example, because he has a large area of land to work. An additional laborer also has a large but slightly smaller marginal product since he is working less land. As more men are hired, the marginal product of each decreases. What wage will the landlord pay his labor force? If we assume each worker receives the same wage, note that under free competition a

SEC. 6-6   Applications From Economics

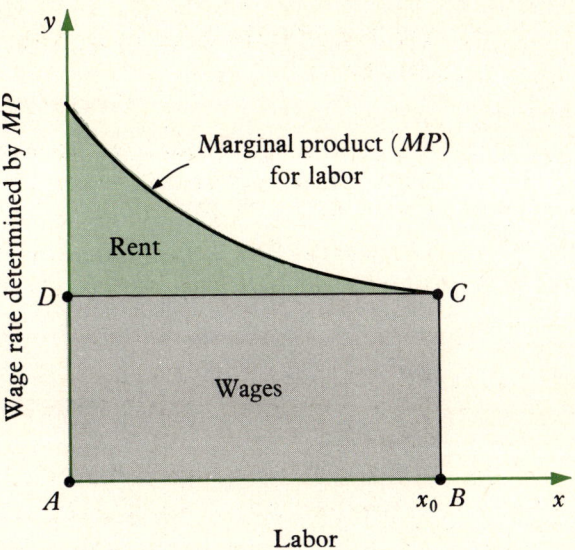

FIGURE 6-18

landlord would never pay a man more than his individual marginal product. So the labor force is paid the wage of the last man hired, since he has the lowest marginal product. Thus in Fig. 6-18, if $y = MP(x)$ is the marginal product of labor and x_0 is the amount of the labor force, then $MP(x_0)$ is the wage paid each labor unit. Total wages then is the area of the rectangle *ABCD*. Economists then refer to the excess marginal productivity produced by the first man, the second, and all up until the last as the landlord's *rent*. This concept of rent can be thought of as the marginal product of the land. Thus marginal productivities help determine competitive pricing of inputs as well as end products.

EXAMPLE 6-25

Suppose the marginal physical product $MP(x)$ for a group of dockworkers is given by

$$MP(x) = 10(x+4)^{-1/2} = P'(x) \quad \text{for } x \in [0, \infty)$$

where x is measured in hundreds of dockworkers and $P(x)$, the productivity function, is measured in tons of cargo unloaded per month. The wage rate for each worker can be determined directly from MP, so we can simply consider MP to be the wage rate (for example, MP may be measured in dollars per hour). If the amount of dockworkers hired is $x = 5$, then the total wages paid (see Fig. 6-19) is

$$MP(5) \cdot 500 = \tfrac{10}{3} \cdot 500 = \tfrac{5000}{3} \text{(dollars/hr)}$$

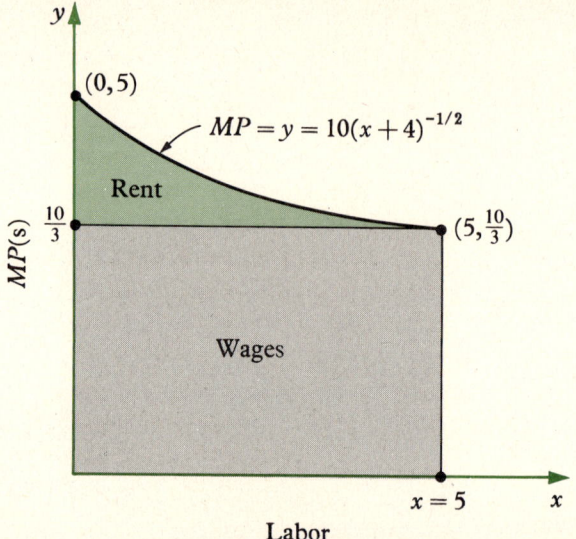

FIGURE 6-19

Then the rent accrued from excess marginal product produced by the labor force is

$$\int_0^5 MP(x)\,dx - \tfrac{50}{3} = \int_0^5 10(x+4)^{-1/2}\,dx - \tfrac{50}{3}$$

$$= 20(x+4)^{1/2}\Big|_0^5 - \tfrac{50}{3}$$

$$= 60 - 40 - \tfrac{50}{3}$$

$$= \tfrac{10}{3}$$

In this example each worker gets paid $MP(5) = \tfrac{10}{3}$ dollars per hr and the landlord saves $\tfrac{10}{3}$ dollars per hr per 100 workers, i.e., the landlord has paid out $\tfrac{50}{3}(100) = \tfrac{5000}{3}$ dollars per hr, yet the marginal product produced by the workers is $\tfrac{5000}{3} + \tfrac{10}{3}(100) = \tfrac{6000}{3}$ dollars per hr.

Another pair of economic concepts best explained by calculus are *consumers' surplus* and *producers' surplus*. Consider the demand function $y = D(x)$ and the supply function $y = S(x)$ where x is the price of a unit of goods, $D(x)$ is the amount of goods requested by consumers at price x, and $S(x)$ is the amount of goods that producers are willing to supply at price x. In a free competitive market, the actual selling price for the goods will be that price x_0 such that $D(x_0) = S(x_0)$. The value x_0 is often referred to as the

SEC. 6-6   **Applications From Economics**

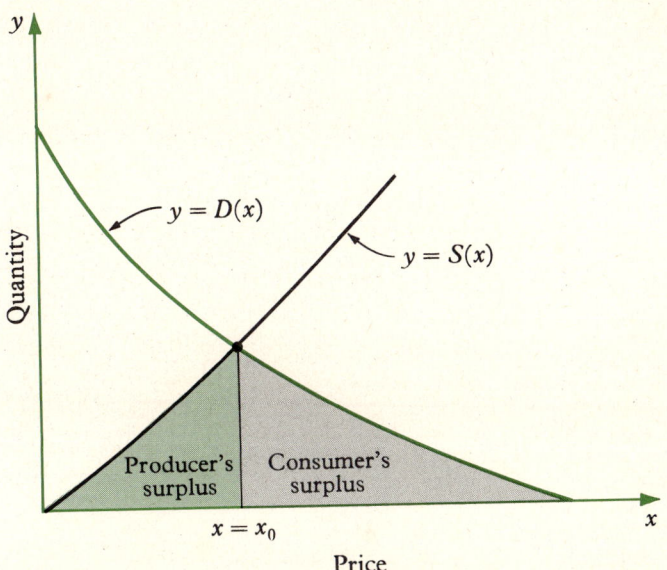

FIGURE 6-20

equilibrium price. Note that some consumers would be willing to pay a higher price for the goods; that is, at price $x_0 + h$, there usually is a positive demand $D(x_0 + h)$, as illustrated in Fig. 6-20. The consumers' surplus is defined to be that area under $y = D(x)$ bounded by $x = x_0$ and the x-axis.

EXAMPLE 6-26

Suppose $D(x) = (x-4)^2$ for $x \in [0, 4]$ and $x_0 = 2$. Then the consumers' surplus (Fig. 6-16) is defined to be

$$\int_2^4 (x-4)^2 \, dx = \frac{(x-4)^3}{3} \Big|_2^4 = \frac{8}{3}$$

Similarly, the producers who were willing to supply the goods at a price below x_0 have gains. The sum total of their gain, called the producers' surplus, is defined to be the area of the region bounded by $y = S(x)$, the x- and y-axes, and $x = x_0$.

EXAMPLE 6-27

Suppose $S(x) = x^2 + 2x$ for $x \in [0, 4]$ and $x_0 = 2$. Then the producers' surplus (Fig. 6-22) is defined to be

$$\int_0^2 (x^2 + 2x) \, dx = \frac{x^3}{3} + x^2 \Big|_0^2 = \frac{8}{3} + 4 = \frac{20}{3}$$

The concepts of consumers' surplus measure the benefit we gain

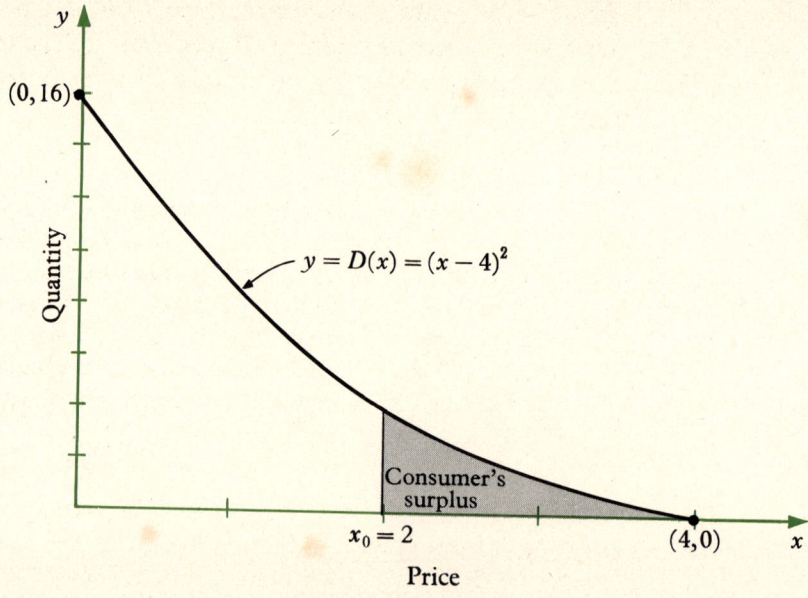

FIGURE 6-21

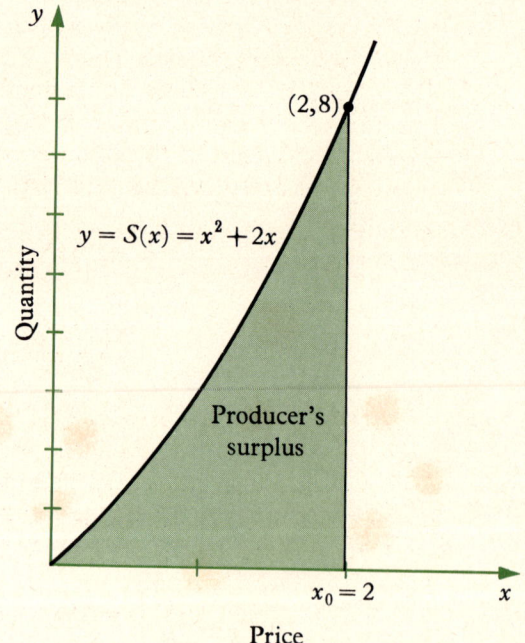

FIGURE 6-22

from the power of buying at low prices and not having to pay the price upon which a ruthless monopolist may insist. Such concepts as consumers' surplus and producers' surplus also help us make social decisions as illustrated in the next example.

EXAMPLE 6-28

A new community is debating whether to self-impose a substantial tax in order to improve their property value, say, to install street lights. For simplicity, suppose the community has 1000 users of the improvement and they are all alike in that they will be taxed the same dollar amount and receive the same utility from the improvement. If they contact an assortment of contractors they will get various prices for various jobs. Suppose they find that the contractors are willing to supply the street lights according to the schedule

$$S(x) = x^2 + 9x$$

where x is the price and $S(x)$ takes into account not only the quantity of street lights but the quality of construction and maintenance.

Through compromise, the debate in city council results in the majority agreeing to a schedule that the community is willing to pay:

$$D(x) = (x-9)^2 = x^2 - 18x + 81$$

Here we assume that x is measured in $100,000 units, while $S(x)$ and $D(x)$ are measured in units of 100 street lights. Letting $D(x) = S(x)$, we see that the equilibrium price x_0 is $x_0 = 3$. Thus each individual must pay a tax of $300,000/1000 = $300 for the improvement. Is it worth it to him? Since there are no revenues coming in from this project, the only "profit" that can be measured from the improvement is the increased utility achieved, thus one must consider the consumers' surplus given by

$$\int_3^9 (x-9)^2 \, dx = \frac{(x-9)^3}{3} \bigg|_3^9 = 72$$

or

$$\frac{\$7{,}200{,}000}{1000} = \$7{,}200$$

Hence the consumers' surplus is greater than the dollar amount spent on the improvement, and hence the project is worthwhile, or "profitable". Thus consumers' surplus can help make decisions where the intangible quantity of utility is the only profit enjoyed. Note that if the value of x_0 turned out to be 8, then the consumers' surplus would be

$$\int_8^9 (x-9)^2 \, dx = \frac{(x-9)^3}{3} \bigg|_8^9 = \frac{1}{3}$$

and the project would not be profitable.

EXERCISE 6-6

1. For the following choices of the marginal productivity function $MP(x)$ for x units of labor and for the choices of x_0, that quantity of labor actually hired by the landlord, find the total wages paid by the landlord and his rent.
   (a) $MP(x) = 5(x+9)^{-1/2}$, $x_0 = 7$
   (b) $MP(x) = 10(x+1)^{-1/3}$, $x_0 = 7$
   (c) $MP(x) = 8(4x+7)^{-1/3}$, $x_0 = 5$

2. Note that each marginal productivity function $MP(x)$ in Problem 1 is a decreasing function in the specified interval $[0, x_0]$. Explain why this must be true in order for the function $MP(x)$ to reflect an actual real world situation. Note also that the productivity function $P(x) = \int MP(x)\,dx$ is increasing in $[0, x_0]$. Why must this be necessarily so?

3. For the following choices of the demand function $y = D(x)$ and supply function $y = S(x)$ for an economic model under pure competition, determine the equilibrium price x_0, the consumers' surplus, and the producers' surplus.
   (a) $D(x) = x^2 - 12x + 36$, $S(x) = x^2 + 6x$
   (b) $D(x) = x^2 - 10x + 25$, $S(x) = x^2 + 5x$

7

Exponential and Logarithmic Functions

7-1 THE DEFINITION OF THE NATURAL LOGARITHM FUNCTION

In this section we will utilize much of our knowledge of calculus to define a function in terms of area and hence in terms of the definite integral. Our initial discussion will be reminiscent of the first part of Sec. 5-2.

Consider the area under the curve $y = 1/t$ from $t = 1$ to $t = 2$, that is, the area of the region bounded by $y = 1/t$, $t = 1$, $t = 2$, and the t-axis. The reason that we use the variable t in place of x will soon become apparent.

We know that the area of this region is a specific number, let's call it ln 2.* It is an irrational number, so to find the magnitude of ln 2, we look for a decimal approximation. First note from Fig. 7-1 that the area representing ln 2 is about two-thirds of one square unit.

To get a better approximation, we must use a finely partitioned graph paper. In Fig. 7-2, each subdivision or block is 1/64th of a square unit. Counting all blocks which lie wholly within the area under $y = 1/t$ and counting one-half a block for those through which the curve passes, we get 44 blocks. Hence ln 2 is about $44/64 \simeq 0.69$. An even better approximation to ln 2 is 0.69315.

Using the definite integral notation, we write

$$\ln 2 = \int_1^2 \frac{1}{t}\, dt \simeq 0.69315$$

---

* If you have read Sec. 1-6 where we intuitively introduced the logarithmic function $f(x) = \ln x$, you will see that this number ln 2 is simply the functional value of $f(x) = \ln x$ at $x = 2$.

The area bounded by $y = \frac{1}{t}$, $t = 1$, $t = 2$, *and* $y = 0$

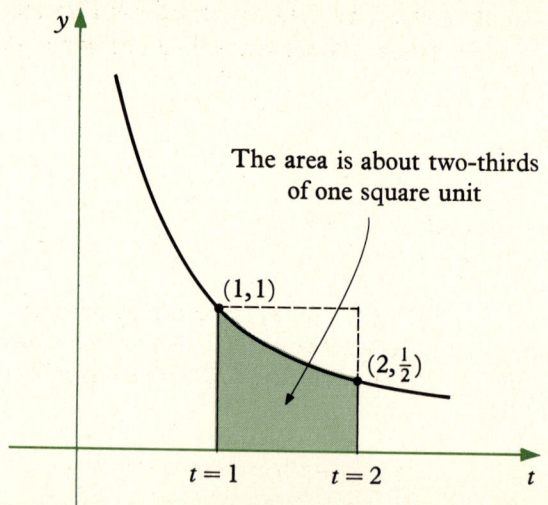

FIGURE 7-1

The area under $y = \frac{1}{t}$ *from* $t = 1$ *to* $t = 2$ *is about* $\frac{44}{64}$

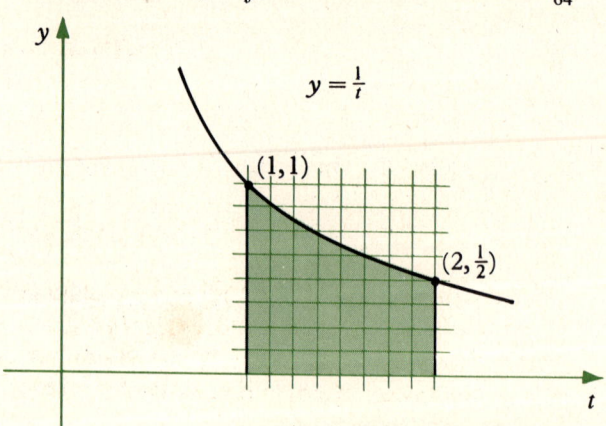

FIGURE 7-2

This discussion could be carried out using any positive number x in place of 2 and then we would say that the area under $y = 1/t$ from $t = 1$ to $t = x$ is $\ln x$. We are now in a position to define the function $y = \ln x$.

 Definition 7-1 The function $y = \ln x$, called the *natural logarithm function*, is defined by

$$\ln x = \int_1^x \frac{1}{t}\, dt$$

for all $x > 0$.

EXAMPLE 7-1

Using the same procedure outlined above for finding an approximation to $\ln 2$, one can use Fig. 7-3 to find a rough approximation to $\ln 3$. A good approximation to $\ln 3$ is 1.09861. (See Sec. 6-2, Ex. 6-5.)

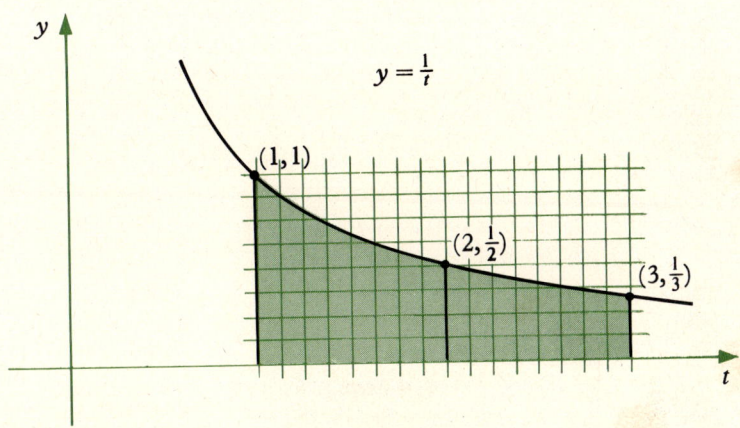

The area under $y = \frac{1}{t}$ *from* $t = 1$ *to* $t = 3$ *is about* $\frac{70}{64}$

FIGURE 7-3

EXAMPLE 7-2

From the definition of $y = \ln x$, one sees that $\ln 1 = 0$ since

$$\ln 1 = \int_1^1 \frac{1}{t}\, dt = 0$$

from Section 5-2, Property 1. Also from the definition of $y = \ln x$, if $0 < a < 1$, then $\ln a < 0$, whereas if $a > 1$, then $\ln a > 0$. Since $y = 1/t$ is unbounded at $t = 0$, $y = \ln x$ is not defined for any nonpositive real number. Hence the domain of $y = \ln x$ is $(0, \infty)$.

The graph of $y = \ln x$ together with a table of values is given in Fig. 7-4. From the graph of $y = \ln x$, or directly from the definition, one sees that

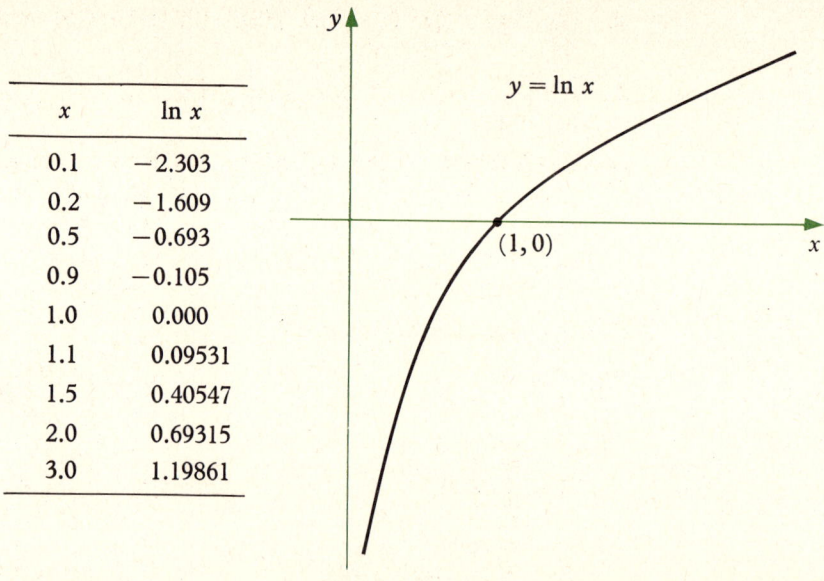

x	$\ln x$
0.1	−2.303
0.2	−1.609
0.5	−0.693
0.9	−0.105
1.0	0.000
1.1	0.09531
1.5	0.40547
2.0	0.69315
3.0	1.19861

FIGURE 7-4

$y = \ln x$ is an increasing function, that is, if $a < b$, then $\ln a < \ln b$. Thus $y = \ln x$ is a one-to-one function, that is, $\ln a = \ln b$ implies that $a = b$.

Refer to Table B-1 of natural logarithms in Appendix B. It lists the values of $\ln x$ for $0 < x \leqslant 99$. The values of $\ln x$ for x greater than 99 can be found by using the properties of $y = \ln x$ presented in the next section.

EXAMPLE 7-3

From Table B-1, one sees that $\ln 4 \cong 1.38629$, $\ln 4.5 \cong 1.50408$ and $\ln 4.58 \cong 1.52170$. Note that $\ln 4.95 \cong 1.59939$ and $\ln 4.96 \cong 1.60141$. It should be pointed out that these values are approximations, as was mentioned above.

EXERCISE 7-1

1. Use graph paper to find the approximate values of $\ln 4$, $\ln \frac{1}{2}$, $\ln 5$.
2. Graph the functions.
   - (a) $y = 2 \ln x$
   - (b) $y = 3 \ln x$
   - (c) $y = \frac{1}{2} \ln x$
   - (d) $y = \ln x + 1$
   - (e) $y = \ln x + 2$
   - (f) $y = \ln 2x$
   - (g) $y = \ln 3x$
   - (h) $y = \ln x + \ln 2$
   - (i) $y = \ln x + \ln 3$
   - (j) $y = \ln x^2$
3. Let $f(t) = \int_1^x 2t\, dt$. Find $f(1), f(2), f(3), f(4)$.

SEC. 7-2   Properties of $y = \ln x$

4. Let $f(t) = \int_1^x t^3 \, dt$. Find $f(1)$, $f(2)$, $f(3)$, $f(4)$, and another formula for $f(t)$.
5. Let $f(t) = \int_1^x (t^2 + t^3) \, dt$. Find $f(1)$, $f(2)$, $f(3)$, $f(4)$, and another formula for $f(t)$.
6. Use Table B-1 in the Appendix to find the natural logarithm of the numbers.
   - (a) 5
   - (b) 10
   - (c) 10.8
   - (d) 9.01
   - (e) 9.08
   - (f) 11
   - (g) 21
   - (h) 60
7. Evaluate the integrals.
   - (a) $\int_1^4 \dfrac{1}{t} \, dt$
   - (b) $\int_1^5 \dfrac{1}{t} \, dt$
   - (c) $\int_1^{8.04} \dfrac{1}{t} \, dt$
   - (d) $\int_1^{9.04} \dfrac{1}{t} \, dt$
   - (e) $\int_1^{10} \dfrac{1}{t} \, dt$
   - (f) $\int_1^{25} \dfrac{1}{t} \, dt$
8. Evaluate the integrals.
   - (a) $\int_3^4 \dfrac{1}{t} \, dt$
   - (b) $\int_3^5 \dfrac{1}{t} \, dt$
   - (c) $\int_{0.5}^9 \dfrac{1}{t} \, dt$
   - (d) $\int_{11}^{12} \dfrac{1}{t} \, dt$

7-2 PROPERTIES OF $y = \ln x$

From the Fundamental Theorem of Calculus, one sees that $D_x \ln x = 1/x$, for if $D_t F(t) = 1/t$ then $\int_1^x 1/t \, dt = F(t)\big|_1^x = F(x) - F(1)$. Thus $D_x \ln x = D_x \int_1^x 1/t \, dt = D_x (F(x) - F(1)) = D_x F(x) = 1/x$.

We record this important fact by

  Rule D5 $D_x \ln x = 1/x$

which was first presented in Sec. 3-2.

When $f(x)$ is a differentiable function of x and $y = \ln f(x)$ then the Chain Rule (Sec. 3-4) yields

  Rule D10 $D_x \ln f(x) = \dfrac{1}{f(x)} D_x f(x)$

221

EXAMPLE 7-4

If $g(x) = \ln(x^3 + x^2)$, then $g(x) = \ln f(x)$ where $f(x) = x^3 + x^2$ is a differentiable function of x. By Rule D10, we have

$$D_x \ln(x^3 + x^2) = \frac{1}{x^3 + x^2} D_x(x^3 + x^2) = \frac{1}{x^3 + x^2}(3x^2 + 2x)$$

$$= \frac{3x^2 + 2x}{x^3 + x^2} = \frac{3x + 2}{x^2 + x}$$

EXAMPLE 7-5

If $f(x) = \ln \frac{x^2}{x+1}$, then

$$D_x \ln \frac{x^2}{x+1}$$

$$= \frac{1}{x^2/(x+1)} D_x \frac{x^2}{x+1} = \frac{x+1}{x^2} \cdot \frac{2x(x+1) - x^2}{(x+1)^2} = \frac{x+1}{x^2} \cdot \frac{x^2 + 2x}{(x+1)^2}$$

$$= \frac{x^2 + 2x}{x^2(x+1)} = \frac{x+2}{x^2 + x}$$

From Rule D5 we get the antiderivative formula

Rule I4 $\quad \int \frac{1}{x} dx = \ln|x| + C$

which was first mentioned in Table 4-1 in Sec. 4-7.

EXAMPLE 7-6

To evaluate $\int 1/(x+1)\,dx$ one uses the substitution $u = x + 1$ so that $du = dx$ and

$$\int \frac{1}{x+1} dx = \int \frac{1}{u} du = \ln|u| + C = \ln|x+1| + C$$

EXAMPLE 7-7

To evaluate $\int x/(1+x^2)\,dx$ one uses the substitution $u = 1 + x^2$ so that $du = 2x\,dx$ and

$$\int \frac{x}{1+x^2} dx = \int \frac{1/2\,du}{u} = \frac{1}{2} \ln|u| + C = \frac{1}{2} \ln(1+x^2) + C$$

EXAMPLE 7-8

$$\int_{-1}^{-2} \frac{1}{x} dx = \ln|x| \Big|_{-1}^{-2}$$

$$= \ln|-2| - \ln|-1|$$

$$= \ln 2 - \ln 1$$

$$= \ln 2$$

$$\cong 0.69315$$

The interest in logarithmic functions stems in part from their usefulness as a computational tool, which is derived from the following properties.

Rule L1 $\ln ab = \ln a + \ln b$

Rule L2 $\ln a^r = r \ln a$

Rule L3 $\ln a/b = \ln a - \ln b$

Before we prove these properties, we will use a few examples to demonstrate their usefulness.

EXAMPLE 7-9

To find $\ln 100$ from Table B–1 we write $100 = 10^2$ and hence

$$\ln 100 = \ln 10^2 = 2 \ln 10 \qquad \text{(Rule L1)}$$

$$\cong 2(2.30259) \qquad \text{(From Table B–1)}$$

$$= 4.60518$$

EXAMPLE 7-10

To find $\ln 250$, we write $250 = 2.5 \times 10^2$. A number expressed in this form, that is, as the product of a number between 0 and 10 and a power of 10, is said to be expressed in *scientific notation*. Then

$$\ln 250 = \ln(2.5 \times 10^2)$$

$$= \ln 2.5 + \ln 10^2 \qquad \text{(Rule L1)}$$

$$= \ln 2.5 + 2 \ln 10 \qquad \text{(Rule L2)}$$

$$\cong .91629 + 2(2.30259)$$

$$= .91629 + 4.60518$$

$$= 5.52147$$

Using the method described in Example 7-10, that is, by expressing the number in scientific notation, the natural logarithm of any positive number can be found.

The following example will indicate how logarithms are used as a computation aid.

EXAMPLE 7-11

We compute $a = (63,100)^{1/2}(.00212)^{3/5}$. We first note that $\ln a = \ln(63,100)^{1/2} + \ln(.00212)^{3/5}$ from Rule L1. From Rule L1 and L2 we compute $\ln(63,100)^{1/2}$ and $\ln(.00212)^{3/5}$.

CHAP. 7   Exponential and Logarithimc Functions

$$\ln(63{,}100)^{1/2} = \tfrac{1}{2}\ln 63{,}100 \qquad \text{Rule L2}$$
$$= \tfrac{1}{2}\ln(6.31 \times 10^4)$$
$$= \tfrac{1}{2}\ln 6.31 + \tfrac{1}{2}\ln 10^4 \qquad \text{Rule L1}$$
$$= \tfrac{1}{2}\ln 6.31 + 2\ln 10 \qquad \text{Rule L2}$$
$$\cong \tfrac{1}{2}(1.84214) + 2(2.30259)$$
$$= 0.92107 + 4.60518$$
$$= 5.52625$$

$$\ln(.00212)^{3/5} = \tfrac{3}{5}\ln .00212 \qquad \text{Rule L2}$$
$$= \tfrac{3}{5}\ln(2.12 \times 10^{-3})$$
$$= \tfrac{3}{5}\ln 2.12 + \tfrac{3}{5}\ln 10^{-3} \qquad \text{Rule L1}$$
$$= \tfrac{3}{5}\ln 2.12 + (-\tfrac{9}{5})\ln 10 \qquad \text{Rule L2}$$
$$\cong \tfrac{3}{5}(0.75142) + (-\tfrac{9}{5})(2.30259)$$
$$= 3(0.15028) - 9(0.46052)$$
$$= 0.45084 - 4.14468$$
$$= -3.69384$$

Thus $\ln a = 5.52625 - 3.69384 = 1.83241$. From Table B-1 we get that $a = 6.25$. Note that in Table B-1, $\ln 6.25 = 1.83258$ so that the number a is actually little less than 6.25, but when rounded off, we can say that $a = 6.25$.

PROOF OF RULE L1   We want to show that

$$\ln ab = \ln a + \ln b$$

Let $g(x) = \ln ax$ for the constant a. Then

$$g'(x) = D_x \ln ax$$
$$= \frac{1}{ax} D_x\, ax$$
$$= \frac{1}{ax}(a) = \frac{1}{x}$$
$$= D_x \ln x$$

Thus $\ln ax$ and $\ln x$ have the same derivative. By Theorem 4-5, they differ only by a constant, so that there exists a constant c such that $\ln ax = c + \ln x$. By letting $x = 1$ we get $\ln a = c + \ln 1 = c$, since

$\ln 1 = 0$. Hence $c = \ln a$ and thus $\ln ax = \ln a + \ln x$. By letting $x = b$ we have the desired result, namely, $\ln ab = \ln a + \ln b$.

PROOF OF RULE L2 We want to show that
$$\ln a^r = r \ln a$$
If we let $f(x) = \ln x^r$, then
$$f'(x) = \frac{1}{x^r} D_x x^r$$
$$= \frac{1}{x^r}(rx^{r-1}) = \frac{r}{x}$$

which is equal to the derivative of $r \ln x$. Hence $\ln x^r$ and $r \ln x$ differ only by a constant. Thus there exists a c such that $\ln x^r = r \ln x + c$. Letting $x = 1$ yields $c = 0$, and therefore $\ln x^r = r \ln x$. By letting $x = a$, we have $\ln a^r = r \ln a$.

PROOF OF RULE L3 We want to show that
$$\ln \frac{a}{b} = \ln a - \ln b$$

We can write $\ln a/b = \ln ab^{-1} = \ln a + \ln b^{-1} = \ln a - \ln b$ by Rules L1 and L2 respectively.

EXERCISE 7-2

1. Find $f'(x)$.
   (a) $f(x) = 2\ln x$
   (b) $f(x) = \ln(x^2 + 1)$
   (c) $f(x) = \ln(3x^2 + x)$
   (d) $f(x) = \ln(x^3 + 5x)$
   (e) $f(x) = x^2 + x\ln(x^3 + 1)$
   (f) $f(x) = (\ln x)^2$
   (g) $f(x) = x\ln x - x$
   (h) $f(x) = \ln \ln x$

2. Find $f'(x)$.
   (a) $f(x) = \dfrac{x}{\ln x}$
   (b) $f(x) = \dfrac{\ln x}{x}$
   (c) $f(x) = \dfrac{x + \ln x}{x^2 + 1}$
   (d) $f(x) = (\ln x)^3$
   (e) $f(x) = (x + \ln x)^2$
   (f) $f(x) = \ln x^2 + (\ln x)^2$

3. Evaluate.

(a) $\displaystyle\int \frac{2}{x}\,dx$

(b) $\displaystyle\int \frac{dx}{x+1}$

(c) $\displaystyle\int \frac{dx}{2x+5}$

(d) $\displaystyle\int \frac{x\,dx}{1+x^2}$

(e) $\displaystyle\int \frac{x^2\,dx}{1+x^3}$

(f) $\displaystyle\int \frac{dx}{x\ln x}$

4. Evaluate.

(a) $\displaystyle\int \frac{x\,dx}{3+2x^2}$

(b) $\displaystyle\int \frac{x\,dx}{x+3}$

(c) $\displaystyle\int \frac{(x+2)\,dx}{x^2+4x}$

(d) $\displaystyle\int \frac{\ln x\,dx}{x}$

5. Find the natural logarithm.
   (a) 120
   (b) 125
   (c) 3,500
   (d) 3.5×10^5
   (e) 0.0012
   (f) 0.0765
   (g) 0.00912
   (h) 6.12×10^{-5}

6. Use logarithms to calculate the numbers.

(a) $\displaystyle\frac{(216)(81)}{62}$

(b) $\displaystyle\frac{(31.6)(3,700)}{.0081}$

7. Let a, b, and c be numbers such that $\ln a = .6$, $\ln b = .7$, and $\ln c = 1.8$. Find the given expressions.
   (a) $\ln ab$
   (b) $\ln bc$
   (c) $\ln abc$
   (d) $\ln(ab/c)$
   (e) $\ln a^2$
   (f) $\ln(ab)^2$
   (g) $\ln a^{\ln b}$

7–3 DEFINITION OF THE NUMBER e

In Sec. 1-5 we introduce the irrational number e by simply saying that it is an important number whose decimal equivalent is a nonterminating, nonrepeating one given by

$$e \cong 2.7182818285$$

The student is encouraged to think of the numerical equivalent of e as "about 2.7", just as π is thought of an "about 22/7". In this manner, expressions such as e^2 take on a more concrete numerical value, namely e^2 is "about $(2.7)^2 = 7.29$."

The number e plays a significant role in many diverse mathematical discussions and it appears in many different forms. In this section we present one definition of e and derive one other formulation.

To motivate the definition of e, let us recall a few properties of $y = \ln x$. First note that the graph of $y = \ln x$ passes through the point $(1, 0)$ since $\ln 1 = 0$. Since $D_x \ln x = 1/x$, we see that $\ln x$ is a differentiable function. Since the domain of $y = \ln x$ is $(0, \infty)$, we see that $D_x \ln x > 0$ for all x in $(0, \infty)$ and thus by Theorem 4-3 $y = \ln x$ is increasing in $(0, \infty)$. In Sec. 7-1 we noted that $\ln 3 > 1$. Hence the graph of $y = \ln x$ must intersect the line $y = 1$ (see Fig. 7-5). In algebraic terms, this means that there is a solution to the equation $\ln x = 1$.

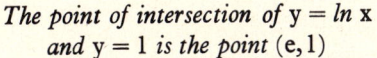

The point of intersection of $y = \ln x$ *and* $y = 1$ *is the point* $(e, 1)$

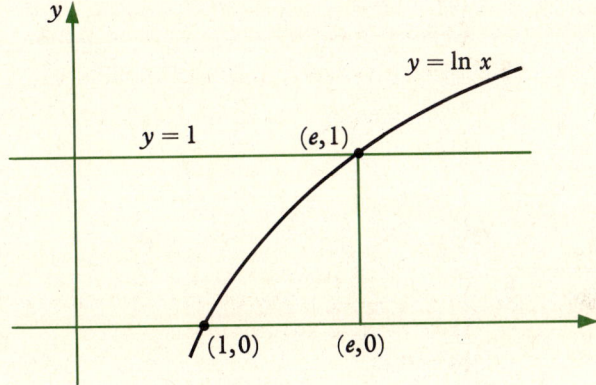

FIGURE 7-5

Definition 7-2 The solution of the equation $\ln x = 1$ is defined to be the number e, that is, e is that number such that $\ln e = 1$. (See Fig. 7-6.)

We now present another common formulation of the number e. Often the following theorem is given as the definition of e. In that case, the fact that $\ln e = 1$ might then be proved from that definition. In short, the two expressions of e are equivalent, each follows from the other. In the statement and proof of the theorem, we restrict the values of x to $(0, \infty)$ for convenience.

The number e *is defined by* $\int_1^e \frac{1}{t}\,dt = 1$, *that is, the area under* $y = \frac{1}{t}$ *from* $t = 1$ *to* $t = e$ *is* 1

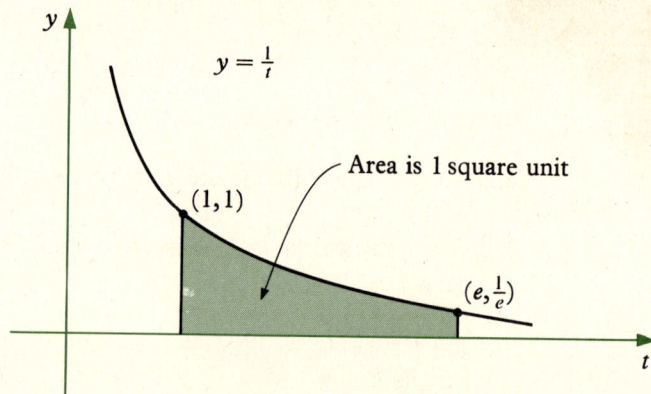

FIGURE 7-6

Theorem 7-1 $e = \lim_{x \to 0} (1+x)^{1/x}$

PROOF We first show that

$$\lim_{x \to 0} \ln(1+x)^{1/x} = 1$$

By Rule L2 we have $\ln(1+x)^{1/x} = (1/x)\ln(1+x)$. Utilizing L'Hôpital's Rule yields

$$\lim_{x \to 0} \ln(1+x)^{1/x} = \lim_{x \to 0} \frac{1}{x} \ln(1+x)$$

$$= \lim_{x \to 0} \frac{\ln(1+x)}{x} \quad \text{(algebraic manipulation)}$$

$$= \lim_{x \to 0} \frac{1/(1+x)}{1} \quad \text{(L'Hôpital's Rule)}$$

$$= \lim_{x \to 0} \frac{1}{1+x}$$

$$= 1$$

Since $\ln e = 1$, we can write

$$\lim_{x \to 0} \ln(1+x)^{1/x} = \ln e$$

Intuitively, this means that as the values of x get very close to 0, $\ln(1+x)^{1/x}$ gets very close to $\ln e$. Hence, as the values of x get very

close to 0, $(1+x)^{1/x}$ gets very close to e. Therefore,
$$\lim_{x \to 0} (1+x)^{1/x} = e$$
as desired.

EXERCISE 7-3

1. Use the approximation of e as "about 2.7" to approximate e^2, e^3, e^4, $e^{1/2}$, e^{-1}, e^{-2}, πe.
2. Graph the points $(1,e)$, $(e,1)$, $(2,e^2)$, $(3,e^3)$, $(-1,e^{-1})$, $(-2,1/e^2)$.
3. Find a value of x.
    (a) $\ln x = 2$
    (b) $\ln x = 3$
    (c) $\ln x = -1$
    (d) $\ln e^2 = x$
    (e) $\ln e^3 = x$
4. Calculate $(1+x)^{1/x}$ for $x = 1, 0.1, 0.01, 0.001$.

7-4 THE DEFINITION OF THE EXPONENTIAL FUNCTION $y = \exp x$

Let us refer back to Ex. 7-11. We found that $\ln a = 1.83241$. We then found a by looking in the body of Table B-1, finding the entry which was closest to 1.83241 (which is the number 1.83258) and then concluding that $a \cong 6.25$ when rounded off. This process of finding the number corresponding to a given logarithm is referred to as finding the antilogarithm and the number found is called the antilogarithm. Thus the antilogarithm of 2.30259 is 10. The notation *exp*, read exponential, is often used to represent the antilogarithm. Hence $\exp 2.30259 \cong 10$ and $\exp 1.83258 \cong 6.25$. Therefore, if $\ln a = b$, then $\exp b = a$.

Definition The *exponential function* $y = \exp x$ is defined by $y = \exp x$ if and only if $\ln y = x$.

Since $\ln y$ is defined only for positive numbers y, we see that $\exp x$ is always positive. Note that $\exp 0 = 1$ since $\ln 1 = 0$, and $\exp 1 = e$ since $\ln e = 1$. One can also see from the definition of $\exp x$ that
$$\exp(\ln x) = x \qquad (x > 0)$$
$$\ln(\exp x) = x$$

Because of the above identities, the two functions $y = \ln x$ and $y = \exp x$ are called *inverses* of each other.

We are now in a position to discover the true identity of the exponential function. We will now show that

$$\exp x = e^x$$

Let $y = e^x$. Then $\ln y = \ln e^x = x \ln e = x \cdot 1 = x$. But $\ln(\exp x) = x$ and hence $y = \exp x$. Thus $\exp x = e^x$ as desired.

Note that the two notations $\exp x$ and e^x can be used interchangeably. Generally "exp" is preferred if the expression in the exponent is complicated, such as $\exp((x^2 + \ln x)/x)$.

EXAMPLE 7-12

One can often use the above identities to find decimal equivalents of certain irrational numbers, such as $e^{\sqrt{2}}$. We write

$$\begin{aligned}
e^{\sqrt{2}} &= \exp(\ln e^{\sqrt{2}}) & \\
&= \exp(\sqrt{2} \ln e) & (Rule\ L2) \\
&= \exp \sqrt{2} & (\ln e = 1) \\
&\cong \exp 1.41 & (\sqrt{2} \cong 1.41) \\
&\cong 4.055 &
\end{aligned}$$

The graph of $y = e^x$ is given in Fig. 7-7 together with a table of values. A more extensive table of values of e^x is given in Table B-2 in the Appendix.

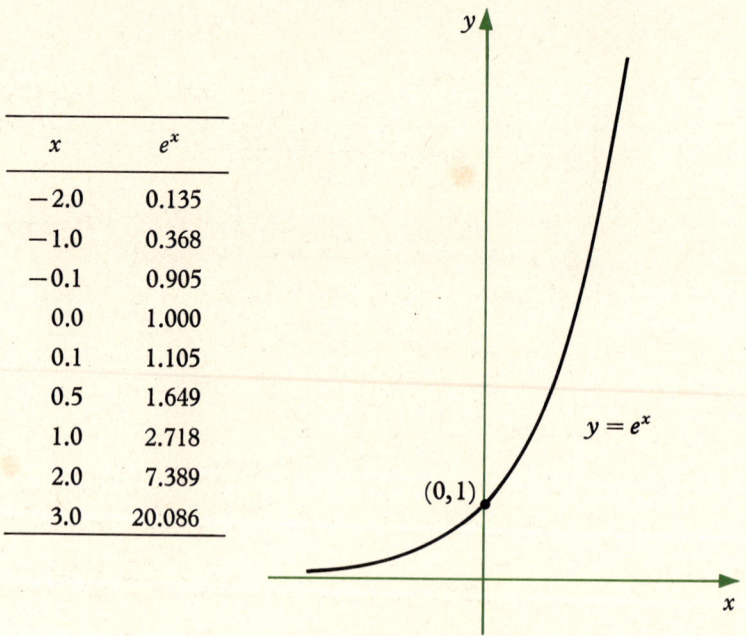

x	e^x
−2.0	0.135
−1.0	0.368
−0.1	0.905
0.0	1.000
0.1	1.105
0.5	1.649
1.0	2.718
2.0	7.389
3.0	20.086

FIGURE 7-7

One immediately sees from the graph that $y = e^x$ is an increasing, one-to-one function which is always positive.

EXERCISE 7-4

1. Find $\exp x$ for the given values of x.
   - (a) 0.14
   - (b) 0.24
   - (c) 0.80
   - (d) 1.1
   - (e) 2.5
   - (f) 3.0
   - (g) 3.8
   - (h) 8
   - (i) 6
   - (j) 12

2. Graph the functions.
   - (a) $y = 2e^x$
   - (b) $y = 3e^x$
   - (c) $y = e^x + 1$
   - (d) $y = 2e^x - 3$
   - (e) $y = e^{2x}$

3. Find approximations to the decimal equivalent to the irrational numbers.
   - (a) $2^{\sqrt{2}}$
   - (b) 3^e
   - (c) $\pi^{\sqrt{2}}$
   - (d) e^π
   - (e) π^e

7-5 PROPERTIES OF $y = e^x$

We will first set out to find $D_x e^x$. In Section 7-2 we mentioned Rule D10, which states that $D_x \ln f(x) = (1/f(x)) D_x f(x)$. If we let $f(x) = \exp x$ in Rule D10, we get

$$D_x \ln(\exp x) = \frac{1}{\exp x} D_x \exp x$$

But $\ln(\exp x) = x$ and so $D_x \ln(\exp x) = D_x x = 1$. Hence

$$1 = \frac{1}{\exp x} D_x \exp x$$

And thus

$$D_x \exp x = \exp x$$

In other words, $D_x e^x = e^x$. We record this property of $y = e^x$ as Rule D4, $D_x e^x = e^x$.

Incorporating Rule D4 into the Chain Rule, we get

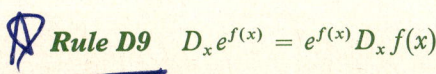

Rule D9 $\quad D_x e^{f(x)} = e^{f(x)} D_x f(x)$

CHAP. 7 Exponential and Logarithmic Functions

EXAMPLE 7-13
Consider the function $y = e^{x^2+3x} = \exp(x^2+3x)$. If we let $f(x) = x^2+3x$ and apply Rule D9 we get

$$D_x \exp(x^2+3x) = \exp(x^2+3x) D_x(x^2+3x) = \exp(x^2+3x)(2x+3)$$

EXAMPLE 7-14
Consider the function $y = e^{x^2} \ln x$. To find y', we use the Product Rule and get

$$D_x(e^{x^2} \ln x) = e^{x^2} D_x \ln x + (\ln x) D_x e^{x^2} = e^{x^2}\left(\frac{1}{x}\right) + (\ln x) e^{x^2} D_x x^2$$

$$= x^{-1} e^{x^2} + 2x e^{x^2} \ln x = e^{x^2}(x^{-1} + 2x \ln x)$$

Rewriting Rule D9 as an antiderivative formula yields

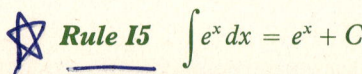
Rule I5 $\int e^x \, dx = e^x + C$

EXAMPLE 7-15
Consider $\int e^{5x} \, dx$. By letting $u = 5x$, we get $du = 5 \, dx$ so that

$$\int e^{5x} \, dx = \int e^u \left(\frac{1}{5}\right) du$$

$$= \frac{1}{5} \int e^u \, du$$

$$= \frac{1}{5} e^u + C$$

$$= \frac{1}{5} e^{5x} + C$$

EXAMPLE 7-16
Consider $\int x e^{x^2} \, dx$. By letting $u = x^2$ we get $du = 2x \, dx$ and so

$$\int x e^{x^2} \, dx = \int e^u \left(\frac{1}{2} du\right)$$

$$= \frac{1}{2} \int e^u \, du$$

$$= \frac{1}{2} e^u + C$$

$$= \frac{1}{2} e^{x^2} + C$$

The graph of $y = e^x$ *and* $y = \ln x$
are mirror images through $y = x$

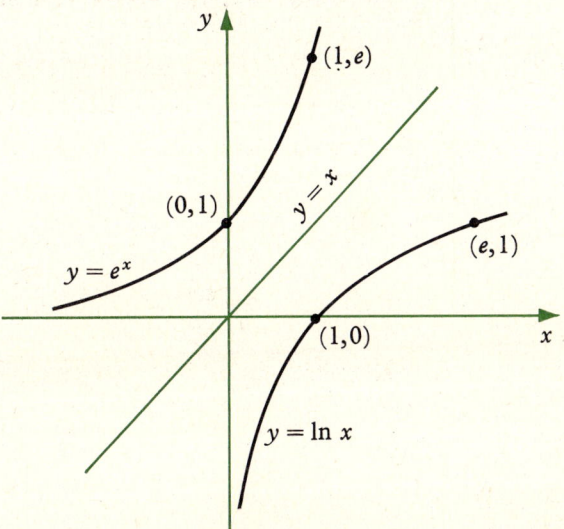

FIGURE 7-8

It is interesting to graph the functions $y = \ln x$ and $y = e^x$ in the same figure. This is done in Fig. 7-8. Also inserted in the figure is the line $y = x$. Note that the graphs of $y = \ln x$ and $y = e^x$ are "mirror images" of each other through the line $y = x$. Algebraically speaking, if the point (a,b) is on one graph then the point (b,a) is on the other. This is another method of stating that the two functions are inverses of each other.

Since $y = e^x$ is an exponential function, it satisfies the properties associated with the laws of exponents, namely

$$e^x e^y = e^{x+y}$$

$$(e^x)^y = e^{xy}$$

$$e^{-x} = 1/e^x$$

If we use the notation $\exp x$ these properties are written in the following form.

$$(\exp x)(\exp y) = \exp(x+y)$$

$$(\exp x)^y = \exp xy$$

$$\exp(-x) = \frac{1}{\exp x}$$

EXERCISE 7-5

1. Find $D_x y$.
   (a) $y = 2e^x + x$
   (b) $y = 5e^{3x}$
   (c) $y = xe^x$
   (d) $y = x^2 e^{x^3}$
   (e) $y = xe^{x^2} + e^{-2x}$
   (f) $y = (x+e^{3x})^{1/2}$
   (g) $y = \ln(x+e^x)$
   (h) $y = \ln xe^x$
   (i) $y = e^x \ln x$
   (j) $y = e^{x^2} \ln(x+e^{5x})$

2. Evaluate the antiderivatives.
   (a) $\int e^{5x} dx$
   (b) $\int (e^{10x} + e^{-x}) dx$
   (c) $\int xe^{2x^2} dx$
   (d) $\int x^2 e^{3x^3} dx$
   (e) $\int (1+2x) e^{x+x^2} dx$
   (f) $\int xe^{2x} dx$

3. Evaluate the integrals.
   (a) $\int_0^1 e^{3x} dx$
   (b) $\int_0^1 (e^{6x} + x) dx$
   (c) $\int_1^2 xe^{x^2} dx$
   (d) $\int_0^2 x^2 e^{x^3} dx$
   (e) $\int_0^1 xe^{1+x^2} dx$
   (f) $\int_0^1 xe^{1+x} dx$

7-6 OTHER BASES

Common logarithms, that is, logarithms to the base 10, written $\log_{10}$ or simply log are defined by

$$y = \log_{10} x \text{ if and only if } 10^y = x$$

The natural logarithms can be thought of as the logarithm to the base e because

$$y = \ln x \text{ if and only if } e^y = x$$

The natural question which arises is whether the two logarithms are related. To answer this question, consider $x = 10^y$. Taking the natural logarithm of each side of this equation yields $\ln x = \ln 10^y = y \ln 10$. But $y = \log_{10} x$ since $x = 10^y$. Hence

$$\ln x = (\log_{10} x) \ln 10$$

Thus to determine the common logarithm of a number, the formula is given by

$$\log_{10} x = \frac{\ln x}{\ln 10} \cong \frac{\ln x}{2.3}$$

For example, $\log_{10} 8.5 = \frac{\ln 8.5}{\ln 10} \cong \frac{2.14007}{2.3} \cong 0.93$

If we replace the number 10 with b in the above discussion we see that

$$\log_b x = \frac{\ln x}{\ln b}$$

for $0 < b \neq 1$. By differentiating each side of the above equation with respect to x we get the following rule.

Rule D11 $D_x \log_b x = \dfrac{1}{x \ln b}$ for $0 < b \neq 1$

Incorporating this rule into the Chain Rule we have

Rule D12 $D_x \log_b u = \dfrac{1}{u \ln b} D_x u$ for $0 < b \neq 1$ and u a function of x

For example,

$$D_x \log_8 (x^2 + 5x) = \frac{2x + 5}{(x^2 + 5x) \ln 8}$$

and

$$D_x \log_{10} (3x^2 + e^x) = \frac{6x + e^x}{(3x^2 + e^x) \ln 10}$$

just remember formulas that were used in homework assignment

Consider the function $y = b^x$ for $0 < b \neq 1$. To find $D_x y$ we write

$$y = b^x$$
$$\ln y = \ln b^x = x \ln b$$

and differentiate implicity to get

$$D_x \ln y = D_x x \ln b$$
$$\frac{y'}{y} = \ln b$$
$$y' = y \ln b$$

Hence we have

Rule D13 $D_x b^x = b^x \ln b$ for $0 < b \neq 1$

Incorporating this rule into the Chain Rule yields

CHAP. 7   Exponential and Logarithmic Functions

☆ **Rule D14** $D_x b^u = b^u \ln b \, D_x u$ for $0 < b \neq 1$ and u a function of x

For example, we have $D_x 8^{x^2} = 2x 8^{x^2} \ln 8$ and

$$D_x 10^{3x^4+x} = (12x^3+1) 10^{3x^4+x} \ln 10$$

We can also rewrite this differentiation rule as a rule of integration.

☆ **Rule I5** $\displaystyle\int b^x \, dx = \frac{b^x}{\ln b} + C$

EXERCISE 7-6

1. Differentiate the functions.
   - (a) $y = \log_5(x+5)$
   - (b) $y = \log_7(x^3+x)$
   - (c) $y = 5\log_{10}(x^2+3x) + \log_8(4x+1)$
   - (d) $y = 5^{x-1}$
   - (e) $y = 8^{x^3}$
   - (f) $y = 10^{3x^2+x}$
   - (g) $y = 5^x 8^{x-1}$
   - (h) $y = 5^x \log_8 x + 8^x \log_5 x$
   - (i) $y = \log_{10} 5^{x^2} + x 5^{x^2}$

2. Integrate the functions.
   - (a) $y = 5^x$
   - (b) $y = 8^{x-1}$
   - (c) $y = x 7^{x^2}$
   - (d) $y = \dfrac{\log_7 x}{x}$
   - (e) $y = \dfrac{\log_5(x^2+x)}{2x+1}$

3. Find $\log_{10} x$ for the given numbers.
   - (a) 5
   - (b) 20
   - (c) 36
   - (d) 102

8

Elementary Differential Equations

8–1 DIFFERENTIAL EQUATIONS REVISITED

A vast number of physical phenomena can best be described by equations involving rates of change, or derivatives. An equation giving information about the derivative of a function is called a *differential equation*. Examples of differential equations are:

1. $y' + y = 0$
2. $y' = 2x$
3. $y' + y - 5x = 1$
4. $y'' + yy' = 5$

The number of the highest derivative that appears is called *the order of the differential equation*. Hence the order of each of the first three equations above is one, whereas the order of the fourth equation is two. In this chapter we concern ourselves mainly with first-order differential equations.

To *solve* a differential equation is to find all functions which when substituted into the equation yield an *identity*, that is, an expression that is true for every value in the domain of the function. Thus, one solution to the differential equation $y' = 2x$ is $y = x^2$, but from Theorem 4-5 we know that $y = x^2 + C$ for any constant C is also a solution. Theorem 4-5 is a fundamental tool in the solution of differential equations, and so we recall it here.

Theorem 4-5 *If f and g are two functions that are differentiable in the open interval (a, b) and continuous at a and b, and if $f'(x) = g'(x)$ for all $x \in (a, b)$, then $f(x) = g(x) + C$ for some real number C and for every $x \in (a, b)$.*

Therefore, if f and g have the same derivative, then there is a constant C such that $f(x) = g(x) + C$.

EXAMPLE 8-1

Psychologists often use mathematical models to describe how excitation levels $E(t)$ depend upon the time t it takes to reach a goal. It is reasonable to assume that the rate of increase of the excitation level is easier to measure than $E(t)$ itself. Suppose that a psychologist conducting an excitation level experiment discovers from empirical data that

$$E'(t) = bt^{1/2}$$

for $t \in [0, 10]$, where t is measured in seconds and $E(t)$ is measured in units derived from an excitation level chart, and where b is some constant. Then

$$\int E'(t)\,dt = E(t) = \int bt^{1/2}\,dt = \frac{2b}{3} t^{3/2} + C$$

Thus

$$E(t) = \frac{2b}{3} t^{3/2} + C$$

for some constant C. If it is measured that the subject is at excitation level E_0 when the experiment starts, i.e., when $t = 0$, then $E(0) = E_0 = C$, and we have

$$E(t) = \frac{2b}{3} t^{3/2} + E_0$$

As illustrated in Ex. 8-1, it is often the case in physical problems that one particular value of the function is known; thus in Ex. 8-1, we knew the value $E(0)$. Such a value is called an *initial condition* or a *boundary value*. The solution of a first-order differential equation involves one arbitrary constant, because of Theorem 4-5. Such a solution is called a *general solution*. A solution obtained by giving particular values to the arbitrary constants, as we did in Ex. 8-1, is called a *particular solution*.

EXAMPLE 8-2

Consider the differential equation

$$y'(x) = 2x + x^{1/2}$$

We have
$$y(x) = \int y'(x)\, dx = \int (2x + x^{1/2})\, dx = x^2 + \tfrac{2}{3}x^{3/2} + C$$

Thus the general solution is $y(x) = x^2 + \tfrac{2}{3}x^{3/2} + C$. If we assign a value to C, such as $C = 5$, we have a particular solution $y(x) = x^2 + \tfrac{2}{3}x^{3/2} + 5$.

EXAMPLE 8-3

A common model in anatomy as well as other disciplines concerns a solution moving along a thin tubular membrane and a solute diffusing across the tube wall. The problem is to find the equation that expresses the concentration $C(x)$ of the solute at a point x of the membrane. Fick's law states that the amount of solute diffusing out across the tube wall is given by the product of the wall surface area, a permeability constant k for the given solution, and the average concentration difference over a given short interval. Given the assumptions that the membrane is a right circular cylinder of radius r and length L and that the average concentration difference can be approximated by $A - Bx^2$ where A and B are constant depending upon the velocity of the solution and other characteristics of the membrane, we get from Fick's law that
$$C'(x) = 2\pi rLk(A - Bx^2)$$
and so
$$C(x) = \int C'(x)\, dx$$
$$= 2\pi rLk \int (A - Bx^2)\, dx$$
$$= 2\pi rLk \left(Ax - \frac{Bx^3}{3} \right) + C_1$$

If we further assume that $C(0) = 0$, we have
$$C(x) = 2\pi rLk \left(Ax - \frac{Bx^3}{3} \right)$$

EXAMPLE 8-4

When the temperature of a body is raised, many different effects may be noticed. For example, gases expand, solids and liquids usually expand, but in rare cases they may contract, thermocouples develop an electromotive force, and many other phenomena occur. In the theory of thermodynamics one studies energy changes involved in the flow of heat and performance of work in a reaction. Van't Hoff discovered that the constant k of the rate of a reaction in which one mole of substance takes part is a function of the temperature T in degrees Kelvin given by
$$D_T \ln k = CT^{-2}$$
for some constant C that depends upon the heat freed or absorbed in the

reaction as well as upon the particular gas involved in the reaction. To solve for k in the above differential equation, we integrate.

$$\int D_T \ln k \, dT = \int CT^{-2} \, dT = C \int T^{-2} \, dT$$

$$\ln k = -CT^{-1} + C_1$$

$$k = e^{-CT + C_1} = C_2 e^{-CT}$$

where $C_2 = e^{C_1}$.

EXERCISE 8-1

1. Determine the order of the differential equations.

   (a) $\dfrac{dy}{dx} + xy^2 = 0$

   (b) $\dfrac{d^2y}{dx^2} + xy^2 = 0$

   (c) $\dfrac{dy}{dx} + y^3 = 0$

   (d) $y' + y + x = 0$

   (e) $(y')^2 + y' + y = x^2$

2. Show that the given functions $y = f(x)$ are solutions to the corresponding differential equations.
   (a) $y = x^2$, $y' - 2x = 0$
   (b) $y = x^3 + x - 1$, $y' - 3x^2 = 1$
   (c) $y = e^x$, $y' = y$
   (d) $y = e^{-3x}$, $y'' + 2y' - 3y = 0$
   (e) $y = e^{x^2}$, $y' - 2xy = 0$
   (f) $y = 1 - e^{-t}$, $y' = 1 - y$

3. Find the general solution and two particular solutions to the differential equations.
   (a) $y' = 3t^2 - 1$
   (b) $y' - 4t = e^t$
   (c) $y' = t^2(t^3 - 1)^{1/2}$
   (d) $y'' = t^3$
   (e) $y'' = t^{1/2} + t$
   (f) $e^{x^2} y' = x$

4. Verify that the function $y = ae^x - bxe^x$ is a solution to the differential equation $y'' - 2y' + y = 0$.

5. A psychologist predicts that a certain test he is going to administer will have an excitation level $E(t)$ depending upon the time t it takes to reach the goal of the test, which is governed by the differential equation

$$E'(t) = bt^{2/3} + C$$

for constants b and C. Find $E(t)$ where $E(0) = E_0$.

8-2 SEPARATION OF VARIABLES

A differential equation of the form

$$f(y)\frac{dy}{dx} = g(x)$$

can be solved in a straightforward manner, provided that the functions $f(y)$ and $g(x)$ can be easily integrated. If a differential equation has the above form, we say that it has separated variables.

To illustrate our method of solution of a differential equation of this type, we present the following example

EXAMPLE 8-5

Consider the differential equation

$$y^2 y' = x$$

If we make the substitution of variable $u = y(x)$, and then formally find $du = y' dx$, as we did in Chap. 5, we can write

$$\int y^2 y' = \int u^2 \, du$$

$$= \int x \, dx + C$$

Therefore

$$\frac{u^3}{3} = \frac{x^2}{2} + C$$

$$y(x) = u = \sqrt[3]{\frac{3x^2}{2} + C_1}$$

for the constant of integration $C_1 = 3C$. The reader can easily check to show that this is actually a solution to the given equation.

Thus if we are confronted with the differential equation

$$f(y) y' = g(x)$$

we let $u = y(x)$, then $du = y' dx$, and so

$$f(y) y' \, dx = f(u) \, du = g(x) \, dx$$

Now if $f(u)$ ahd $g(x)$ are easily integrated we can then solve for $u = y(x)$. In practice it simplifies matters simply to treat dy/dx as the ratio of dy over dx, which in fact it is not, but then the equation

$$f(y)\frac{dy}{dx} = g(x)$$

becomes
$$f(y)\,dy = g(x)\,dx$$
upon "multiplying" the equation by dx, and thus we can immediately write
$$\int f(y)\,dy = \int g(x)\,dx$$
Let us emphasize that this technique is merely a symbol manipulation which is justified for the sake of expediency, or, in other words, because it works.

EXAMPLE 8-6

Consider the equation
$$\frac{dy}{dx} = 3x^2 y$$
We write $(1/y)\,dy = 3x^2\,dx$ and so
$$\int \frac{dy}{y} = \int 3x^2\,dx$$
$$\ln y = x^3 + C$$
But
$$y = e^{\ln y}$$
so, by substitution
$$y = e^{x^3 + C}$$
Thus
$$y = C_1 e^{x^3}$$
where $C_1 = e^C$.

EXAMPLE 8-7

Recall Ex. 4-39 where we introduced the concept of a "growth equation", given by
$$\frac{dy}{dx} = ky$$
Hence
$$\int \frac{1}{y}\,dy = \int k\,dx$$
$$\ln y = kx + C$$
But
$$y = e^{\ln y}$$
so by substitution
$$y = e^{kx + C} = C_1 e^{kx}$$
where $C_1 = e^C$.

EXAMPLE 8-8

Bacteria reproduce by simple division, and thus the rate of change of a bacteria population $DN(t)$ is proportional to the existing population $N(t)$ where t is measured in hours, and hence

$$D_t N(t) = kN(t)$$

As in the previous example, $N(t) = C_1 e^{kt}$. Suppose after 3 hr there are ten thousand bacteria present and at the end of 4 hr, there are twenty thousand. Then

$$N(3) = C_1 e^{3k} = 10^4$$
$$N(4) = C_1 e^{4k} = 2 \times 10^4$$

Dividing the second expression by the first yields $e^k = 2$, so that $k = \ln 2$ and hence

$$C_1 = 10^4 e^{-3 \ln 2} = 10^4 e^{\ln 2^{-3}} = 10^4 e^{\ln(1/8)} = \frac{10^4}{8}$$

Therefore

$$N(t) = \frac{10^4}{8} e^{t \ln 2} = \frac{10^4}{8} (e^{\ln 2})^t = \frac{10^4}{8} 2^t$$

EXAMPLE 8-9

When considering blood flow through the aorta, the diastole phase of the heartbeat is the period of relaxation of the heart, while the systole phase is the period of contraction. During the diastole the volume V of the aorta decreases as blood flows away from the heart into blood vessels. Poiseuille's law states that $V'(t)$ is proportional to the pressure P, given by

$$V'(t) = \frac{-1}{w} P(t)$$

where w is a constant depending upon the viscosity of the blood, length, and radius of the aorta. Biologists also assume that P is proportional to V, i.e., that there exists a constant k such that $P = kV$. Thus

$$\frac{dP}{dt} = -\frac{k}{w} P(t)$$

By separating variables and integrating we get that

$$P(t) = P_0 e^{-(k/w)t}$$

where P_0 is the pressure when $t = 0$, i.e., where $P_0 = P(0)$. Thus

$$V'(t) = -\frac{P_0}{w} e^{-(k/w)t}$$

and

$$V(t) = \frac{P_0}{k} (e^{-(k/w)t} - 1) + V_0$$

where
$$V(0) = V_0$$

EXAMPLE 8-10

In constructing a meaningful model of ecology, many scientists, notably Lotka and Volterra, focussed attention on the concept of *biological association* in which they assumed that an ecological community consisted of a set of species, each of which influenced the others in some way, but none of which were influenced by a species outside the association. While no such community can adhere to this supposition exactly, the ecologist cannot be far wrong when he studies a community by assuming this model. Suppose N is the population size of a given species in the community. It is often the case that data on the rate of change with respect to time t of the population, dN/dt, are much more accessible than finding N directly. For example, if the species finds itself alone in the community, it will increase in numbers geometrically. If the species is a primary producer, such as a plant that depends upon no other species for its food, the rate of change of N will increase, whereas it will decrease if the species is a prey, as a hare, who is the prey of a number of predators.

Hence in a simplified sense, we can assume that dN/dt is proportional to N, so that

$$\frac{dN}{dt} = aN(t)$$

where a is a constant, positive or negative, depending upon the characteristics of the species and the community. However, for a finite population N the elements of even the same species must compete with each other to some extent, as trees reaching for sunlight or seals inhabiting crowded rocks. This tends to decrease dN/dt. It is therefore reasonable to modify the change in N by a factor of $K-N$ where K is the equilibrium value or the maximum population of the species which the community can maintain. Thus we assume

$$\frac{dN}{dt} = aN(K-N)$$

for constants a and K.

We can now use the method of separation of variables to solve for $N(t)$. Dividing by $N(K-N)$ and formally multiplying by dt, we have

$$\frac{dN}{N(K-N)} = a\,dt$$

and

$$\int \frac{dN}{N(K-N)} = \int a\,dt = at + d_1$$

where d_1 is a constant of integration.

To find the antiderivative of $1/N(K-N)$ we utilize the technique of integration called partial fractions. We write the fraction $1/N(K-N)$ as the sum of the two fractions

$$\frac{A}{N} + \frac{B}{K-N}$$

where A and B are to be determined. Thus we write

$$\frac{1}{N(K-N)} = \frac{A}{N} + \frac{B}{K-N} = \frac{AK - AN + BN}{N(K-N)}$$

We equate the numerators.

$$AK - AN + BN = 1 = AK + (B-A)N$$

Hence we have

$$AK = 1 \quad \text{or} \quad A = \frac{1}{K}$$

$$A - B = 0 \quad \text{so} \quad B = \frac{1}{K}$$

Thus

$$\int \frac{dN}{N(K-N)} = \int \frac{dN}{KN} + \int \frac{dN}{K(K-N)}$$

$$= \frac{1}{K} \ln N - \frac{1}{K} \ln(K-N)$$

$$= -\frac{1}{K} \ln \frac{K-N}{N}$$

Hence we have

$$\ln \frac{K-N}{N} = -Kat - Kd_1$$

Considering each side as an exponent of e and writing $d = e^{-Kd_1}$

$$e^{\ln(K-N)/N} = \frac{K-N}{N} = de^{-Kat}$$

Solving for N we get

$$N = \frac{K}{1 + de^{-Kat}}$$

You should now check to show that N is actually a solution to the differential equation

$$\frac{dN}{dt} = aN(K-N)$$

EXAMPLE 8-11

In sociology, the study of the diffusion of information is of fundamental importance. Sociologists assume that the spread of rumors and messages throughout a population is predictable. Such an assumption allows for the use of probability models to establish laws of interaction. It is reasonable to assume that the rate of diffusion is proportional to the frequency of contact between those who have received the message and those who have not. Hence if $N(t)$ is the proportion of the population who have received the message and $1-N(t)$ the proportion who have not, then

$$\frac{dN(t)}{dt} = kN(t)[1-N(t)]$$

for some constant k. Separating variables, we get

$$\int \frac{dN}{N(1-N)} = \int k\, dt$$

As in the previous example, we use the method of partial fractions to write

$$\frac{1}{N(1-N)} = \frac{1}{N} + \frac{1}{1-N}$$

and hence

$$\int \frac{dN}{N(1-N)} = \int \frac{dN}{N} + \int \frac{dN}{1-N} = \int k\, dt$$

$$\ln N - \ln(1-N) = kt + C$$

$$\ln \frac{N}{1-N} = kt + C$$

$$\frac{N}{1-N} = e^{kt+C}$$

$$= C_0 e^{kt}$$

$$N(t) = \frac{C_0 e^{kt}}{1 + C_0 e^{kt}}$$

where $C_0 = e^C$.

EXERCISE 8-2

1. Find the general solution of the differential equations.
   (a) $y' = 2y$
   (b) $y' = xy$
   (c) $y' = x^2 y$
   (d) $yy' = 2$
   (e) $yy' = x$
   (f) $y^2 y' = x$
   (g) $(y^2 + 1)y' = x^2 - 2$
   (h) $y' = \dfrac{x - x^3}{y - y^3}$
   (i) $y' = e^{x-y}$
   (j) $y' = xe^y$
   (k) $yy' = x^2 e^{y^2}$
   (l) $y' = xy - x^2 y$

2. Find particular solutions for the differential equations with the given boundary value.
   (a) $y' = y^2 x, y(1) = 1$
   (b) $yy' = xe^y, y(0) = 1$
   (c) $(x^2+1)y' - xy = 0, y(1) = 3$

3. Differential equations can be used to describe the spread of a disease model that we have discussed previously. Consider the following model. A portion of a community has been infected by a disease. The health officials hypothesize that if the number of infected members of the community is $x(t)$ at time t, then the rate at which $x(t)$ changes is proportional to the number already infected $x(t)$, since the disease is communicable. But in a finite population P, as $x(t)$ increases, there are fewer people susceptible to the disease, so that the rate $x'(t)$ is diminished by a factor of $P - x(t)$. Thus it is reasonable to assume that

$$x'(t) = ax(t)(P - x(t))$$

for constants a and P. Find $x(t)$.

4. The growth of a species depends not only upon the availability of food but also upon the rate of reproduction. A familiar growth model to ecologists assumes that the population $N(t)$ at time t of a certain species has a rate of change that depends upon the birth rate, which is usually assumed to be proportional to the number of encounters between males and females at time t, i.e., $k_1 N(t)^2$ for some constant of proportionality k_1, and the death rate, which is usually assumed to be proportional to $N(t)$, i.e., $k_2 N(t)$ for some constant k_2. Thus we have

$$N'(t) = k_1 N(t)^2 - k_2 N(t)$$

Find $N(t)$.

5. An autocatalytic reaction is one in which one substance changes into a new substance in such a way that the amount itself of the new substance $x(t)$ present at time t increases the rate of the reaction. Often it is assumed that the rate of formation dx/dt of the new substance is given by

$$\frac{dx}{dt} = kx(a - x)$$

where k is a constant and a is the initial amount of the substance. Find $x(t)$.

8–3 INTEGRATING FACTORS

Unfortunately there is no one technique which can be used to solve every first-order differential equation. The theory behind the solution of first-order equations involves classifying these equations into sets of equations, each having its own technique of solution. Hence there are those equations which can be solved by separation of variables, and in this section we discuss those equations of the form

$$y' + p(x)\, y = q(x)$$

which can be solved by multiplying the equation by a function called an *integrating factor*. Given the above differential equation, the integrating factor needed is

$$f(x) = e^{\int p(x)\, dx}$$

We shall show why this procedure works after we present an example designed to motivate the method. Let us illustrate the technique with an example.

EXAMPLE 8-12

Consider the equation

$$y' + 2xy = x$$

We have $p(x) = 2x$, or $\int p(x)\, dx = x^2$, and so our integrating factor is e^{x^2}. If we multiply our differential equation by e^{x^2}, we get

$$y' e^{x^2} + 2xy e^{x^2} = x e^{x^2}$$

The key step in this technique is to notice now that

$$\frac{d}{dx}(y e^{x^2}) = y' e^{x^2} + 2xy e^{x^2}$$

which is the left-hand side of the above equation. Thus, equating the last two expressions,

$$\frac{d}{dx}(y e^{x^2}) = x e^{x^2}$$

$$\int \frac{d}{dx}(y e^{x^2})\, dx = \int x e^{x^2}\, dx$$

$$y e^{x^2} = \tfrac{1}{2} e^{x^2} + C$$

$$y = Ce^{-x^2} + 1/2$$

which is the general solution to the above equation.

Thus, in general, the key step in solving the differential equation in the form

$$y' + p(x)\, y = q(x)$$

whose integrating factor is $f(x) = e^{\int p(x)\, dx}$ is to note that the left-hand side of the equation

$$y' f(x) + f(x)\, p(x)\, y = f(x)\, g(x)$$

can be expressed as
$$\frac{d}{dx}[yf(x)]$$
Thus we set
$$\frac{d}{dx}[yf(x)] = f(x)\,g(x)$$
and hence we must solve
$$yf(x) = \int \frac{d}{dx}[yf(x)]\,dx = \int f(x)\,g(x)\,dx$$
And thus the solution is
$$y = \frac{1}{f(x)} \int f(x)\,g(x)\,dx$$
Let us now consider a few more examples.

EXAMPLE 8-13

Consider the equation
$$y' + y = 1$$
Then $p(x) = 1$, $\int p(x)\,dx = x$ and our integrating factor is e^x. Multiplying the above equation by e^x yields
$$y'e^x + ye^x = e^x$$
and
$$e^x = \frac{d}{dx}(ye^x)$$
Thus
$$\int \frac{d}{dx}(ye^x)\,dx = \int e^x\,dx$$
$$ye^x = e^x + C$$
$$y = Ce^{-x} + 1$$

EXAMPLE 8-14

Biologists describe the operation of a muscle by a mathematical model that takes into account the contraction and extension of the muscle when flexed. Thus the force exerted by the muscle is the sum of the two forces, contraction and extension. The muscle will stretch according to the laws of elasticity, discussed in the previous chapter, and thus the force F_1 due to extension is proportional to the length l of the extension, so that
$$F_1 = k_1 l$$

for a constant of proportionality k_1. The force F_2 of contraction is a force of friction which is proportional to the rate of contraction with respect to time. Hence

$$F_2 = k_2 \frac{dl}{dt}$$

where k_2 is a constant of proportionality and F_2 is the force resisting the friction. If F is the total force exerted by the muscle, then

$$F = F_1 + F_2 = k_1 l + k_2 \frac{dl}{dt}$$

where here we are assuming the muscle is in isometric contraction where the total force F is a constant, and where the contractible part of the muscle shrinks while the elastic part stretches so that the tension $F_1 = al$ in the muscle increases without motion of the limb. To find the tension F_1, we write the equation in the form

$$l' + \frac{a}{b} \cdot l = \frac{F}{b}$$

where we make the substitutions $a = k_1$ and $b = 1/k_2$.

The integrating factor is $e^{(a/b)t}$ and thus

$$\frac{d}{dt}(le^{(a/b)t}) = l'e^{(a/b)t} + \frac{a}{b}le^{(a/b)t}$$

$$= \frac{F}{b}e^{(a/b)t}$$

$$\int \frac{d}{dt}(le^{(a/b)t})\,dt = \int \frac{F}{b}e^{(a/b)t}\,dt$$

$$le^{(a/b)t} = \frac{F}{a}e^{(a/b)t} + C$$

$$l = \frac{F}{a} + Ce^{-(a/b)t}$$

and thus

$$F_1 = al = F + Cae^{-(a/b)t}$$

EXAMPLE 8-15

The following model can be used to describe applications involving the addition of one solution to another, whether it be water added to a salt solution, liquid waste being added to a stream, or pollutants added to the atmosphere. Consider a biologist studying the infusion of glucose to the blood stream at a steady rate. The concentration $c(t)$ of glucose in the body at time t depends not only upon the amount of glucose admitted per volume of the liquids of the body V, but also the amount of glucose expended by bodily functions, which is usually assumed to be proportional to the change in the concentration of the glucose. Hence it is assumed that

$$\frac{dc}{dt} = \frac{A}{V} - kc$$

where k is a constant of proportionality, A and V are constant, and the units are milligrams per minute. If we rearrange the above equation to get

$$c' + kc = \frac{A}{V}$$

we see that we can use the integrating factor e^{kt} to solve for $c(t)$. Multiplying by e^{kt} we get

$$e^{kt}c' + e^{kt}kc = \frac{A}{V}e^{kt} = \frac{d}{dt}(ce^{kt})$$

and thus

$$ce^{kt} = \int \frac{A}{V}e^{kt} = \frac{A}{kV}e^{kt} + C$$

$$c(t) = Ce^{-kt} + \frac{A}{kV}$$

EXAMPLE 8-16

In psychology, an important model of learning called the *Hullian model* assumes that on each trial of an experiment, the *strength* of the correct response is increased; that is, because the subject has learned more about the experiment from previous trials, the probability $P(t)$ that a correct response will be given at time t increases as t increases. It is assumed in this model that the change in the probability, $P'(t)$, depends not only upon $P(t)$ but on a certain threshold C which $P'(t)$ cannot exceed. This is a reasonable assumption since the knowledge accrued from previous trials can only increase one's chances of success by a certain given amount. Thus the Hullian model of learning can be described by the differential equation

$$P'(t) = C - aP(t)$$

for constants C and a. To solve for $P(t)$, we rearrange the equation and multiply the equation by the integrating factor e^{at} and so

$$e^{at}P'(t) + ae^{at}P(t) = \frac{d}{dt}[e^{at}P(t)] = Ce^{at}$$

and thus

$$e^{at}P(t) = \int Ce^{at}\,dt = \frac{C}{a}e^{at} + k$$

for the constant of integration k. Hence

$$P(t) = \frac{C}{a} + ke^{-at}$$

EXERCISE 8-3

1. Find the general solution to the differential equations.
   - (a) $y' + y = 1$
   - (b) $y' - y = x$
   - (c) $y' - y = e^{-x}$
   - (d) $y' + xy = 2x$
   - (e) $y' + x^2 y = 3x^2$
   - (f) $y' + (x^2 + 1)y = x^2 + 1$
   - (g) $y' - (y/x) = x^3 - 5x$
   - (h) $y' = y + 5x$
   - (i) $y' = x^2(y + 2)$
   - (j) $y' = \frac{1}{x} - \frac{y}{x \ln x}$

2. Find a particular solution to the differential equations with the given boundary value.
   - (a) $y' = 2 - y$, $y(1) = 1$
   - (b) $y' = x^3 - yx$, $y(1) = 1$
   - (c) $y' - (y/x) = x^2 - 1$, $y(1) = 1$
   - (d) $y' - (y/x^2) = 5x^{-2}$, $y(1) = 3$
   - (e) $y' = xe^{-x^2} - 2xy$, $y(0) = 1$

3. A psychologist predicts that the test he is going to administer will be best described by a Hullian model where

$$P'(t) = tC - atP(t)$$

where $P(t)$ is the probability of a correct response at time t. Find $P(t)$.

4. Another psychologist predicts that his test will be best described by

$$P'(t) = t^2 C - at^2 P(t)$$

where $P(t)$ is the probability of a correct response at time t. Find $P(t)$.

9

Functions of Several Variables

9-1 EXAMPLES AND GRAPHS

The mathematical models we have presented thus far have been of a comparatively simple nature. For expediency, we have assumed that the concept or quantity we wished to study depended upon only one variable quantity while all other factors remained constant. In real-life situations, however, it is often the case that one quantity is a function of two or more independent variables. Consider the following illustrations.

1. Your grade in this course will depend upon many variables, including perhaps three or four exam grades.
2. In economics, the procurement lead time of a supply item will depend upon the time to process an acquisition and to deliver and perhaps the availability of storage space, as well as upon many other factors.
3. The total population of a given species depends upon its rate of reproduction, the availability of food, its death rate, and so on.
4. Total production of a firm depends upon the amount of labor and of land utilized.

Just as we found graphing functions of one variable an invaluable aid, so must we make extensive use of the graphs of functions of two variables. To this end, we construct a coordinate system in three-dimensional space. Consider three mutually perpendicular coordinate axes

CHAP. 9   Functions of Several Variables

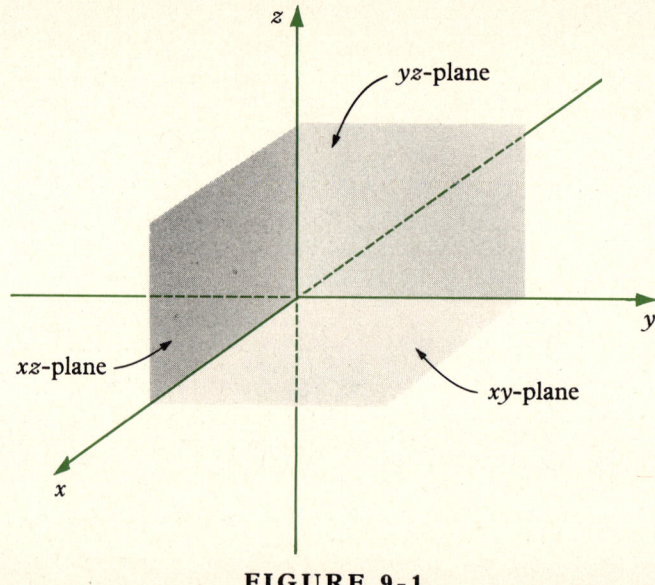

FIGURE 9-1

intersecting at a common origin. It is customary to label one axis, the x-axis, as that line pointing straight out from the paper. Then the y-axis is the horizontal line in the paper and the z-axis is the vertical line, as in Fig. 9-1.

Each pair of coordinate axes determines a *coordinate plane*. The xy-plane is that plane determined by the x- and y-axes. It is perpendicular to the paper. Similarly the yz-plane is the plane of the paper and the xz-plane is a vertical plane that is perpendicular to the paper. The xy-plane can be identified with the coordinate plane we have studied previously. The positively directed y-axis is to the right and the positively directed z-axis is vertical.

This coordinate system establishes a one-to-one correspondence between the set of all ordered triples (x,y,z) where x, y, and z are real numbers and the set of all points in three-dimensional space. We locate the point corresponding to the triple (a,b,c) by moving a units along the x-axis, then moving b units in the xy-plane parallel to the y-axis, then c units directly up or down depending upon whether c is positive or negative. An alternate method of viewing the location of the point (a,b,c) is to first plot the point (a,b) in the xy-plane as we have done earlier, and then move up or down c units. In the obvious way, every point in space corresponds to a unique triple.

EXAMPLE 9-1

Let us plot the points $(3,0,0)$, $(1,2,3)$, $(0,3,-2)$, $(-1,-4,-1)$, $(-4,4,3)$,

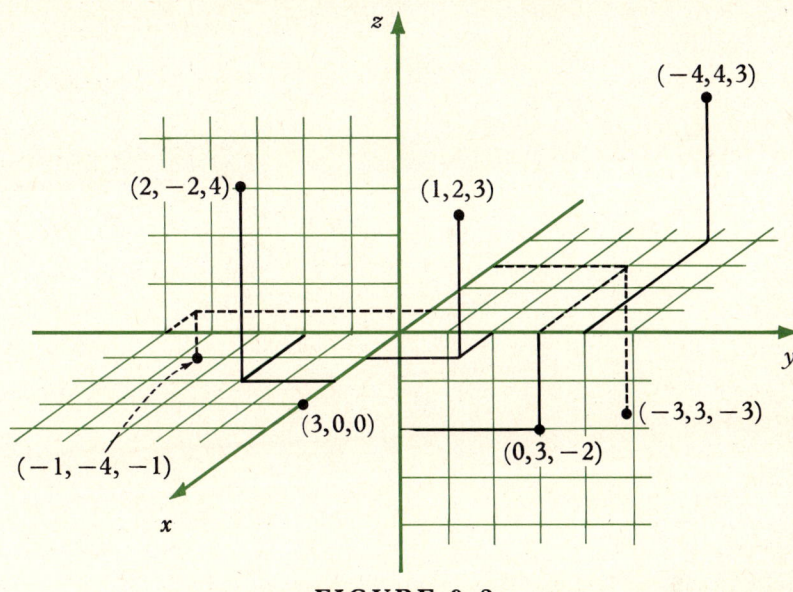

FIGURE 9-2

$(-3,3,-3)$, and $(2,-2,4)$. It is necessary to draw a few reference lines tracing a path to the point in order to get a perspective of the position of the point relative to the coordinate axes. We plot these points in Fig. 9-2.

In order to draw useful graphs in three-dimensional space one must develop skill in perspective drawings. Such a skill is indispensable in solving problems concerning the calculus of higher dimensions and its applications. A satisfactory drawing of a solid object must show clearly that it is three-dimensional. It may be helpful to visualize the coordinate axis system as the corner of a room where the *xy*-plane is the floor and the *xz*-plane and the

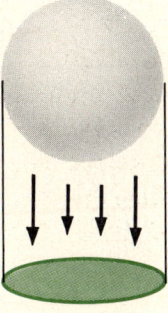

FIGURE 9-3

yz-plane are the walls. For example, if you project a sphere onto a two-dimensional piece of paper, you get a circle or an ellipse, depending upon the angle from which you are viewing the plane of the paper (see Fig. 9-3).

It is often useful to draw dotted "hidden" lines such as those in Fig. 9-4 to indicate depth and shape of solids on a two-dimensional surface. If one thinks of the solid as positioned in a coordinate system, then the "hidden" lines can be thought of as the intersection of a plane which is perpendicular to one of the coordinate planes of the solid. This intersection

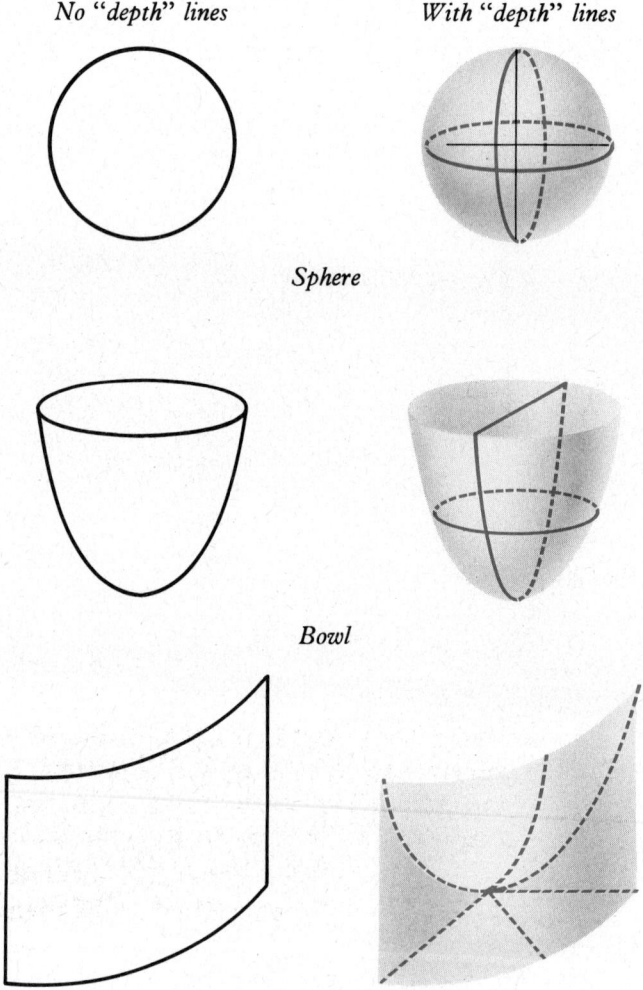

FIGURE 9-4

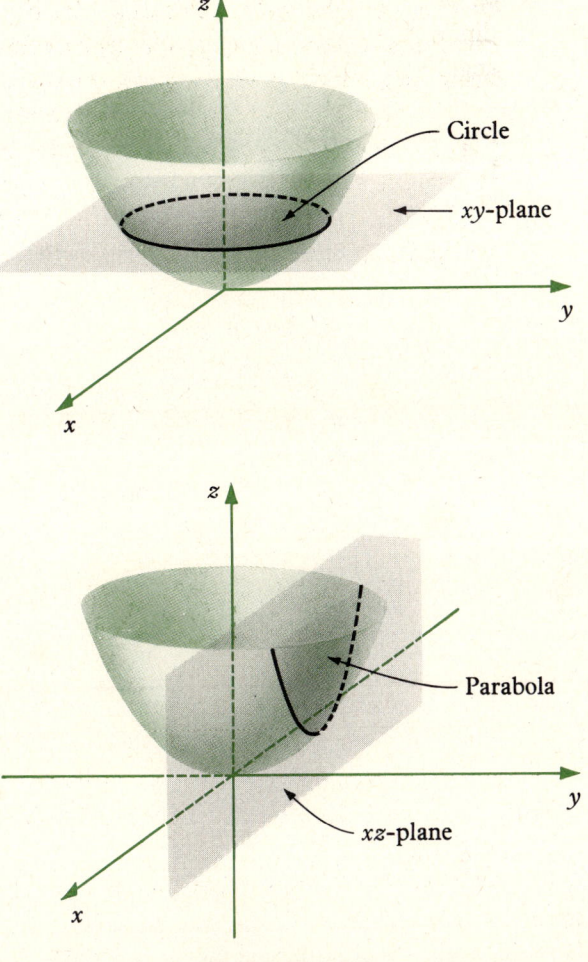

FIGURE 9-5

is then a two-dimensional curve in that plane. Consider the bowl-shaped region in Fig. 9-5. If we cut the region with a plane parallel to the *xy*-plane, the resulting intersection is a circle. If we cut the region with a plane perpendicular to the *xy*-plane, the intersection is a parabola.

It is precisely this type of graphing technique which will play a fundamental role in our discussion of functions of more than one variable.

EXERCISE 9-1

1. Describe three physical quantities that are functions of more than one variable.

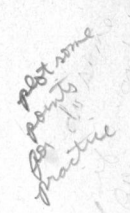

2. Plot the points.
   - (a) (1,0,0)
   - (b) (−1,0,0)
   - (c) (8,0,1)
   - (d) (6,0,2)
   - (e) (−1,0,−3)
   - (f) (0,1,0)
   - (g) (0,−5,0)
   - (h) (5,1,0)
   - (i) (−1,3,0)
   - (j) (0,0,1)
   - (k) (0,0,−5)
   - (l) (1,1,1)
   - (m) (1,−1,−1)
   - (n) (2,3,−6)
   - (o) (−2,8,−3)
   - (p) (6,−2,1)

3. Verify that the distance d between the two point P_1 (x_1, y_1, z_1) and P_2 (x_2, y_2, z_2) is given by

$$d = \sqrt{(x_2-x_1)^2 + (y_2-y_1)^2 + (z_2-z_1)^2}$$

by using the Pythagorean theorem twice. Find the distance between the pairs of points.
   - (a) (2,1,7) and (−1,0,3)
   - (b) (−1,2,0) and (0,1,1)
   - (c) (6,5,2) and (1,3,2)
   - (d) (5,−1,−2) and (1,0,−6)

4. Use Problem 3 to show that the point

$$\left(\frac{x_1+x_2}{2}, \frac{y_1+y_2}{2}, \frac{z_1+z_2}{2}\right)$$

is the midpoint of the line segment joining the points (x_1, y_1, z_1) and (x_2, y_2, z_2).

5. Which of the statements are true concerning three-dimensional space?
   - (a) Two intersecting lines determine a unique plane.
   - (b) A point and a line determine a unique plane.
   - (c) Three points not on a line determine a unique plane.
   - (d) Two intersecting planes determine a unique line.
   - (e) Infinitely many planes contain the same line.
   - (f) Any plane divides three-dimensional space into two sections.
   - (g) Two planes always intersect.
   - (h) Two lines always intersect.
   - (i) A line perpendicular to two intersecting lines is perpendicular to the plane containing them.
   - (j) Three planes may intersect in one and only one point. (*Hint*: Think of the origin in our coordinate system.)
   - (k) One and only one line perpendicular to a given plane passes through a given point. (*Hint*: Think of a plumb line.)
   - (l) Lines perpendicular to the same plane are parallel.
   - (m) Four points exist that are not in any single plane.

6. Draw a picture of a cube. Note that if you project the cube squarely on the paper, your picture will look like a square, not a cube. Be sure that parallel

edges of equal length are drawn. Also note that without "hidden" lines being drawn, your cube could be taken to be the picture of the corner of a room. You have the makings of an optical illusion.

7. Make a three-dimensional sketch of each shape.
   (a) A sphere
   (b) A right circular cylinder
   (c) A pyramid
   (d) A triangular prism
   (e) Two concentric spheres
   (f) Two concentric cylinders
   (g) A cube with a plane passing through two diagonals
   (h) A doughnut
   (i) A headlight
   (j) A square building with a dome-shaped roof

8. For each of the sketches in Problem 7, position the figure in a coordinate system, then cut the figure with planes perpendicular to each coordinate plane and describe the intersection of the figure and the plane.

9-2 DEFINITION OF A FUNCTION OF TWO VARIABLES

By a *function f* of *two variables* is meant a rule of correspondence between the elements of a set D of ordered pairs (x,y) of real numbers, called the *domain* of f, and the set of real numbers. This definition is in keeping with our definition of a function in Chap. 1, yet now we take the domain to be a set of ordered pairs. Customarily, functions of two variables are written as a formula in the form $z = f(x,y)$ where x and y are considered to be independent variables and z a dependent variable.

EXAMPLE 9-2

Consider the function

$$z = f(x,y) = x + 2y$$

Now $f(0,0) = 0$ so that $(0,0)$ corresponds to zero according to the function f. Similarly $f(1,-5) = 1+2(-5) = -9$ so that $(1,-5)$ corresponds to -9.

It is always assumed that the domain of $f(x,y)$ encompasses all those ordered pairs for which $f(x,y)$ is defined. Thus the domain of $f(x,y) = x+2y$ is the set of *all* ordered pairs of real numbers.

EXAMPLE 9-3

Consider

$$f(x,y) = \sqrt{1-(x^2+y^2)}$$

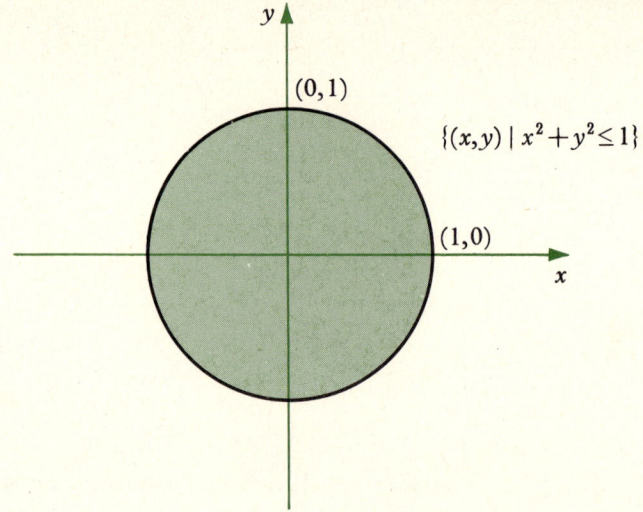

FIGURE 9-6

Note that if $x^2 + y^2 > 1$, then $f(x,y)$ is not defined. Hence the domain of f is

$$\{(x,y) \mid x^2 + y^2 \leq 1\}$$

whose graph is given in Fig. 9-6.

EXAMPLE 9-4

The familiar formula for the volume of a right circular cylinder is

$$V = V(r,h) = \pi r^2 h$$

which is a function of r and h. Here, of course $r > 0$ and $h > 0$, so the domain of V is

$$\{(r,h) \mid r > 0 \text{ and } h > 0\}$$

Hence a function of two variables $z = f(x,y)$ can be pictured as a machine that generates a real number $f(x,y)$ from two separate "inputs" x and y, or treated as an ordered pair as in Fig. 9-7. Think of a machine that takes two separate strands of twine and twists them into a single rope.

To graph the function $z = f(x,y)$, we think of the domain as a subset of the xy-plane in the three-dimensional coordinate system described in Sec. 9-1. Then the point $(x,y,f(x,y))$ is a point on the graph of the function. Hence we plot $(x,y,0)$ in the xy-plane and then locate (x,y,z) where $z = f(x,y)$, which is z units above (or below) the point $(x,y,0)$. Hence the graph of $z = f(x,y)$ is simply $\{(x,y,z) \mid z = f(x,y)\}$.

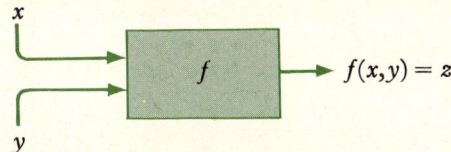

FIGURE 9-7

EXAMPLE 9-5

Consider the origin of our coordinate system to be at the center of a city and the xy-plane to be the ground, so that any point (x,y) is $\sqrt{x^2+y^2}$ miles from the center of the city. Let the positively directed x-axis denote the southerly direction and the positively directed y-axis point the easterly direction. Let the function $f(x,y)$, given by

$$f(x,y) = 200 + 2x^2 + y^2$$

denote the height in feet of the cloud cover over the point (x,y). Hence the cloud cover is $f(0,0) = 200$ ft over the center of the city. The point

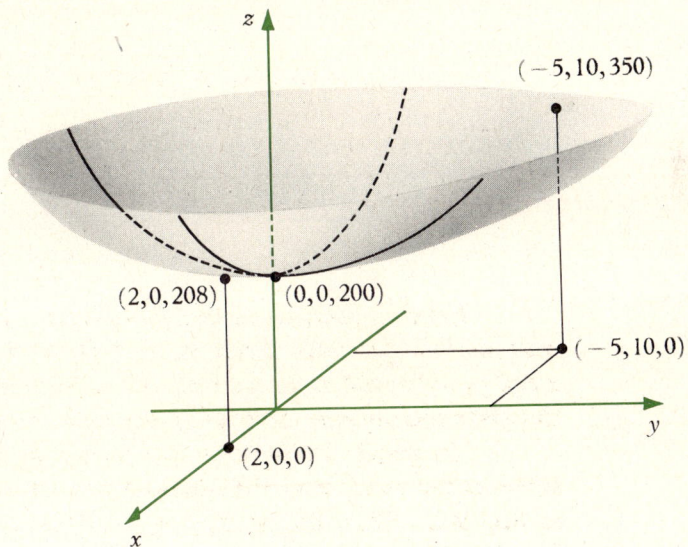

$$f(x,y) = 200 + 2x^2 + y^2$$

FIGURE 9-8

(2,0), which is 2 miles directly south of the city, has cloud cover $f(2,0) = 200 + 8 = 208$ ft overhead. The point 5 miles west and 10 miles south, i.e., $(-5,10)$, has a cloud cover $f(-5,10) = 200 + 50 + 100 = 350$ ft. Hence the points $(0,0,200)$, $(2,0,208)$, and $(-5,10,350)$ are on the graph of $f(x,y)$. This graph is pictured as a surface (or cloud cover) over the xy-plane, as given in Fig. 9-8.

There is no easy method for graphing functions $z = f(x,y)$. The problem entails working algebraically with the formula $z = f(x,y)$ to determine what type of figures one gets when one intersects the graph of $z = f(x,y)$ with the planes mentioned in the last section. Once these

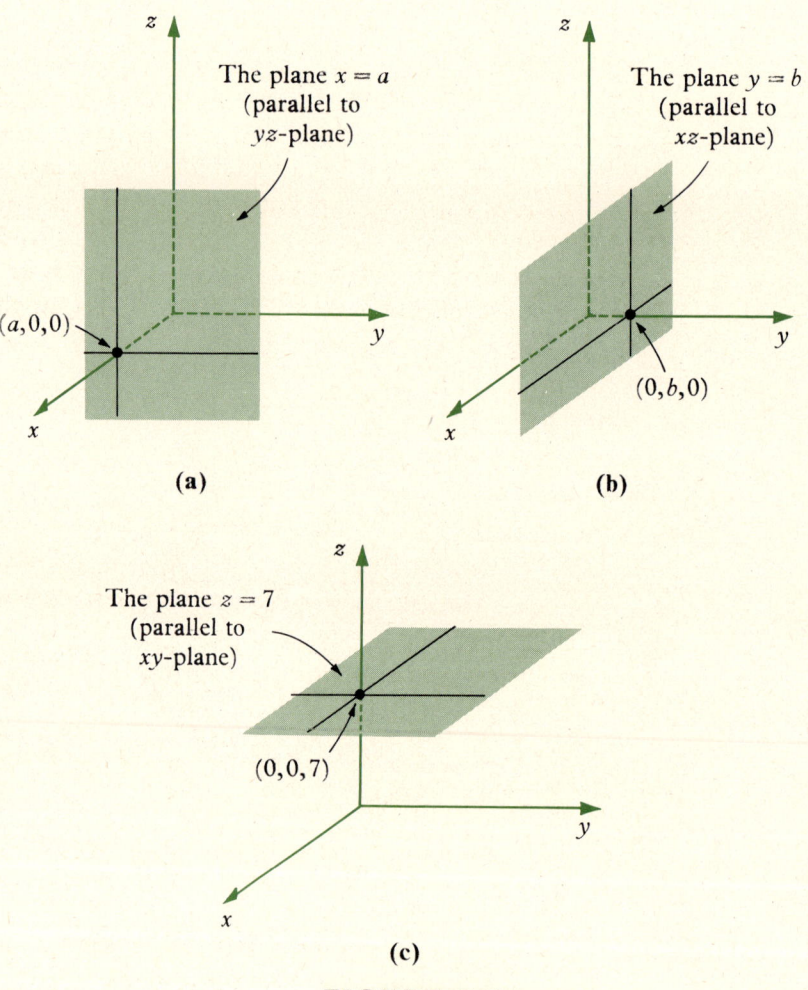

FIGURE 9-9

intersections are determined one can formulate a perspective figure as done in Sec. 9-1. The planes used to section the surface in question are those planes which are parallel to one of the coordinate axes. The reader should verify that:

1. The graph of the equation $x = a$ is a plane parallel to the yz-plane.
2. The graph of the equation $y = b$ is a plane parallel to the xz-plane.
3. The graph of the equation $z = c$ is a plane parallel to the xy-plane.

EXAMPLE 9-6

> The graph of $z = 0$ is $\{(x,y,0)\}$, which is the xy-plane. The graph of $z = 7$ is $\{(x,y,7)\}$, which is a plane parallel to the xy-plane and 7 units above it (see Fig. 9-9c).

The intersection of a surface $z = f(x,y)$ with a coordinate plane is called a *trace*. Therefore a trace is a two-dimensional figure in one of the coordinate planes whose equation is obtained by solving the equations $z = f(x,y)$ and the equation of the coordinate plane simultaneously. A *section* of a surface $z = f(x,y)$ is the intersection of the surface with a plane that is parallel to one of the coordinate planes. Thus a section is a two-dimensional figure in one of the planes $x = a$, $y = b$, or $z = c$. Its formula is obtained by solving the equation $z = f(x,y)$ and the equation of the plane simultaneously.

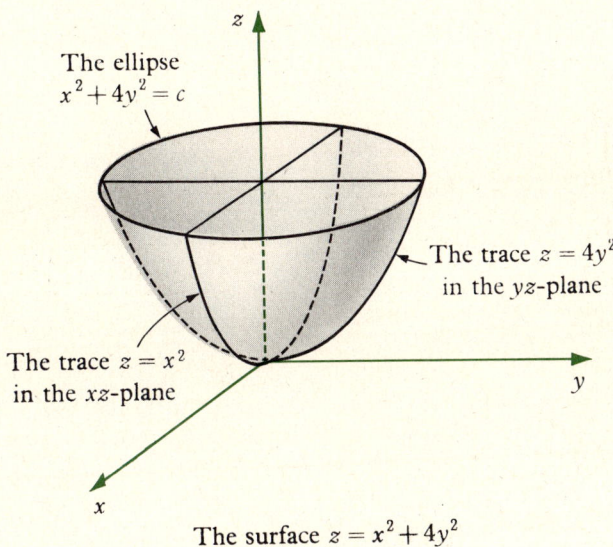

FIGURE 9-10

EXAMPLE 9-7

To sketch the graph of the function

$$z = f(x,y) = x^2 + 4y^2$$

we note that the trace in the yz-plane is the parabola

$$z = 0 + 4y^2 = 4y^2$$

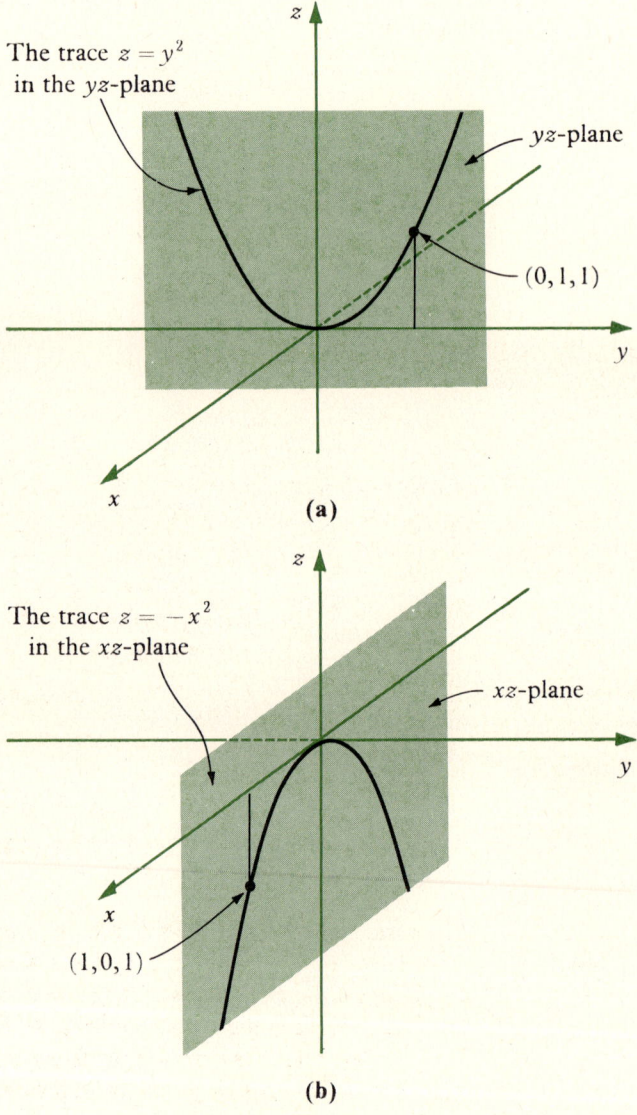

FIGURE 9-11

while the trace in the xz-plane is

$$z = x^2 + 4 \cdot 0 = x^2$$

If we section the surface by a plane $z = c$, we get the ellipse

$$z = c = x^2 + 4y^2$$

This surface is a bowl-shaped surface called an *elliptic paraboloid* and is sketched in Fig. 9-10.

EXAMPLE 9-8

Consider the function

$$z = f(x,y) = y^2 - x^2$$

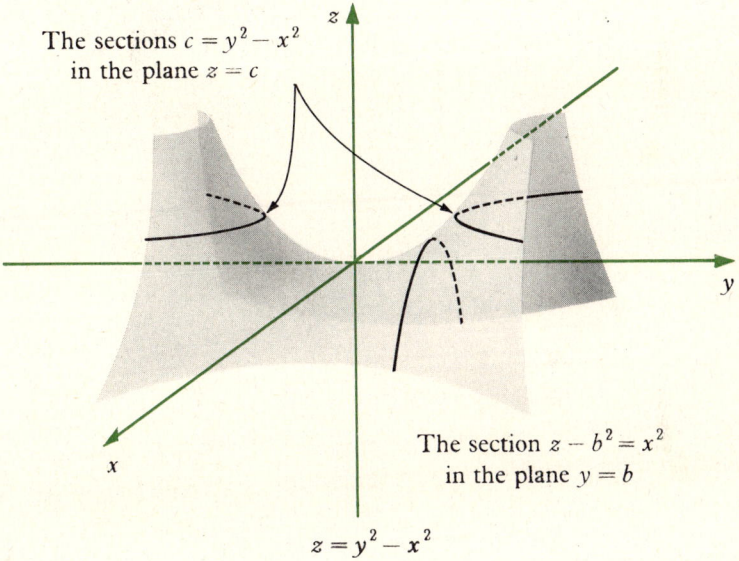

$z = y^2 - x^2$

FIGURE 9-12

The trace in the yz-plane, obtained by substituting $x = 0$ into $z = y^2 - x^2$, is the parabola $z = y^2$, whose graph is given in Fig. 9-11a. The trace in the xz-plane, obtained by substituting $y = 0$ into $z = y^2 - x^2$, is the parabola $z = -x^2$, whose graph is given in Fig. 9-11b.

Sections of the surface by planes $z = c$ yield hyperbolas $y^2 - x^2 = c$ and sections by planes $y = b$ yield parabolas. Putting this information together we can draw a reasonable sketch of this surface and obtain the saddle-shaped surface in Fig. 9-12.

EXERCISE 9-2

1. Given the function $f(x,y) = 3x - y$ and $g(x,y) = x^2 + y^2$, find the following.
   - (a) $f(0,0)$
   - (b) $f(1,2)$
   - (c) $f(-1,8)$
   - (d) $f(6,\frac{3}{2})$
   - (e) $f(2,-7)$
   - (f) the domain of f
   - (g) $g(0,0)$
   - (h) $g(1,1)$
   - (i) $g(-1,3)$
   - (j) $g(\frac{3}{2},-4)$
   - (k) $g(-6,-1)$
   - (l) the domain of g

2. The equation of the sphere having a radius of one unit and its center at the origin is $x^2 + y^2 + z^2 = 1$. Is it possible to represent this sphere as a function $z = f(x,y)$?

3. Give a geometrical test using lines perpendicular to the xy-plane to determine whether a given surface is a function $z = f(x,y)$. Recall the geometrical test given in Chap. 1 to determine whether a two-dimensional curve is a function.

4. Suppose the cloud cover of a point (x,y) that is $\sqrt{x^2 + y^2}$ miles from the center of a city located at the origin of our coordinate system is given by

$$f(x,y) = 100 + 3x^2 + 2y^2$$

   What is the cloud cover of the points $(1,5)$, $(2,-6)$, $(-1,-3)$? Does the height of the cloud cover increase more rapidly as one proceeds out along the positively directed x-axis or the positively directed y-axis, or the line $y = x$ in the xy-plane? Sketch the graph of f.

5. Find the domain of the functions.
   - (a) $z = x + y$
   - (b) $z = \dfrac{1}{xy}$
   - (c) $z = \dfrac{1}{(x-1)(y-2)}$
   - (d) $z = x - 1$
   - (e) $z = \sqrt{4 - (x^2 + y^2)}$

6. Graph the planes.
   - (a) $x = 1$
   - (b) $y = 2$
   - (c) $z = 5$
   - (d) $x = -3$
   - (e) $y = -1$
   - (f) $z = \sqrt{5}$
   - (g) $x + y = 1$
   - (h) $x - z = 2$
   - (i) $2x - y + z = 1$

7. Graph the functions by finding their traces and as many sections as needed for a reasonable perspective sketch.
   - (a) $z = x + y$
   - (b) $z = 2x^2 + 3y^2$
   - (c) $z = 200 + 2x^2 + y^2$
   - (d) $z = \sqrt{1 - (x^2 + y^2)}$
     (Note: $x^2 + y^2 + z^2 = 1$ and $z \geq 0$)
   - (e) $z = \sqrt{4 - x^2 - y^2}$
   - (f) $z = \sqrt{x^2 + y^2}$
   - (g) $z = 2y^2 - x^2$ (see Ex. 9-8)
   - (h) $z = x^2 - 2y^2$
   - (i) $z = e^x$

9-3 LIMITS AND CONTINUITY

We alter the usual approach to the discussion of limits and continuity by first discussing the property of a function of two variables being *continuous* at the point (x_0, y_0). We feel that continuity has a firmer intuitive base than the concept of a limit, and once the reader has a good grasp of continuity, the concept of a limit follows quite easily.

Geometrically or intuitively, by a function $z = f(x,y)$ being continuous at the point (x_0, y_0) we mean that the surface is "smooth", or at least contains no holes or gaps, for all the points sufficiently close to (x_0, y_0). Thus, there exists a circle in the xy-plane with center at (x_0, y_0) and of radius r, where r may be taken to be some small positive number, such that the surface of $z = f(x,y)$ immediately above (or below) the points within this circle is "smooth", in the sense mentioned above. In Fig. 9-13 we've selected such a circle and illustrated that portion of the surface which is immediately above the disk within the circle in the xy-plane. Perhaps the reader would do well to picture an infinite cylinder perpendicular to the xy-plane defined by the above-mentioned circle. Then that portion of the surface which we are considering is the set of all points on the graph whose corresponding points in the domain, or in the xy-plane, is within this circle.

EXAMPLE 9-9

The function

$$z = f(x,y) = 2$$

is a plane parallel to the xy-plane and two units above it. If (x_0, y_0) is any point in the xy-plane and r is any positive real number, then the surface above the disk of radius r and center (x_0, y_0) is the disk in the plane $z = 2$ with center $(x_0, y_0, 2)$ and radius r, which certainly fits our intuitive description of a "smooth" surface. Thus the function $z = 2$ is continuous at every point (x_0, y_0) in the xy-plane.

EXAMPLE 9-10

The function

$$z = f(x,y) = \frac{2(x^2+y^2)}{x^2+y^2}$$

is identical with the function $z = 2$ except when $x^2 + y^2 = 0$, i.e., when $x = 0 = y$, the origin. Hence the graph of $z = 2(x^2+y^2)/(x^2+y^2)$ is the plane $z = 2$ except that it has a hole at the point $(0,0,2)$. Hence this function is not continuous at $(0,0)$, whereas it is continuous at every other point in the xy-plane.

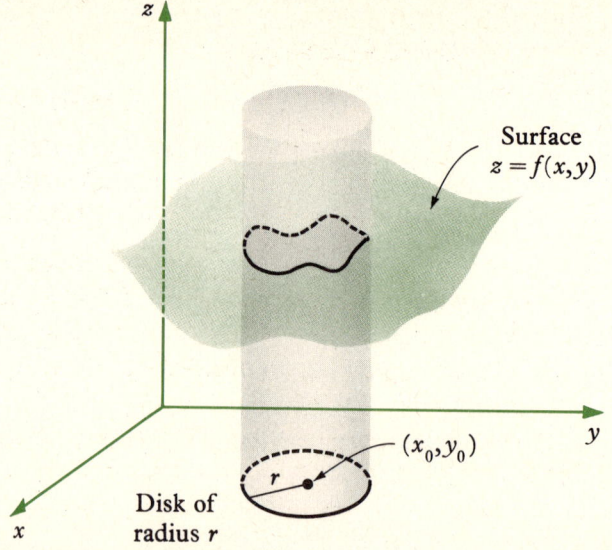

FIGURE 9-13

A more exact yet perhaps less intuitive way of describing the property of a function $z = f(x,y)$ being continuous at a point (x_0, y_0) is to say that if (x,y) is sufficiently close to (x_0, y_0), then $f(x,y)$ is sufficiently close to $f(x_0, y_0)$. We say that (x,y) is within r units of (x_0, y_0) if (x,y) is a point in the disk with center (x_0, y_0) and radius r; i.e., that (x,y) is an element of the set

$$\{(x,y) \mid (x-x_0)^2 + (y-y_0)^2 < r^2\}$$

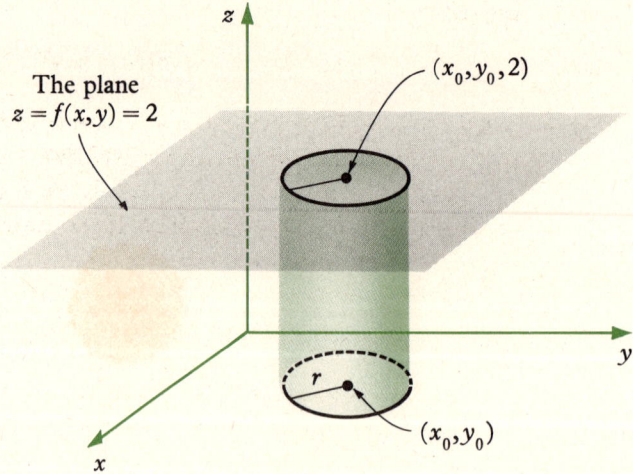

FIGURE 9-14

EXAMPLE 9-11

Consider the function

$$z = f(x,y) = x^2 + y^2$$

We claim that $f(x,y)$ is continuous at $(0,0)$. Suppose (x,y) is close to $(0,0)$, say, within r units for some small positive number r. Then $x^2 + y^2 < r^2$, so $f(x,y)$ is very close to $f(0,0) = 0$, in fact within r^2 units of 0.

To present the precise definition of continuity, it is necessary to formulate our intuitive concept of "close" into algebraic terms. Furthermore, to guarantee that $f(x,y)$ is close to $f(x_0,y_0)$ whenever (x,y) is close to (x_0,y_0), we must approach the problem in a different direction from that presented in Ex. 9-11.

 **Definition** A function

$$z = f(x,y)$$

is said to be continuous at the point (x_0,y_0) if given any $d > 0$, there exists an $r > 0$ such that

$$|f(x,y) - f(x_0,y_0)| < d$$

whenever (x,y) is within r units of (x_0,y_0).

When we speak of "close" we are actually talking about a limit. To say that $z = f(x,y)$ is continuous at (x_0,y_0) is to say that the limit of $f(x,y)$ as (x,y) approaches (x_0,y_0) is the number $f(x_0,y_0)$. To say that a number L is the limit of $f(x,y)$ as (x,y) approaches (x_0,y_0), which we indicate by

$$\lim_{(x,y) \to (x_0,y_0)} f(x,y) = L$$

is to say, intuitively, that $f(x,y)$ is close to L whenever (x,y) is close to (x_0,y_0). But we have just formulated this type of concept in the previous definition, so now if we replace $f(x_0,y_0)$ by L in that definition, we have the definition of the limit of a function of two variables.

 **Definition** We define

$$\lim_{(x,y) \to (x_0,y_0)} f(x,y) = L$$

to mean that for every $d > 0$ there exists $r > 0$ such that

$$|f(x,y) - L| < d$$

whenever (x,y) is within r units of (x_0,y_0) and $(x,y) \neq (x_0,y_0)$.

It may be the case that $f(x_0,y_0)$ does not exist or that $f(x_0,y_0)$ does exist but is not equal to L, even though

$$\lim_{(x,y)\to(x_0,y_0)} f(x) = L$$

does exist.

EXAMPLE 9-12
Again consider the function

$$z = f(x,y) = \frac{2(x^2+y^2)}{x^2+y^2}$$

discussed in Ex. 9-10. The graph has a hole at the point $(0,0,2)$. But note that if $(x,y) \neq (0,0)$, then $f(x,y) = 2$ and hence

$$\lim_{(x,y)\to(0,0)} \frac{2(x^2+y^2)}{x^2+y^2} = 2$$

even though $f(0,0)$ does not exist.

EXAMPLE 9-13
Consider the function

$$z = f(x,y) = \frac{x}{|x|}$$

To graph this function, note that $f(x,y)$ is the plane $z = 1$ if $x > 0$, i.e., if we are to the right of the xz-plane, while $f(x,y)$ is the plane $z = -1$ if $x < 0$,

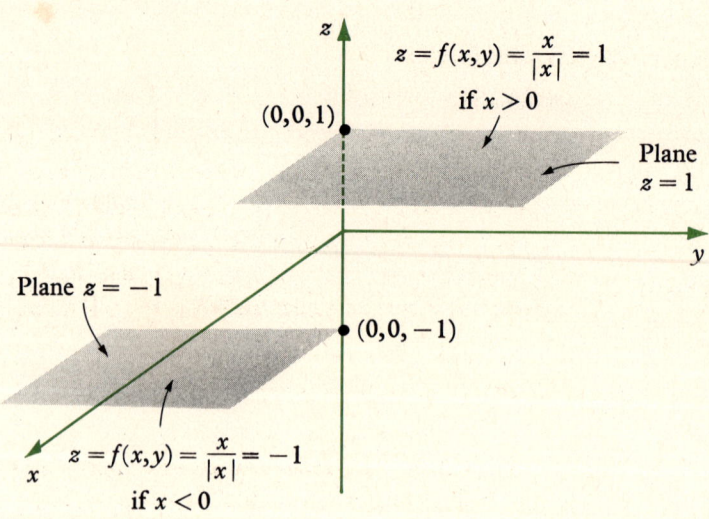

FIGURE 9-15

i.e., if we are to the left of the xz-plane. This graph is given in Fig. 9-15. Consider any point $(x_0,0)$ on the x-axis. If

$$D = \{(x,y) \mid (x-x_0)^2 + y^2 < r^2\}$$

is the disk with center $(x_0,0)$ and radius r, no matter how small we choose r, it is not the case that the functional values of all the points in D are close to any one number L. For, if $(a,b) \in D$ and $a > 0$, then $f(a,b) = 1$, while if $a < 0$, then $f(a,b) = -1$, and no matter how small r is, each type of point exists in D. In fact, there always must exist points in D, namely those on the x-axis, such that $f(x,y)$ does not exist. Hence there is no number L that satisfies our definition, so $\lim_{(x,y)\to(x_0,y_0)} f(x,y)$ does not exist.

EXERCISE 9-3

1. Graph the disks in the xy-plane.
   (a) $\{(x,y) \mid x^2 + y^2 < 2\}$
   (b) $\{(x,y) \mid (x-1)^2 + y^2 < 4\}$
   (c) $\{(x,y) \mid (x-1)^2 + (y-3)^2 < 9\}$
   (d) $\{(x,y) \mid (x-3)^2 + (y+1)^2 < 1\}$
2. Sketch the disk $D = \{(x,y) \mid (x-1)^2 + (y-2)^2 < 4\}$ in the xy-plane with a three-dimensional coordinate system. In the same graph sketch the disk in the plane $z = 5$ with center $(1,2,5)$ and radius 2. That is, graph the function $z = 5$ and sketch that portion of $z = 5$ immediately above D.
3. Use the definition of continuity to show that the functions are continuous at the point $(0,0)$. Sketch each function.
   (a) $z = 2$
   (b) $z = x + y$
   (c) $z = x^2$
   (d) $z = x^2 + y^2$
4. Explain why $z = 2x/|x|$ is not continuous at $(0,0)$.
5. Explain why $\lim_{(x,y)\to(0,0)} 2x/|x|$ does not exist.
6. Show that $\lim_{(x,y)\to(0,0)} (x-y)/(x+y)$ does not exist by looking at points on the x-axis and then at points on the y-axis which are close to $(0,0)$.
7. Show that $\lim_{(x,y)\to(0,0)} xy/(x^2+y^2)$ does not exist by looking at points on the x-axis and points on the line $x = y$ which are close to $(0,0)$.

9-4 PARTIAL DERIVATIVES

Just as the rate of change of a function of one variable is the underlying fundamental concept of differential calculus in two dimensions, so too is rate of change of functions of two or more variables basic to the study of calculus in higher dimensions. The situation for functions of two or more variables differs greatly from functions of one variable. Consider a function $z = f(x,y)$ of two variables. Ideally, we want to measure the rate of change of $z = f(x,y)$ in each possible direction from a point (a,b) in the domain of f. That is, as we move away from (a,b) in the xy-plane in some direction, we want to determine the rate at which f is increasing or decreasing.

For our purposes, it will suffice to consider only those two directions parallel to the x- and y-axes. If we restrict ourselves to one such direction, say, the direction parallel to the x-axis, by investigating the rate of change of $z = f(x,y)$ at the point (a,b) we are graphically considering the section of the surface determined by the plane passing through (a,b) and parallel to the xz-plane, that is, the plane $y = b$. Hence we are dealing with a two-dimensional curve defined by the intersection of $z = f(x,y)$ and $y = b$, that is, the function of one variable, $z = f(x,b)$. But we already have the means at hand to study the rate of change of such a function, namely, the derivative of $z = f(x,b)$ at $x = a$.

This number will then be called the *partial derivative* of $z = f(x,y)$ with respect to x at the point (a,b).

Definition The *partial derivative* of $z = f(x,y)$ *with respect* to x at the point (a,b) is defined by

$$\left.\frac{\partial f}{\partial x}\right|_{(a,b)} = \lim_{x \to a} \frac{f(x,b) - f(a,b)}{x - a}$$

Similarly, the partial derivative of $z = f(x,y)$ with respect to y at the point (a,b) is defined by

$$\left.\frac{\partial f}{\partial y}\right|_{(a,b)} = \lim_{y \to b} \frac{f(a,y) - f(a,b)}{y - b}$$

Other notations for the partial derivative of $z = f(x,y)$ are

$$\left.\frac{\partial f}{\partial x}\right|_{(a,b)} = f_x(a,b) = \frac{\partial f}{\partial x}(a,b) = f_1(a,b)$$

$$\left.\frac{\partial f}{\partial y}\right|_{(a,b)} = f_y(a,b) = \frac{\partial f}{\partial y}(a,b) = f_2(a,b)$$

We say that $z = f(x,y)$ is *differentiable* at (a,b) if $f_x(a,b)$ and $f_y(a,b)$ exist.

Figure 9-16 gives a graphical description of the numbers $f_x(a,b)$ and $f_y(a,b)$ as the slopes of the tangent lines to the surface $z = f(x,y)$ at the point $(a,b,f(a,b))$ in the directions parallel to the x- and y-axes, respectively.

SEC. 9-4   **Partial Derivatives**

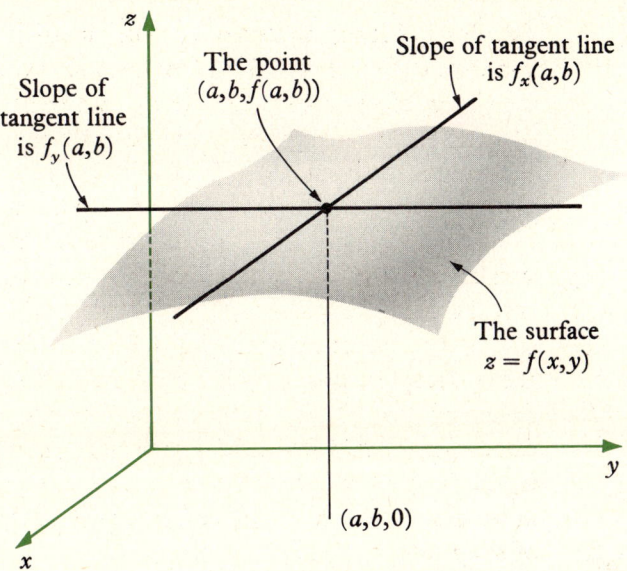

FIGURE 9-16

Of course we can compute the partial derivatives of $f(x,y)$ at an arbitrary point (x,y), and then the partial derivatives $f_x(x,y)$ and $f_y(x,y)$ are functions of x and y. The procedure used to calculate f_x and f_y is a relatively easy one. To find $f_x(x,y)$ from the formula $f(x,y)$, we treat y as a constant and consider the formula $f(x,y)$ as representing a function of the one variable x, and then differentiate with respect to x.

EXAMPLE 9-14

Consider the function

$$z = f(x,y) = x^2 + xy$$

Then

$$\frac{\partial f}{\partial x} = \frac{\partial}{\partial x}(x^2 + xy)$$

$$= \frac{\partial}{\partial x}(x^2) + \frac{\partial}{\partial x}(xy)$$

$$= 2x + y\frac{\partial x}{\partial x}$$

$$= 2x + y$$

Thus $f_x(x,y) = 2x + y$.

273

Similarly, to calculate $f_y(x,y)$ from the formula $f(x,y)$ we hold x constant and differentiate with respect to y.

EXAMPLE 9-15
Consider the function
$$z = f(x,y) = x + xy + y^3$$
Then
$$\frac{\partial f}{\partial y} = \frac{\partial}{\partial y}(x+xy+y^3)$$
$$= \frac{\partial}{\partial y}(x) + \frac{\partial}{\partial y}(xy) + \frac{\partial}{\partial y}(y^3)$$
$$= 0 + x\frac{\partial}{\partial y}(y) + 3y^2$$
$$= x + 3y^2$$

A few more examples are provided to help clarify this procedure.

EXAMPLE 9-16
Consider the function
$$z = f(x,y) = x^2 + xy + xy^2$$
Then
$$f_x(x,y) = 2x + y + y^2 \quad \text{and} \quad f_y(x,y) = x + 2xy$$
Then
$$f_x(1,2) = 2 + 2 + 4 = 8 \quad \text{while} \quad f_y(-1,3) = -1 + 2(-1)(3) = -7$$

EXAMPLE 9-17
Consider the function
$$z = f(x,y) = x^2 y^3 + xe^y$$
Then
$$f_x(x,y) = 2xy^3 + e^y \quad \text{and} \quad f_y(x,y) = 3x^2 y^2 + xe^y$$

EXAMPLE 9-18
Consider the function
$$z = f(x,y) = xe^{xy} + x \ln y$$
Then
$$f_x(x,y) = e^{xy} + xye^{xy} + \ln y \quad \text{and} \quad f_y(x,y) = x^2 e^{xy} + \frac{x}{y}$$

EXAMPLE 9-19

Consider the function

$$z = f(x,y) = \frac{x^2 y}{\sqrt{x+ye^x}}$$

Then

$$f_x(x,y) = \frac{2xy\sqrt{x+ye^x} - x^2 y(\frac{1}{2})(x+ye^x)^{-1/2}(1+ye^x)}{x+ye^x}$$

$$= \frac{xy(4x+4ye^x - x - xye^x)}{2(x+ye^x)^{1/2}(x+ye^x)}$$

$$= \frac{3x^2 y + 4xy^2 e^x - x^2 y^2 e^x}{2(x+ye^x)^{3/2}}$$

$$f_y(x,y) = \frac{x^2\sqrt{x+ye^x} - x^2 y(\frac{1}{2})(x+ye^x)^{-1/2} e^x}{x+ye^x}$$

$$= \frac{(x+ye^x)^{-1/2}(x^3 + x^2 ye^x - \frac{1}{2}x^2 ye^x)}{x+ye^x}$$

$$= \frac{2x^3 + x^2 ye^x}{2(x+ye^x)^{3/2}}$$

Just as we can define higher-order derivatives of functions of one variable, so too can we define higher-order partial derivatives. If $z = f(x,y)$, then the four second partial derivatives of f are

$$f_{xx}(x,y) = \frac{\partial^2 f}{\partial x\, \partial x} = \frac{\partial^2 f}{\partial x^2} = \frac{\partial}{\partial x}\left(\frac{\partial f}{\partial x}\right)$$

$$f_{yx}(x,y) = \frac{\partial^2 f}{\partial x\, \partial y} = \frac{\partial}{\partial x}\left(\frac{\partial f}{\partial y}\right)$$

$$f_{xy}(x,y) = \frac{\partial^2 f}{\partial y\, \partial x} = \frac{\partial}{\partial y}\left(\frac{\partial f}{\partial x}\right)$$

$$f_{yy}(x,y) = \frac{\partial^2 f}{\partial y\, \partial y} = \frac{\partial}{\partial y}\left(\frac{\partial f}{\partial y}\right)$$

Similarly one can define higher-order partial derivatives such as

$$f_{yxx}(x,y) = \frac{\partial}{\partial x}\left[\frac{\partial}{\partial x}\left(\frac{\partial f}{\partial y}\right)\right]$$

For all of the functions which we deal with, we have $f_{xy} = f_{yx}$.

EXAMPLE 9-20

Consider the function
$$z = f(x,y) = x^2y + xe^{xy}$$
Then
$$f_x(x,y) = 2xy + e^{xy} + xye^{xy} \quad \text{and} \quad f_y(x,y) = x^2 + x^2e^{xy}$$
Thus
$$f_{xx}(x,y) = 2y + ye^{xy} + ye^{xy} + xy^2e^{xy}$$
$$f_{yx}(x,y) = 2x + 2xe^{xy} + x^2ye^{xy}$$
$$f_{xy}(x,y) = 2x + xe^{xy} + xe^{xy} + x^2ye^{xy} = f_{xy}(x,y)$$
$$f_{yy}(x,y) = x^3e^{xy}$$

EXERCISE 9-4

1. Find f_x and f_y for the following functions.
   (a) $f(x,y) = 2x + 3y$
   (b) $f(x,y) = 5x + 6xy - 2y$
   (c) $f(x,y) = x^2 + xy + y^2$
   (d) $f(x,y) = x^3 + 2x^2y + xy^2 + 3y^3$
   (e) $f(x,y) = (x^2+1)(y^2+1)$
   (f) $f(x,y) = (x^2+1)^{1/2}(y^2+1)^{1/2}$
   (g) $f(x,y) = e^x \ln y$
   (h) $f(x,y) = ye^{x^2} + x^2 \ln y$
   (i) $f(x,y) = \ln(2e^{x^2} + xy)$
   (j) $f(x,y) = \dfrac{x^2 + xy}{e^{xy} + xy \ln xy}$

2. Use the definition of the partial derivative to find $f_x(1,2)$ where $f(x,y) = 2xy$.
3. Let $g(r,s) = 2r^2s + e^{rs}$ and find $g_s(1,-1)$ and $g_r(0,1)$.
4. Find $f_{xx}, f_{yx}, f_{xy},$ and f_{yy} for the given functions.
   (a) $f(x,y) = 3x^2 + 4xy^2$
   (b) $f(x,y) = x^2ye^y + yx^2e^{xy}$
   (c) $f(x,y) = xy^3 + \ln y(x^2+1)$
   (d) $f(x,y) = \dfrac{x^2 + xy}{e^{xy} + x^2}$

5. Find $f_{xxy}, f_{xyx},$ and f_{yxx} for the given functions.
   (a) $f(x,y) = x^4 + x^3y + x^2y^2 + x^5$
   (b) $f(x,y) = xe^y + x^2y \ln xy$
   (c) $f(x,y) = \dfrac{x + e^{xy}}{x^2}$

6. Show that the function $f(x,y) = x^3 + 5x^2 + e^y$ satisfies the partial differential equation

$$\frac{\partial^2 f}{\partial x\, \partial y} = 0$$

7. Show that the function $f(x,y) = x^3 - 3xy^2$ satisfies the partial differential equation

$$f_{xx} + f_{yy} = 0$$

8. It is possible to define the partial derivative of a function of more than two variables. One simply holds constant all variables except one and then defines the derivative of the resulting function of one variable. Let $w = f(x,y,z)$ be a function of three variables. Define the partial derivative of f with respect to x at the point (a,b,c).

9. Consider the function $f(x,y,z) = x^2 + y^2 + xyz$. To find the partial derivative f_x of f with respect to x, hold y and z constant, and then differentiate with respect to x. Thus $f_x = 2x + yz$. Find f_y and f_z.

10. Find f_x, f_y, and f_z.
    (a) $f(x,y,z) = x^2 y + xyz + yz^2$
    (b) $f(x,y,z) = 3x^2 yz - 4y^2 z^3$
    (c) $f(x,y,z) = xye^{yz}$
    (d) $f(x,y,z) = \dfrac{x^2 + e^{yz}}{xz^2 + \ln xyz}$

9-5 MAXIMA AND MINIMA

Many applications of functions of more than one variable involve finding a value of the function which is a maximum or minimum. As per our earlier discussions in this chapter, we shall center our attention on functions $z = f(x,y)$ of two variables because we have a firm geometrical base, i.e., three-dimensional sketches, which yield good intuitive descriptions of the concepts under discussion. The results we obtain can easily be generalized to pertain to functions of more than two variables.

Geometrically, a relative maximum of a function $z = f(x,y)$ resembles a mountain peak on the surface, whereas a relative minimum looks like a valley. One can also picture a dome as representing a maximum and a bowl as representing a minimum.

We can describe relative extrema of functions of two variables analytically by the following definition.

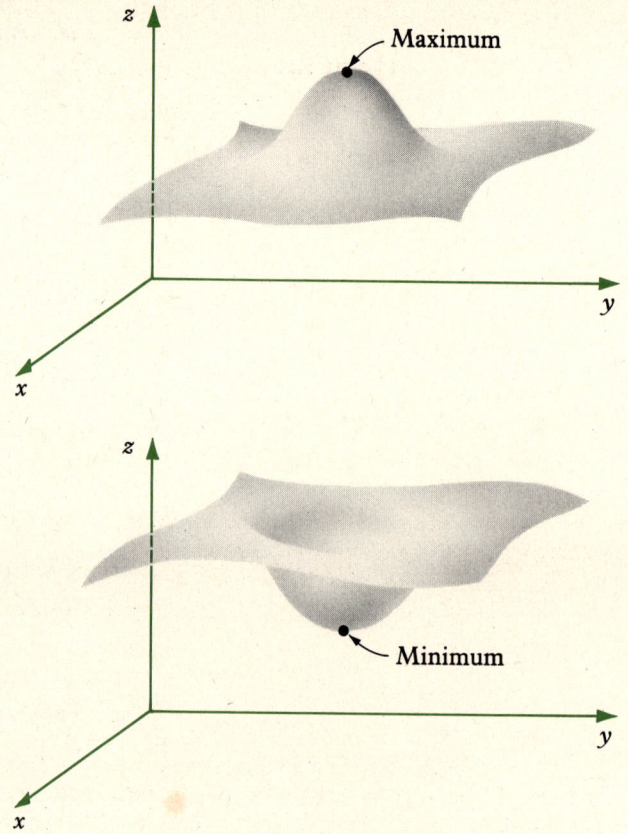

FIGURE 9-17

Definition A function $z = f(x,y)$ has a *relative maximum* [*minimum*] at the point (a,b) if

$$f(a,b) \geq f(x,y) \qquad [f(a,b) \leq f(x,y)]$$

for all points (x,y) within a distance r of (a,b), for some $r > 0$.

To say that (x,y) is within a distance r of (a,b) we mean (x,y) is within the circle with radius r and center at (a,b).

EXAMPLE 9-21

The function

$$z = f(x,y) = x^2 + 4y^2$$

was discussed in Ex. 9-11 and its graph given in Fig. 9-10. It is clear from the graph that the function has a minimum value of $f(0,0) = 0$ at the point

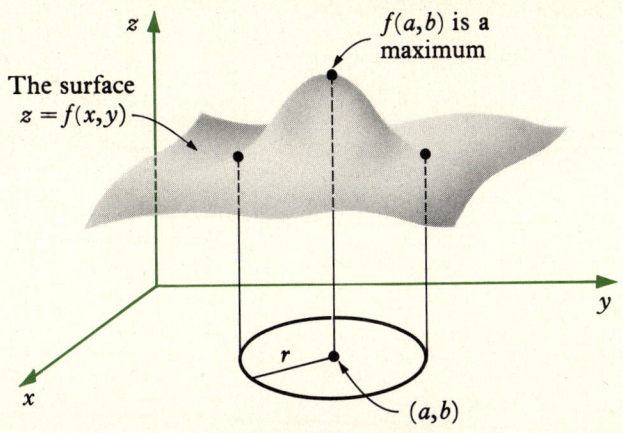

FIGURE 9-18

(0,0). This fact is also clear from the definition. That is, if (x,y) is *any* point, then $x^2 + 4y^2 \geq 0$, and thus, in particular, we can choose r to be any positive real number and have $f(0,0) = 0 \leq x^2 + 4y^2 = f(x,y)$ for all points (x,y) within the circle of radius r with center $(0,0)$.

EXAMPLE 9-22

The function
$$f(x,y) = 200 + 2x^2 + y^2$$
was discussed in Ex. 9-5 and its graph given in Fig. 9-8. We have that $f(x,y)$ has a minimum value of $f(0,0) = 200$ at the point $(0,0)$.

EXAMPLE 9-23

Recall the function
$$f(x,y) = y^2 - x^2$$
mentioned in Ex. 9-8 and its graph given in Fig. 9-12. At the point $(0,0)$ the function has a "saddle point", which is neither a maximum nor a minimum. No matter how small a radius r is chosen for a circle about $(0,0)$ for points $(0,y)$ on the y-axis with $|y| < r$ we have $f(0,y) = y^2 > 0$, and for points $(x,0)$ on the x-axis with $|x| < r$ we have $f(x,0) = -x^2 < 0$. Hence $f(0,0) = 0$ is not an extremum.

Definition A function $z = f(x,y)$ has an *absolute maximum* [*minimum*] at the point (a,b) if
$$f(a,b) \geq f(x,y) \qquad [f(a,b) \leq f(x,y)]$$
for all points (x,y) in the domain of $f(x,y)$.

As before, we refer to a relative maximum (or minimum) simply as a maximum (or minimum) or as an extremum.

Our process for finding those points for which $z=f(x,y)$ has an extremum is analogous to the techniques we used in dealing with functions of one variable, namely, we first use the partial derivatives to find candidates for extrema, and then use other means to determine which of these candidates actually do yield extrema.

Consider the graphical description given in Figs. 9-17 and 9-18. The tangent lines to the surface at the extremum point are parallel to the xy-plane. In particular the tangent lines that are parallel to the xz-plane and the yz-plane are parallel to the xy-plane and therefore have a slope equal to zero. In the last section we discovered that the partial derivatives of the function measure the slopes of these tangent lines. Hence, it seems reasonable to expect that if $z=f(x,y)$ has an extremum at (a,b) and if $z=f(x,y)$ is differentiable at (a,b), then $f_x(a,b)=f_y(a,b)=0$. The case where $z=f(x,y)$ is not differentiable at (a,b) is discussed in Exercise 9-6. Such functions do not often arise in applications so we are not considering them here.

We now present the theorem that allows us to find candidates for extrema, namely, those points (a,b) such that $f_x(a,b)=f_y(a,b)=0$. These are our candidates because only at these points can $z=f(x,y)$ have an extremum. However, as we see in Ex. 9-24, it may well happen that $f_x(a,b)=f_y(a,b)=0$ for a differentiable function, but that (a,b) is not a maximum or minimum.

Theorem 9-1 *If $z=f(x,y)$ has a maximum or a minimum at the point (a,b) and if $f_x(a,b)$ and $f_y(a,b)$ exist, then $f_x(a,b)=f_y(a,b)=0$.*

EXAMPLE 9-24
Consider the function
$$z = f(x,y) = x^2 + y^2$$
We have $f_x(x,y) = 2x$ and $f_y(x,y) = 2y$. We have seen that f has a minimum at $(0,0)$ and that $f_x(0,0) = f_y(0,0) = 0$.

EXAMPLE 9-25
Consider the function
$$z = f(x,y) = 2x^3 + 3x^2 - 12x + y^2 - y + 2$$
To find candidates for extrema we find f_x and f_y. Therefore
$$f_x(x,y) = 6x^2 + 6x - 12 = 6(x^2 + x - 2) = 6(x+2)(x-1)$$
while
$$f_y(x,y) = 2y - 1$$

If we set $f_x = 0$ and $f_y = 0$, we get $x = -2, 1$ and $y = \frac{1}{2}$. Hence our candidates for extrema are $(-2, \frac{1}{2})$ and $(1, \frac{1}{2})$. If this function has a maximum or a minimum, it must occur at $(-2, \frac{1}{2})$ or $(1, \frac{1}{2})$.

The next example illustrates that one cannot conclude that the function $z = f(x,y)$ has an extremum at (a,b) simply because $f_x(a,b) = 0 = f_y(a,b)$.

EXAMPLE 9-26

Consider the function

$$z = f(x,y) = y^2 - x^2$$

Then $f_x(x,y) = -2x$ and $f_y(x,y) = 2y$. Hence $f_x(0,0) = 0$ and $f_y(0,0) = 0$; so $(0,0)$ is a candidate for an extremum. However, in Ex. 9-23 we noted that this function does not have a maximum or a minimum at $(0,0)$. The point $(0,0)$ is a "saddle point" of the graph.

Often the problem of determining whether a candidate for an extremum is actually a maximum or a minimum is difficult. One method is to first find the candidate by solving $f_x = 0$ and $f_y = 0$ and then analyzing some properties of the function itself. We utilized this method when investigating the functions in Examples 9-24, 9-25, and 9-26. The next example further illustrates this technique. Another method, called *Lagrange multipliers* is discussed in Sec. 9-6. Still another method is given in Theorem 9-2 after Example 9-27.

EXAMPLE 9-27

A rectangular box with a top is to made out of 150 sq in. of paper. If V is the volume of the box, we want to find the dimensions of the box which yield the maximum volume. If we let x and y be the lengths of the sides and z the height of the box (in inches), then the surface area of the box is the sum of the areas of the four sides and the top and bottom, which is $2xy + 2yz + 2xz$ (see Fig. 9-19).

This surface area is to be 150 sq in., so we have $2xy + 2yz + 2xz = 150$; thus $xy + yz + xz = 75$. If we solve for z, we get

$$z = \frac{75 - xy}{x + y}$$

The volume V is the *length* × *width* × *height*, and so $V = xyz$. Therefore

$$V = xy\left(\frac{75 - xy}{x + y}\right) = \frac{75xy - x^2y^2}{x + y}$$

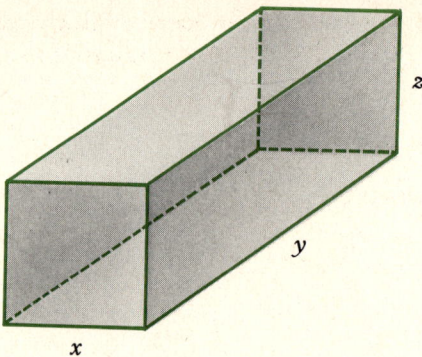

FIGURE 9-19

We now find the partial derivatives and get

$$V_x = \frac{(x+y)(75y-2xy^2) - (75xy-x^2y^2)}{(x+y)^2} = \frac{75y^2 - x^2y^2 - 2xy^3}{(x+y)^2}$$

$$V_y = \frac{(x+y)(75x-2x^2y) - (75xy-x^2y^2)}{(x+y)^2} = \frac{75x^2 - x^2y^2 - 2x^3y}{(x+y)^2}$$

Letting $V_x = 0$ and $V_y = 0$ yields

$$y^2(75 - x^2 - 2xy) = 0$$

and

$$x^2(75 - y^2 - 2xy) = 0$$

From the statement of the problem, since x, y, and z are lengths, x, y, and z must be greater than zero. Hence we can dispense with the y^2 and x^2 in the above equations and consider

$$75 - x^2 - 2xy = 0$$

and

$$75 - y^2 - 2xy = 0$$

Note that if we subtract these equations, we get

$$x^2 - y^2 = 0$$

Thus $x = \pm y$. Again, since x and y are positive, we have $x = y$. Substituting $x = y$ in the latter equation yields $75 - 3x^2 = 0$ and so $x = 5 = y$. Our only candidate for an extremum is the point (5,5). From the statement of the problem, it is clear that such a largest box does exist, and so the function must have a maximum at this point. If $x = y = 5$, then $z = (75 - 25)/10 = 5$, and so the largest box is a cube and its volume is 125 cu in.

The second derivative test for extrema for functions of one variable has an analogous test for functions of two variables. The *second partial derivative test* is more difficult to state and certainly more difficult to prove, so we present it in Theorem 9-2 without proof.

Theorem 9-2 Let $z = f(x,y)$ have partial derivatives at all points close to the point (a,b) and suppose that $f_x(a,b) = f_y(a,b) = 0$. To simplify notation we let $A = f_{xx}(a,b)$, $B = f_{yy}(a,b)$, and $C = f_{xy}(a,b)$.
  (I) If $A > 0$ and $AB - C^2 > 0$, then $f(a,b)$ is a minimum.
  (II) If $A < 0$ and $AB - C^2 > 0$, then $f(a,b)$ is a maximum.
  (III) If $AB - C^2 < 0$, then $f(x\ y)$ has a saddle point at (a,b).
  (IV) No conclusion can be immediately reached if $AB - C^2 = 0$ so that further tests must be conducted.

EXAMPLE 9-28

To find the extrema of the function

$$f(x,y) = x^2 - 4x + y^2 - 6y$$

we first find f_x and f_y, then set them equal to zero.

$$f_x(x,y) = 2x - 4$$

$$f_y(x,y) = 2y - 6$$

Hence we set $2x - 4 = 0$ and $2y - 6 = 0$ so that $x = 2$, $y = 3$. Thus $f_x(2,3) = f_y(2,3) = 0$. Now $f_{xx} = 2$, $f_{yy} = 2$, and $f_{xy} = 0$. Hence

$$A = f_{xx}(2,3) = 2$$

$$B = f_{yy}(2,3) = 2$$

$$C = f_{xy}(2,3) = 0$$

Hence $A > 0$ and $AB - C^2 = 2 \cdot 2 - 0 = 4 > 0$. Therefore $f(2,3) = -13$ is a minimum by part (I) of Theorem 9-2.

EXAMPLE 9-29

To find the extrema of the function

$$f(x,y) = x^2 + 4x - 2y^3 + 3y^2$$

we calculate

$$f_x(x,y) = 2x + 4$$
$$f_y(x,y) = -6y^2 + 6y$$
$$f_{xx}(x,y) = 2$$
$$f_{yy}(x,y) = -12y + 6$$
$$f_{xy}(x,y) = 0$$

By setting $f_x = 0$ and $f_y = 0$ we get $2x+4 = 0$ and $-6y^2+6y = 0 = -6y(y-1)$, so that $x = -2$ and $y = 0,1$. Hence the two points $(-2,0)$ and $(-2,1)$ are candidates for extrema. We first consider the point $(-2,0)$. We calculate

$$A = f_{xx}(-2,0) = 2$$
$$B = f_{yy}(-2,0) = 6$$
$$C = f_{xy}(-2,0) = 0$$

Thus $AB - C^2 = 2(6) = 12 > 0$ and $A = 2 > 0$, and so $f(x,y)$ has a minimum at $(-2,0)$ by part (I) of Theorem 9-2.

We now consider the point $(-2,1)$. We calculate

$$A = f_{xx}(-2,1) = 2$$
$$B = f_{yy}(-2,1) = -6$$
$$C = f_{xy}(-2,1) = 0$$

Hence $AB - C^2 = 2(-6) = -12$, and so $f(x,y)$ has a saddle point at $(-2,1)$ by part (III) of Theorem 9-2.

Part (IV) of Theorem 9-2 needs clarification. If $AB - C^2 = 0$, one must investigate functional values of all points close to (a,b). If the functional values are all greater than $f(a,b)$, then $f(a,b)$ is a minimum. If they are all less than $f(a,b)$, then $f(a,b)$ is a maximum. If some are greater and some are less than $f(a,b)$, then $f(x,y)$ has a saddle point at (a,b).

It is often a difficult task to inspect all points close to (a,b). It usually suffices to inspect four points, each in a separate direction from (a,b). The only restriction on the allowable distance away from (a,b) which one can choose for these points is that each chosen point must be closer to (a,b) than the nearest other candidate for extrema.

EXAMPLE 9-30

To find the extrema of the function

$$f(x,y) = x^2 + y^2 + 4xy$$

we calculate
$$f_x(x,y) = 2x + 4y$$
$$f_y(x,y) = 2y + 4x$$
$$f_{xx}(x,y) = 2$$
$$f_{yy}(x,y) = 2$$
$$f_{xy}(x,y) = 4$$

If we set $f_x = 0$ and $f_y = 0$, we get $2x + 4y = 0$ so $x = -2y$ and $2y + 4x = 0$, and so $y = -2x$. Substituting the last equation into $x = -2y$ yields $x = -2(-2x) = 4x$, and hence $3x = 0$, so $x = 0$. Therefore $y = 0$ and the candidate for extrema is the point $(0,0)$. Now
$$A = f_{xx}(0,0) = 2$$
$$B = f_{yy}(0,0) = 2$$
$$C = f_{xy}(0,0) = 4$$

Thus $AB - C^2 = 4 - 4 = 0$. It is therefore necessary to investigate points close to $(0,0)$. If we choose the points $(1,1), (-1,-1), (1,-1),$ and $(-1,1)$ we get $f(0,0) = 0, f(1,1) = 6, f(-1,-1) = 6, f(1,-1) = -2,$ and $f(-1,1) = -2$. Therefore $f(x,y)$ has a saddle point at $(0,0)$.

EXERCISE 9-5

1. Use the definition of a minimum to show that the functions have a minimum at the given point.
   (a) $f(x,y) = x^2 + y^2 + 2$ at $(0,0)$
   (b) $f(x,y) = (x-1)^2 + y^2$ at $(1,0)$
   (c) $f(x,y) = \sqrt{2x^2 + y^2}$ at $(0,0)$
2. Show that the function $f(x,y) = 2x^2 - y^2$ does not have an extremum at $(0,0)$ even though $f_x(0,0) = 0$ and $f_y(0,0) = 0$. At $(0,0)$ this function has a saddle point.
3. Find the points (a,b) such that $f_x(a,b) = 0 = f_y(a,b)$. Test for extrema.
   (a) $f(x,y) = x^2 + 2x - y^2$
   (b) $f(x,y) = x^3 + 3x^2 + y^2 - 4y + 2$
   (c) $f(x,y) = (x^2 + x + y^2)^{1/2}$
4. Find the dimensions of the box with maximum volume if the box is to be made from 36 sq in. of material.
5. Find all extrema for the function $f(x,y) = (x^2 + y^2)^{1/3}$.
6. Find three numbers x, y, and z such that their sum is 30 and their product is a maximum.
7. Consider the surface $f(x,y) = \sqrt{x^2 + y^2}$ whose graph given in the figure below resembles an inverted cone. Use the definition of a minimum to show that this function has a minimum at $(0,0)$. Show that $f_x(0,0)$ and $f_y(0,0)$ do not exist.

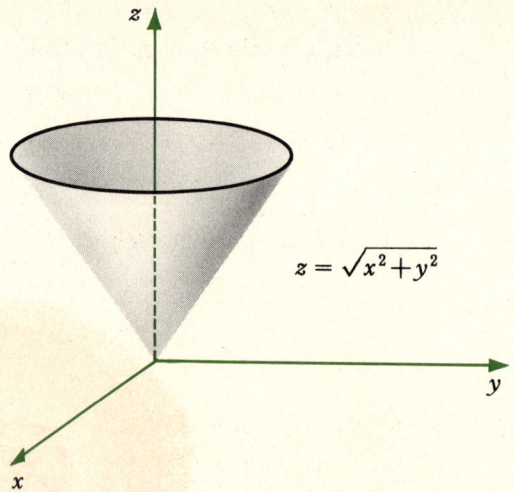

9-6 LAGRANGE MULTIPLIERS

The technique of finding extrema of functions of more than one variable by Lagrange multipliers can be applied to problems that involve maximizing or minimizing a function with an added restriction or constraint. In applications, often this restriction is necessary for the problem to make sense. In the previous section we discussed maximizing the function $V = xyz$ subject to the condition that $2xy + 2yz + 2xz = 150$. Without the restriction, the function $V = xyz$ cannot be maximized.

The great mathematician Joseph Louis Lagrange (1736–1813) developed this technique, which can be expressed for functions of n variables. We first explain the method for functions of two variables and then proceed to functions of three variables.

Consider the function $z = f(x,y)$. We wish to distinguish the extrema for f subject to the constraint $g(x,y) = 0$. Introducing the new variable t, called the *Lagrange multiplier*, we form the new function $h(x,y,t)$, a function of the three variables x, y, and t, defining

$$h(x,y,t) = f(x,y) - tg(x,y)$$

We now find the candidates for extrema of the function h. It can be shown that each extremum value of h is also an extremum of f relative to the constraint $g = 0$. Thus we find the partial derivatives of h, namely h_x, h_y, and h_t and set them equal to zero.

EXAMPLE 9-31

To find the minimum of the function

$$f(x,y) = x^2 + y^2$$

subject to the constraint $0 = g(x,y) = x + 2y - 10$, we form the function

$$h(x,y,t) = x^2 + y^2 - t(x + 2y - 10)$$

and find its partial derivatives and set them equal to zero.

$$h_x = 2x - t = 0$$
$$h_y = 2y - 2t = 0$$
$$h_t = -(x + 2y - 10) = 0$$

Thus $t = 2x$ and $t = y$, so $2x = y$. Substituting this equation into the linear equation $x + 2y - 10 = 0$ yields $5x = 10$, so $x = 2$, and thus $y = 4$. Hence the function $f(x,y) = x^2 + y^2$ has a minimum value of $f(2,4) = 20$ at the point $(2,4)$ subject to the constraint $x + 2y - 10 = 0$. The problem can be illustrated graphically by picturing the graph of the function $f(x,y) = x^2 + y^2$, which resembles an "infinite bowl", intersecting this function with the plane $0 = x + 2y - 10$, i.e., the plane $x = 10 - 2y$ which is perpendicular to the xy-plane, and then finding the minimum value of the two-dimensional curve that is the intersection of these two surfaces.

EXAMPLE 9-32

Consider the problem of determining the minimum distance from the line $3x + 4y - 25 = 0$ to the origin $(0,0)$. The function to be minimized is the distance formula

$$d(x,y) = (x^2 + y^2)^{1/2}$$

subject to the constraint $g(x,y) = 0 = 3x + 4y - 25$. We form the function

$$h(x,y,t) = (x^2 + y^2)^{1/2} - t(3x + 4y - 25)$$

and find its partial derivatives and set them equal to zero

$$h_x = (x^2 + y^2)^{-1/2}x - 3t = 0$$
$$h_y = (x^2 + y^2)^{-1/2}y - 4t = 0$$
$$h_t = -(3x + 4y - 25) = 0$$

From the first two equations we get

$$(x^2 + y^2)^{-1/2} = \frac{3t}{x} = \frac{4t}{y}$$

so that $x = \frac{3}{4}y$. Substitution of this equation into the third yields

$$3(\tfrac{3}{4}y) + 4y = 25$$

so that $y = 4$ and thus $x = 3$. Hence the minimum distance is
$$d(3,4) = (3^2+4^2)^{1/2} = 5$$

To find the extrema of a function $w = f(x,y,z)$ of three variables subject to the constraint $g(x,y,z) = 0$, we form the new function
$$h(x,y,z,t) = f(x,y,z) - tg(x,y,z)$$
and find its extrema.

EXAMPLE 9-33

Consider the function
$$w = f(x,y,z) = x^2 + y^2 + z^2$$
To find the minimum of f subject to the constraint $0 = g(x,y,z) = x-y+z+6$, we form
$$h(x,y,z,t) = x^2 + y^2 + z^2 - t(x-y+z+6)$$
and set the partial derivatives equal to zero,
$$h_x = 2x - t = 0$$
$$h_y = 2y + t = 0$$
$$h_z = 2z - t = 0$$
$$h_t = -(x-y+z+6) = 0$$
and thus $t = 2x = -2y = 2z$ and so $x = -y = z$. From the last equation we get $3x = -6$, $x = -2$, so that $f(-2,2,-2) = 12$ is a minimum value for f subject to the constraint $x-y+z+6 = 0$.

EXAMPLE 9-34

Consider the problem of finding the minimum distance from the plane $x+2y-z+12 = 0$ to the origin $(0,0,0)$. The function to be minimized is the distance formula,
$$d(x,y,z) = (x^2+y^2+z^2)^{1/2}$$
subject to the constraint $g(x,y,z) = x+2y-z+12 = 0$. We form the function
$$h(x,y,z,t) = (x^2+y^2+z^2)^{1/2} - t(x+2y-z+12)$$
Then
$$h_x = (x^2+y^2+z^2)^{-1/2}x - t = 0$$
$$h_y = (x^2+y^2+z^2)^{-1/2}y - 2t = 0$$
$$h_z = (x^2+y^2+z^2)^{-1/2}z + t = 0$$
$$h_t = -(x+2y-z+12) = 0$$

Thus
$$\frac{t}{(x^2+y^2+z^2)^{1/2}} = x = \frac{y}{2} = -z$$

Hence, $2x = y = -2z$ from the first three equations, and from the fourth $6x = -12$, $x = -2$. Thus $y = -4$ and $z = 2$, and the minimum distance is $d(-2,-4,2) = \sqrt{24} = 2\sqrt{6}$.

EXAMPLE 9-35

Suppose that a manufacturing company produces three products. If the company produces x of the first product, y of the second, and z of the third, and if the profit margins of the three products are a, b, and c dollars, respectively, then total profit P from the sale of the three products is given by

$$P(x,y,z) = ax + by + cz$$

The company can adjust production lines somewhat to make varying quantities of each product but owing to factors such as packaging specifications, labor costs, and overhead as well as others, they find that the amounts of each product that they can produce is subject to the condition

$$x + y^2 + z^2 = 90$$

items daily. To maximize P subject to this constraint, we form

$$h(x,y,z,t) = ax + by + cz - t(x+y^2+z^2-90)$$

Then

$$h_x = a - t \qquad h_y = b - 2ty \qquad h_z = c - 2tz$$

and

$$h_t = -(x^2+y^2+z^2-90)$$

Letting each partial derivative equal zero yields $t = a$, so $b = 2ay$ and $c = 2az$; thus from $h_t = 0$ we get

$$x^2 + \frac{b^2}{4a^2} + \frac{c^2}{4a^2} = 90$$

Therefore maximum profits are realized when

$$x = \frac{\sqrt{360a^2 - b^2 - c^2}}{2a} \qquad y = \frac{b}{2a} \qquad z = \frac{c}{2a}$$

EXAMPLE 9-36

The daily production capacity P of a firm is given by

$$P = P(x,y) = x^{1/2}y^{1/2}$$

where x is the number of man hours available daily and y is the amount of capital invested in machinery. Labor costs $3 per man hour. Assuming that there are 260 working days in a year and that the company allocates

$500,000 in capital for wages and machinery, the constraining equations are

$$3 \cdot 260x + y = 500,000$$

$$780x + y = 500,000$$

To find the values of x and y which yield the maximum production capacity, we form

$$F(x,y,\lambda) = x^{1/2}y^{1/2} - \lambda(780x+y-500,000)$$

$$F_x = \tfrac{1}{2}x^{-1/2}y^{1/2} - 780\lambda = 0$$

$$F_y = \tfrac{1}{2}x^{1/2}y^{-1/2} - \lambda = 0$$

$$F_\lambda = 780x + y - 500,000 = 0$$

$$\lambda = \tfrac{1}{2}x^{1/2}y^{-1/2} = (\tfrac{1}{2} \cdot 780)\,x^{-1/2}y^{1/2}$$

$$780x = y$$

$$2y = 500,000$$

$$y = \$250,000$$

Thus to maximize production the firm should spend $250,000 on machinery and $250,000 on wages and hence utilize approximately 320 man hours daily. The maximum value for P subject to the given constraint is

$$P(250,000, 250,000) = 250,000$$

EXAMPLE 9-37

A company manufactures two products, product A and product B. Product A is much more costly to produce than product B. The cost function C is given by

$$C = C(x,y) = x^2 + 2y + 10$$

where x is the production level of product A, y is the production level of product B, both measured in thousands of units, 10 is the fixed cost, and C is measured in thousands of dollars. To determine the minimum cost, the company uses the method of Lagrange multipliers. The constraining condition is that only 10,000 units can be produced, that is,

$$x + y = 10$$

The function (x,y,λ) is given by

$$F(x,y,\lambda) = x^2 + 2y + 10 - \lambda(x+y-10)$$

$$F_x = 2x - \lambda = 0$$

$$F_y = 2 - \lambda = 0$$

$$F_\lambda = x + y - 10 = 0$$

thus $x = 1$ and $y = 9$. Minimum cost results from producing 1000 units of product A and 9000 units of product B. The minimum cost is

$$C(1,9) = 29$$

or $29,000.

EXAMPLE 9-38

The revenue function R in the above problem is given by

$$R = R(x,y) = 8x - y^2 + 6y + 10$$

where x and y are the respective production levels for the company's two products, measured in thousands of units, and where R is measured in thousands of dollars. The profit function P is then given by

$$P = P(x,y) = R(x,y) - C(x,y)$$
$$= 8x - x^2 + 4y - y^2$$

the constraint is that 10,000 total units are to be produced, so that the constraining equation is

$$x + y = 10$$

To determine the production levels for each product which yield maximum profit, we form

$$F(x,y,\lambda) = 8x - x^2 + 4y - y^2 - \lambda(x+y-10)$$
$$F_x = 8 - 2x - \lambda = 0$$
$$F_y = 4 - 2y - \lambda = 0$$
$$F_\lambda = x + y - 10 = 0$$

Hence

$$8 - 2x = 4 - 2y \quad \text{and} \quad x + y = 10$$

which simplify to

$$x - y = 2 \quad \text{and} \quad x + y = 10$$

which imply that

$$x = 6, \ y = 4, \ \lambda = -4$$

Thus to maximize profit, the firm will produce 6000 units of product A and 4000 units of product B. Total profits are then

$$P(6,4) = 48 - 36 + 16 - 16$$
$$= 12$$

or $12,000.

It is not particularly difficult to picture the method of solving constrained extrema in three dimensions, that is, when the objective function $z = f(x,y)$ is a function of two variables and the constraining equation $g(x,y) = 0$ is an equation in x and y. The objective function $z = f(x,y)$ can be graphed in a three-dimensional coordinate system. The equation $g(x,y) = 0$, since it does not contain the variable z, is a generalized cylinder. That is, to obtain the graph of $g(x,y) = 0$, graph $g(x,y)$ first in the xy-plane, and then extend this curve vertically, parallel to the z-axis and hence perpendicular to the xy-plane. If $g(x,y) = 0$ is a line in the xy-plane, then $g(x,y) = 0$ if the plane goes through this line and is perpendicular to the xy-plane. If $g(x,y) = 0$ is a circle in the xy-plane, then $g(x,y) = 0$ is an infinite cylinder perpendicular to the xy-plane.

Since the solution of the constrained extrema problem is a value $f(a,b)$ such that $g(a,b) = 0$, one considers all points of $g(x,y) = 0$ which are also on $z = f(x,y)$. Hence we consider the intersection of $g(x,y) = 0$ with $z = f(x,y)$. This intersection is a curve in three-dimensional space and we seek an extrema of this curve. This is similar to the extrema problem in two-dimensional space.

The methods of solving constrained extrema problems by utilizing the notion of Lagrange multipliers can be applied to functions of more than two variables. In the following model we examine an objective function which is a function of three variables.

EXAMPLE 9-39

To construct a detached garage a construction company charges $3.50 per square foot for the floor, $2.50 per square foot for the ceiling, and $4.00 per square foot for the walls. A customer wants to know the dimensions of the largest garage which can be built for $1800. The function to be maximized is the volume V, given by

$$V = V(x,y,z) = xyz$$

subject to the constraint that the cost C is $1800. The cost C is the sum of the costs of the four walls and the floor and the ceiling, where we assume that the cost of the doors and windows are figured in the same way as regular parts of the walls. In practice, they usually cost a set sum and are figured into the total cost as such. Thus, if one door and two windows are desired, and their fixed cost is $100, the problem would be the same if the customer wanted to spend $1900.

If x is the height, and y and z the length and width measured in feet, then the cost of the four walls is $4(2xy + 2xz)$ dollars while the cost of the ceiling is $2.5yz$ dollars and the cost of the floor is $3.5yz$ dollars.

Hence the constraint is given by
$$C = 1800 = 8xy + 8xz + 6yz$$
or
$$8xy + 8xz + 6yz - 1800 = 0$$

To solve the problem by Lagrange multipliers, we form the function $F = F(x,y,z,\lambda)$

$$F = xyz - \lambda(8xy+8xz+6yz-1800)$$
$$F_x = yz - \lambda(8y+8z) = 0$$
$$F_y = xz - \lambda(8x+6z) = 0$$
$$F_z = xy - \lambda(8x+6y) = 0$$
$$F_\lambda = -(8xy+8xz+6yz-1800) = 0$$

Thus, solving for λ in each equation, we have

$$\lambda = \frac{yz}{8y+8z} = \frac{xz}{8x+6z} = \frac{xy}{8x+6y}$$

By cross multiplying we get

$$yz(8x+6z) = xz(8y+8z)$$
$$xz(8x+6y) = xy(8x+6z)$$

Simplifying these equations yields

$$6yz^2 = 8xz^2$$
$$8x^2z = 8x^2y$$

and thus

$$x = \tfrac{3}{4}y, \; y = z$$

Substituting these values into $F_\lambda = 0$ yields

$$8(\tfrac{3}{4})y^2 + 8(\tfrac{3}{4})y^2 + 6y^2 = 1800$$
$$18y^2 = 1800$$
$$y^2 = 100$$
$$y = 10$$

The solution $y = -10$ is not applicable. Thus

$$x = (\tfrac{3}{4})10 = 7.5$$

the dimensions are $10 \times 10 \times 7.5$.

Lagrange multipliers have a particularly important interpretation in extrema problems. For problems in economics in which the objective function f has the dimensions of a monetary value, e.g., profits, revenue, or cost, and the constraining equation reflects a specific boundary value for the independent variables, then the Lagrange multiplier λ measures the sensitivity of the objective function to changes in the independent variables and hence represents a price, called a shadow price. Thus the Lagrange multiplier affords the economist another important predictor.

If $z = f(x,y)$ is the objective function subject to the constraint $g(x,y) = 0$, then to find an extrema by the method of Lagrange multipliers we form the function $F(x,y,\lambda) = f(x,y) - \lambda g(x,y)$. If $f(a,b)$ is the constrained optimum and $\lambda = c$ is the corresponding value of the Lagrange multiplier, the sign of the real number c provides an indication of the effect of a small change in x or y on the constrained optimum $f(a,b)$. If λ is negative, the constrained optimum will increase if the constant in the constraint is decreased, whereas the constrained optimum will decrease if the constant in the constraint is increased. If λ is positive, the constrained optimum will increase if the constant in the constraint is increased, whereas the constrained optimum will decrease if the constant in the constraint is decreased.

This interpretation of the Lagrange multiplier is examined in the following example.

EXAMPLE 9-40

Consider the profit function P encountered in Example 9-38 which was given by

$$P = P(x,y) = 8x - x^2 + 4y - y^2$$

where x and y are the respective production levels for the company's two products, measured in thousands of units, and where P is measured in thousands of dollars. The constraint is that 10,000 total units are to be produced so that the constraining equation is

$$x + y = 10$$

In Example 9-38 it was determined that the constrained optimum value is $P(6,4) = 12$. Hence the company will make a maximum profit of \$12,000 when it produces $x = 6$ or 6000 units of product A and $y = 4$ or 4000 units of product B. The corresponding value of the Lagrange multiplier is $\lambda = -4$. Since λ is negative, it is expected that if the constant in the constraint is increased, say from 10 to 11, then the constrained optimum will decrease. If the constant is decreased, say from 10 to 9, then the constrained optimum will increase.

Thus if we solve the constrained extrema problem

$$P(x,y) = 8x - x^2 + 4y - y^2$$

subject to the constraint

$$x + y = 11$$

we find that $x = 6.5$ and $y = 4.5$ yield the constrained maximum value of

$$P(6.5, 4.5) = 7.5$$

If we solve the constrained extrema problem

$$P(x,y) = 8x - x^2 + 4y - y^2$$

subject to the constraint

$$x + y = 9$$

we find that $x = 5.5$ and $y = 3.5$ yield a constrained maximum value of

$$P(5.5, 3.5) = 15.5$$

It should be pointed out that this application of the interpretation of the Lagrange multiplier as an indicator is valid only for small changes in the independent variables. In our demonstration above, we chose the fairly large increment of one-tenth, i.e., a change from 10 to 11 and from 10 to 9 so that the increment was $(11-10)/10$, only to illustrate the previous discussion. The values obtained in the discussion are not absolute. Thus it is not meant that if the firm decided to produce 9000 total units then the profit expected is $15,500. In practice, the constraint is usually determined and the objective function is then generated. For a different constraint, a possibly different objective function would be obtained. The effect of the Lagrange multiplier is to indicate that the firm should at least investigate the possibility of reducing the total output, say from $x+y = 10$ to $x+y = 9$. Given that $x+y = 9$, a new profit function would be generated from available data.

EXERCISE 9-6

1. Use the method of Lagrange multipliers to find the extrema of the functions f subject to given constraints g.
   (a) $f(x,y) = x^2 + y^2$, $g(x,y) = x + y - 1 = 0$
   (b) $f(x,y) = x^2 + 2x + y^2 - y$, $g(x,y) = x + 2y = 0$
   (c) $f(x,y) = x^3 + 3x^2 + y^2 - 2y$, $g(x,y) = 3x + 2y + 3 = 0$
   (d) $f(x,y,z) = x^2 + y^2 + z^2$, $g(x,y,z) = 3x - y + z = 0$
   (e) $f(x,y,z) = x^2 + 3y^2 + 5z^2$, $g(x,y,z) = 3x - 4y - z = 0$
   (f) $f(x,y,z) = x^2yz$, $g(x,y,z) = x^2 + y^2 + z^2 - 16 = 0$
2. Find the minimum distance from the line $3x+y-5 = 0$ to the origin $(0,0)$.
3. Find the minimum distance from the line $3x-4y+1 = 0$ to the point $(1,2)$.

4. Find the minimum distance from the plane $3x - y + 2z - 5 = 0$ to the origin $(0,0,0)$.

5. Find the minimum distance from the plane $6x - y - 2z + 3 = 0$ to the point $(1, 0, -1)$.

6. Use the method of Lagrange multipliers to find the dimensions of the box with maximum volume if the box is to be made from 216 sq in. of material.

7. Suppose a company produces three products with profit margins of $3, $4, and $6. If x, y, and z are the amounts of each item produced daily and if production is subject to the condition $x^2 + 2y^2 + 3z^2 = 100$, find the values of x, y, and z such that the total profit is maximized.

8. If P is the profit function and L and S are the sales volume of the large size and small size of a product, calculate those values of L and S which maximize P.
   (a) $P(L,S) = 2L + 5S + 2LS - L^2 - S^2 - 10$
   (b) $P(L,S) = 6L + 2S + LS - L^2 - 3S^2 + 6$
   (c) $P(L,S) = 10L + 3S + 2LS - 2L^2 - S^2 - 15$
   (d) $P(L,S) = 3L - 5S + 3LS - L^2 - 6S^2$

9. Daily production capacity P of a firm is a function of the number of man hours x and the amount of capital invested y. Assuming that there are 260 working days in a year and that the company allocates C dollars in capital for wages and machinery, find the values of x and y which maximize P.
   (a) $P(x,y) = x^{1/2}y^{1/2}$, $C = 100{,}000$
   (b) $P(x,y) = x^{1/2}y^{1/2}$, $C = 800{,}000$
   (c) $P(x,y) = x^{1/3}y^{1/3}$, $C = 100{,}000$
   (d) $P(x,y) = x^{1/3}y^{1/3}$, $C = 800{,}000$

10. If P is the profit function and x and y are the respective production levels for the company's two products, determine the values of x and y which maximize P subject to the given constraint.
    (a) $P(x,y) = 8x - x^2 + 2y - y^2$, $x + y = 10$
    (b) $P(x,y) = 4x - x^2 + 8y - y^2$, $x + y = 5$
    (c) $P(x,y) = 6x - x^2 + 10y - 3y^2$, $x + y = 20$
    (d) $P(x,y) = 2x - x^2 + 16y - 5y^2$, $x + y = 15$

9-7 DOUBLE INTEGRALS

Consider the problem of finding the volume of the solid under the surface $z = f(x,y)$ bounded by the area A in the xy-plane sketched in Fig. 9-20.

In Chap. 5 we saw that the integral of a function $y = f(x)$ from $x = a$ to $x = b$ can be used to solve the problem of finding the area under $y = f(x)$ bounded by the lines $x = a$ and $x = b$. Since finding a volume under a surface is a natural extension of finding the area under a curve, you might

The volume of the solid under the surface $z = f(x,y)$
bounded by the area A

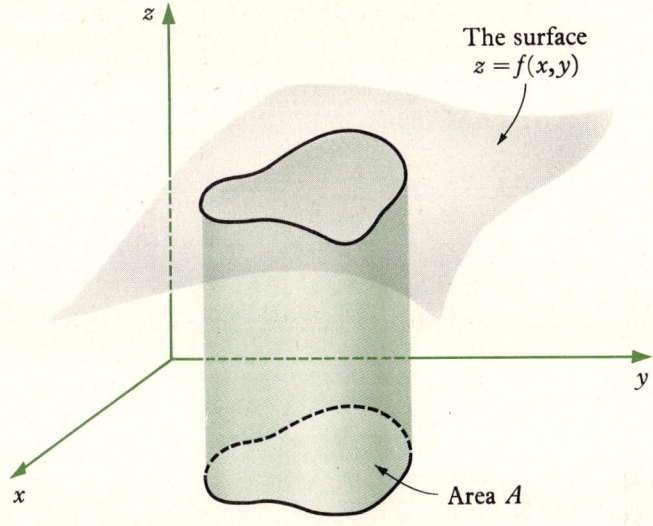

FIGURE 9-20

expect that its solution involves integration. In fact, we must "integrate" twice, so this process is called the *double integral*.

The rigorous definition of the double integral is beyond the scope of this text, but we present an argument designed to convince you that to find the volume under a surface, we must perform integration twice.

Consider the surface $z = f(x,y)$ that lies above the area A in the xy-plane. Suppose A is bounded by the two functions $y = f_1(x)$ and $y = f_2(x)$ as given in Fig. 9-21.

To find an expression for the volume under $z = f(x,y)$ bounded by A, we first section the surface, and hence the solid, by a plane $x = c$ for some constant c. (See Fig. 9-21.) The intersection of the solid with the plane $x = c$ yields an area in the plane $x = c$. This area is the area under a two-dimensional curve (in the plane $x = c$), namely, $z = f(c,y)$, which is a function of the variable y and bounded by the two-dimensional lines (in the plane $x = c$) $z = f_1(c)$ and $z = f_2(c)$. This area is of course

$$\int_{f_1(c)}^{f_2(c)} f(c,y)\, dy$$

If $x = a$ and $x = b$ are the lines that bound the area A, then we can do this procedure for every value c such that $c \in [a, b]$. The volume V of the solid under $z = f(x,y)$ bounded by A is then the "sum" of all these areas. This

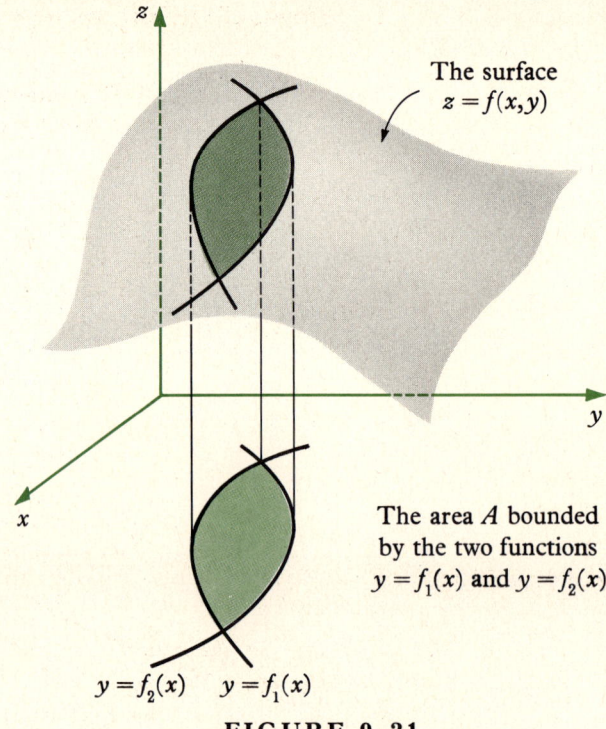

FIGURE 9-21

"sum" leads us to another integration. The function we are to integrate is the "area" function

$$A(x) = \int_{f_1(x)}^{f_2(x)} f(x,y)\, dy$$

That is, for each value of $x \in [a, b]$, we get the cross-sectional area described above. Then the volume V is

$$\int_a^b A(x)\, dx = \int_a^b \left[\int_{f_1(x)}^{f_2(x)} f(x,y)\, dy \right] dx$$

This expression is called the *double integral* of $z = f(x,y)$ over the area A. Note that the limits of integration of the inner integral are *functions* of the one variable x. Moreover, the inner integral was defined only for x constant; therefore one holds x constant while treating $f(x,y)$ as a function of y alone. Then one finds the antiderivative of this function. One then evaluates this antiderivative at $f_2(x)$ and subtracts the evaluation at $f_1(x)$ resulting in a function of the one variable x. This remaining integral can then be evaluated.

It is customary to delete the brackets in the expression for the

The two-dimensional area under $z = f(c,y)$

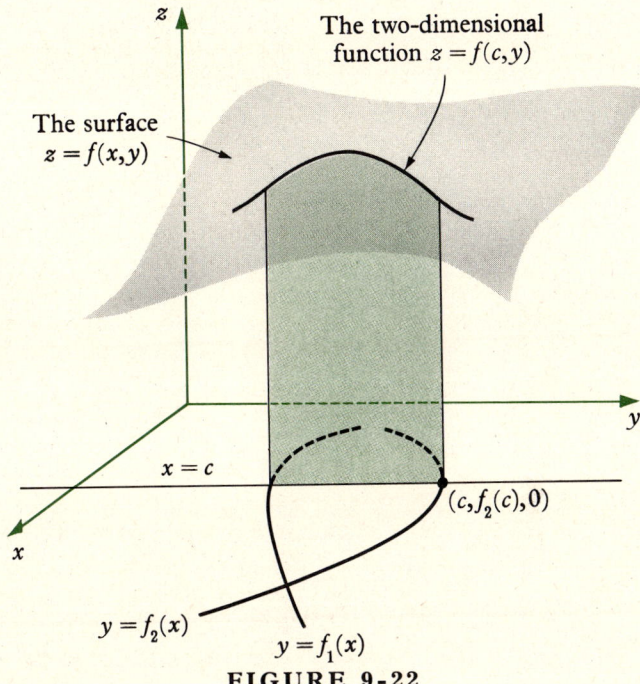

FIGURE 9-22

double integral. We therefore write

$$\int_a^b \left[\int_{f_1(x)}^{f_2(x)} f(x,y) \, dy \right] dx = \int_a^b \int_{f_1(x)}^{f_2(x)} f(x,y) \, dy \, dx$$

Let us illustrate the technique of evaluating the double integral with a few examples.

EXAMPLE 9-41

Consider the double integral

$$\int_1^2 \int_1^x (x+2y+2) \, dy \, dx = \int_1^2 (xy+y^2+2y) \Big|_1^x dx$$
$$= \int_1^2 [x^2 + x^2 + 2x - (x+1+2)] \, dx$$
$$= \int_1^2 (2x^2 + x - 3) \, dx$$
$$= \left(\frac{2x^3}{3} + \frac{x^2}{2} - 3x \right) \Big|_1^2$$
$$= \frac{19}{6}$$

EXAMPLE 9-42

Consider the double integral

$$\int_0^2 \int_0^{x^2} (2xy + e^{x^3}) \, dy \, dx = \int_0^2 (xy^2 + ye^{x^3}) \Big|_0^{x^2} dx$$

$$= \int_0^2 (x^5 + x^2 e^{x^3}) \, dx$$

$$= \frac{x^6}{6} + \frac{e^{x^3}}{3} \Big|_0^2$$

$$= \frac{64}{6} + \frac{e^8}{3} - \frac{1}{3}$$

$$= \frac{31 + e^8}{3}$$

The key to solving volume problems is selecting the proper limits of integration, which are determined from the area A. The functions $y = f_1(x)$ and $y = f_2(x)$ which bound A must first be determined and then the outer limits are the numbers a and b such that the lines $x = a$ and $x = b$ bound A.

EXAMPLE 9-43

Find the volume in the first octant under the function $z = xy$ bounded by the region A in the xy-plane enclosed by $y = x^2$ and $y = x$. Consider the sketch in Fig. 9-23.

The integral is

$$\int_0^1 \int_{x^2}^{x} xy \, dy \, dx = \int_0^1 \frac{1}{2} xy^2 \Big|_{x^2}^{x} dx$$

$$= \int_0^1 \frac{x}{2} (x^2 - x^4) \, dx$$

$$= \frac{1}{2} \int_0^1 (x^3 - x^5) \, dx$$

$$= \frac{1}{2} \left(\frac{x^4}{4} - \frac{x^6}{6} \right) \Big|_0^1$$

$$= \frac{1}{24}$$

SEC. 9-7 Doulbe Integrals

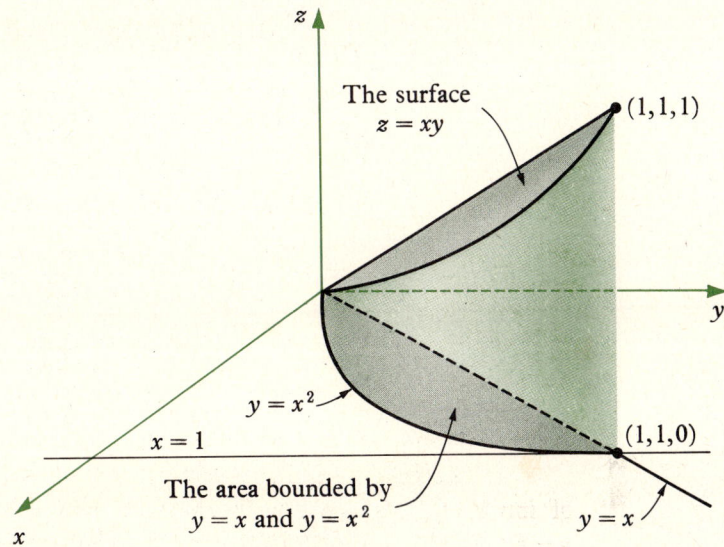

FIGURE 9-23

The volume in the first octant under z = xy bounded by the area enclosed by y = x and y = x²

This volume could also be found by first integrating with respect to x and then y. The limits of integration must also be changed. Thus the volume is

$$\int_0^1 \int_y^{\sqrt{y}} xy \, dx \, dy = \int_0^1 \frac{1}{2} x^2 y \Big|_y^{\sqrt{y}} dy$$

$$= \int_0^1 \frac{y}{2}(y - y^2) \, dy$$

$$= \left(\frac{y^3}{6} - \frac{y^4}{8} \right) \Big|_0^1$$

$$= \frac{1}{24}$$

EXAMPLE 9-44

Find the volume in the first octant under the plane $x + 2y + z - 10 = 0$. The solid is a terahedron formed by the given plane and the three co-ordinate axes. The area A is the triangle formed by the x-axis, y-axis, and the line $x + 2y - 10 = 0$, which is the intersection of the given plane and the xy-plane. Hence $f_1(x) = 0$ and $f_2(x) = \frac{1}{2}(10 - x)$, which are the inner limits of integration. The outer bounds for A are $x = 0$ and $x = 10$. Thus

301

The tetrahedron formed by the plane
$x + 2y + z - 10 = 0$ *and the coordinate axes*

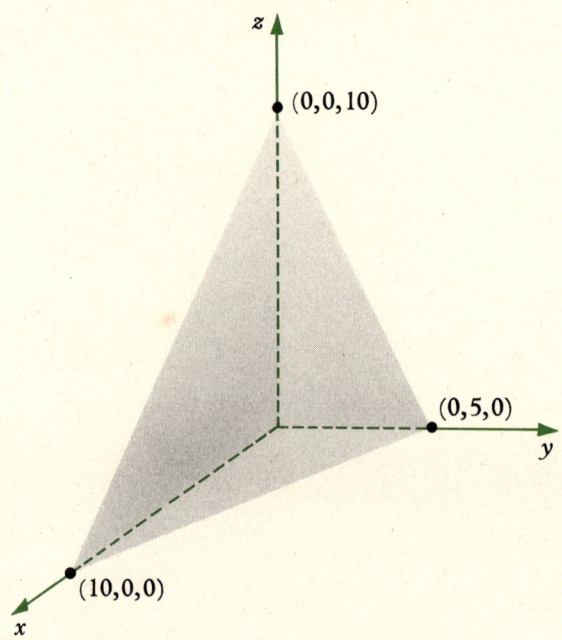

FIGURE 9-24

the volume is given by

$$\int_0^{10} \int_0^{\frac{1}{2}(10-x)} (10-x-2y)\,dy\,dx = \int_0^{10} (10y - xy - y^2) \Big|_0^{\frac{1}{2}(10-x)} dx$$

$$= \int_0^{10} \left[5(10-x) - \frac{x}{2}(10-x) - \frac{(10-x)^2}{4} \right] dx$$

$$= \int_0^{10} \left[25 - 5x + \frac{x^2}{4} \right] dx$$

$$= \left[25x - \frac{5x^2}{2} + \frac{x^3}{12} \right]_0^{10}$$

$$= \frac{250}{3}$$

EXAMPLE 9-45

Find the volume of the solid bounded by the coordinate planes and the surface $z = 4 - 9x^2 - 2y$. To determine A, we must find the intersection

The intersection of the surface $z = 4 - 9x^2 - 2y$
and the xy-plane; $z = 0$ is the parabola
$4 - 9x^2 - 2y = 0$

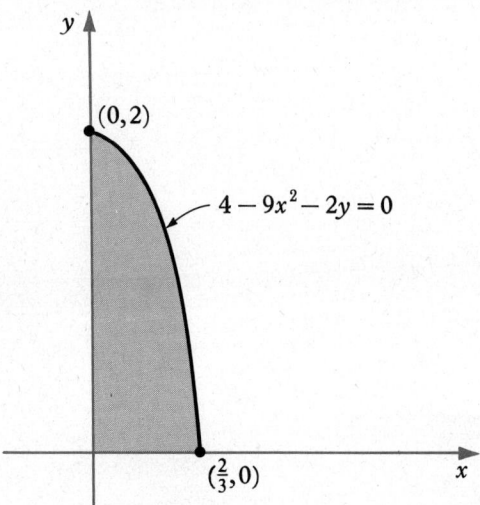

FIGURE 9-25

of the surface with the xy-plane, $z = 0$. This intersection is $0 = 4 - 9x^2 - 2y$, whose graph in the first quadrant in the xy-plane is given in Fig. 9-25.

The integral is given by

$$\int_0^{2/3} \int_0^{2-(9/2)x^2} (4 - 9x^2 - 2y) \, dy \, dx = \int_0^{2/3} (4y - 9x^2 y - y^2) \Big|_0^{2-(9/2)x^2} dx$$

$$= \int_0^{2/3} y(4 - 9x^2 - y) \Big|_0^{2-(9/2)x^2} dx$$

$$= \int_0^{2/3} \left(2 - \frac{9x^2}{2}\right)\left(-\frac{9x^2}{2} + 2\right) dx$$

$$= \int_0^{2/3} \left(\frac{81}{4} x^4 - 18x^2 + 4\right) dx$$

$$= \left(\frac{81}{20} x^5 - 6x^3 + 4x\right) \Big|_0^{2/3}$$

$$= \frac{64}{45}$$

Hence the volume is $\frac{64}{45}$.

EXERCISE 9-7

1. Evaluate the given double integrals.

   (a) $\int_0^1 \int_2^3 (x+y-2)\, dy\, dx$

   (b) $\int_0^1 \int_1^x (x^2+xy)\, dy\, dx$

   (c) $\int_0^2 \int_x^{x^2} (x+y)\, dy\, dx$

   (d) $\int_0^2 \int_1^x (xy-x^2)\, dy\, dx$

   (e) $\int_0^1 \int_0^{1-x^2} y\, dy\, dx$

   (f) $\int_{-1}^1 \int_x^{x^2} (x^2+2xy-3y^2)\, dy\, dx$

   (g) $\int_4^9 \int_0^x \sqrt{x-y}\, dy\, dx$

   (h) $\int_0^1 \int_{x^3}^{x^2} (x^2-xy)\, dy\, dx$

   (i) $\int_0^1 \int_0^x e^{x^2}\, dy\, dx$

   (j) $\int_0^1 \int_0^{3y} \sqrt{x+y}\, dx\, dy$

2. Evaluate the double integrals of the given functions $z = f(x,y)$ over the given regions A.
   (a) $z = x^2+y^2$, A is bounded by $x = 0$, $y = 0$, $x+y = 1$
   (b) $z = xy$, A is bounded by $x = 0$, $y = 0$, $x+2y = 3$
   (c) $z = x^2+2y^2$, A is bounded by $x = 0$, $y = 1$, $y = x^2$
   (d) $z = x^2+2y^2$, A is bounded by $x = 1$, $y = 0$, $y = x^2$

3. Find the volume under the surface $z = xy$ bounded by $y = x^2$ and $y^2 = x$ and the plane $z = 0$.

4. Find the volume under the surface $z = xy^2$ bounded by $y = x$ and $y = x^2$ and the plane $z = 0$.

5. Find the volume in the first octant under the plane $x+y+z-10 = 0$.

6. Find the volume in the first octant under the plane $2x+4y+z+3 = 0$.

7. Find the volume in the first octant under the surface $z = 1-x-y$.

8. Find the volume of the solid bounded by $z = 0$ and $z = 4-x-y$.

9. Find the volume of the solid bounded by $z = 0$ and $z = 10-x-2y$ in the first octant.

10. Find the volume of the solid bounded by $z = x+y$ and $z = 8-x-y$.

9-8 METHOD OF LEAST SQUARES

The applications considered thus far have often assumed that a given function reflects a real-world situation. For example, the marginal analysis model (Ex. 3-6 in Sec. 3-1) assumed that the cost $C(x)$ of making x sets of a hundred units per month was given by $C(x) = 20+0.5x-0.1x^2$. In practice, this function $C(x)$ would often be determined by the manufacturing company

Line 1 *is a good fit while Line* 2 *is a bad fit*

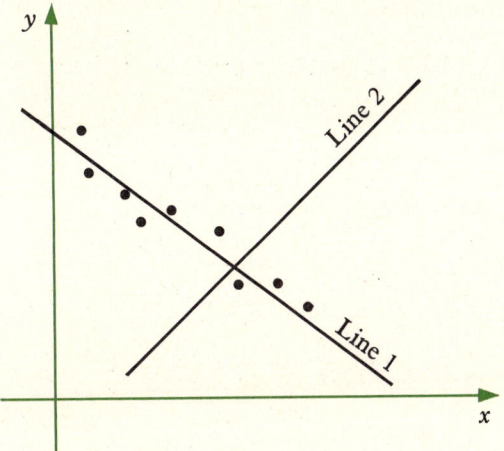

FIGURE 9-26

from previous data compiled from the recent past. Certain values of x and their corresponding cost values would be selected and a curve or function would be chosen which best reflects this data.

In this section, we describe methods of selecting a linear function to best fit a given set of data or points. The problem entails selecting a linear function $f(x) = mx + b$ which passes through points close to a given set of points. What is needed is a definition of what it means to "fit" the data or to "come close" to the given points. Intuitively, one sees that in Fig. 9-26, line 1 is a good fit but line 2 does not fit the data very well.

EXAMPLE 9-46

> Consider the points (0,5), (1,3), (2,4), (3,3) and (4,0). The line $y = -x + 5$ is a good fit but the line $y = x$ is a bad fit. Note that the reason why $y = x$ is a bad fit is that some of the points are quite far away from the line. Even though (3,3) is on $y = x$, (0,5) is five units above the line and (4,0) is four units below the line. However, none of the points are more than one unit away from the line $y = -x + 5$.

Let $(x_1, y_1), (x_2, y_2), \ldots, (x_n, y_n)$ be n points in the plane. The problem is to determine a linear function $f(x) = mx + b$ which is a good fit to these n data points. Note that the x_i and y_i are given numbers, so the task at hand is to determine m and b which are therefore the variables in our discussion.

From Ex. 9-38, it is clear that we must try to minimize the distances from the given points to the line. Here we are speaking of vertical distances. The distance from the point (x_i, y_i) to the point $(x_i, f(x_i))$ is simply

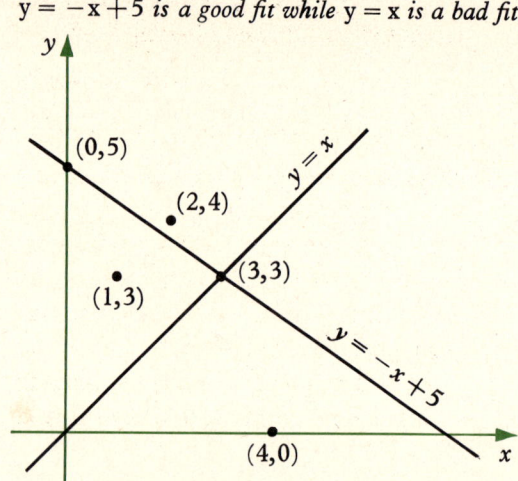

$y = -x + 5$ *is a good fit while* $y = x$ *is a bad fit*

FIGURE 9-27

$|y_i - f(x_i)| = |y_i - mx_i - b|$ since $f(x_i) = mx_i + b$. Hence it would seem reasonable to minimize the sum of these n distances. However, it turns out that there is a slight alteration we can make to this method which yields far superior results.

The method of least squares entails minimizing the sum of the *squares* of the distances, that is, minimizing the quantity

$$(y_1 - f(x_1))^2 + (y_2 - f(x_2))^2 + \cdots + (y_n - f(x_n))^2$$

Note that by squaring the distances, any distance greater than one dramatically affects the sum.

We can now utilize the machinery developed in this chapter by first noting that the above expression is a function of the two variables m and b since the x_i and y_i are given numbers. We can thus define the function

$$\begin{aligned}G(m,b) &= (y_1 - f(x_1))^2 + (y_2 - f(x_2))^2 + \cdots + (y_n - f(x_n))^2 \\ &= (y_1 - mx_1 - b)^2 + (y_2 - mx_2 - b)^2 + \cdots + (y_n - mx_n - b)^2\end{aligned}$$

The problem now becomes one of minimizing the function $G(m,b)$. Thus we find the partial derivatives and set them equal to zero.

$$\begin{aligned}\frac{\partial G}{\partial m} &= 2(y_1 - mx_1 - b)(-x_1) + 2(y_2 - mx_2 - b)(-x_2) + \cdots \\ &\quad + 2(y_n - mx_n - b)(-x_n) = 0\end{aligned}$$

$$\begin{aligned}\frac{\partial G}{\partial b} &= 2(y_1 - mx_1 - b)(-1) + 2(y_2 - mx_2 - b)(-1) + \cdots \\ &\quad + 2(y_n - mx_n - b)(-1) = 0\end{aligned}$$

By multiplying each equation by $-1/2$ and collecting terms, we can rewrite the equations in the following form:

$$(x_1+x_2+\cdots+x_n)m + nb = y_1+y_2+\cdots+y_n$$

$$(x_1^2+x_2^2+\cdots+x_n^2)m + (x_1+x_2+\cdots+x_n)b = x_1y_1+x_2y_2+\cdots+x_ny_n$$

Since the x_i and y_i are all given, the coefficients of m and b are real numbers, as are the expressions on the right-hand side of each equation. Thus, to solve for m and b we must simply solve a system of two equations in two unknowns. We can readily show that the system will have a unique solution unless all the x_i are equal. In the latter case, the line of best fit is certainly the vertical line through the given points.

The line $f(x) = mx + b$ so constructed is often referred to as the "line of regression", especially in the study of statistics.

EXAMPLE 9-47

Consider again the points (0,5), (1,3), (2,4), (3,3), and (4,0). To find the line of best fit, or the line of regression, by the method of least squares we must find the quantities

$$\sum_{i=1}^{5} x_i = x_1 + x_2 + x_3 + x_4 + x_5$$

$$\sum_{i=1}^{5} y_i = y_1 + y_2 + y_3 + y_4 + y_5$$

$$\sum_{i=1}^{5} x_i^2 = x_1^2 + x_2^2 + x_3^2 + x_4^2 + x_5^2$$

$$\sum_{i=1}^{5} x_i y_i = x_1 y_1 + x_2 y_2 + x_3 y_3 + x_4 y_4 + x_5 y_5$$

The following table contains the necessary information.

i	1	2	3	4	5	Σ
x_i	0	1	2	3	4	10
y_i	5	3	4	3	0	15
x_i^2	0	1	4	9	16	30
$x_i y_i$	0	3	8	9	0	20

Hence our equations are

$$10m + 5b = 15$$

$$30m + 10b = 20$$

Dividing the first equation by 5 and the second by 10 yields

$$2m + b = 3$$
$$3m + b = 2$$

Subtracting the first equation from the second yields

$$m = -1$$

and hence

$$b = 5$$

Thus the line of best fit is $f(x) = -x + 5$.

EXAMPLE 9-48

A company determines the following data from past experience relating the cost $c(x)$ to make x units per month where $c(x)$ is measured in thousands of dollars and x in quantities of one hundred.

x	0	1	2	3	4
$c(x)$	7.5	10	10.25	11	11.25

Hence the fixed cost, the cost to make no units, is $c(0) = 7.5$ thousand dollars while it costs $c(4) = 11.25$ thousand dollars to make $x = 4$ hundred units. For cost analysis purposes, the firm wants to determine a function $y = C(x)$ for all values of x. For simplicity, it is assumed that $C(x)$ is a linear function and hence it is of the form $C(x) = mx + b$ for constants m and b. The requisite information is found in the following table

i	1	2	3	4	5	Σ
x_i	0	1	2	3	4	10
c_i	7.5	10	10.25	11	11.25	50
x_i^2	0	1	4	9	16	30
$x_i c_i$	0	10	20.5	33	45	108.5

Hence our equations, with $n = 5$, are given by

$$10m + 5b = 50$$
$$30m + 10b = 108.5$$

Multiplying the first equation by 3 and the second by -1 yields

$$30m + 15b = 150$$
$$-30m - 10b = -108.5$$

Adding the equations gives us

$$5b = 41.5$$

$$b = 8.3$$

Substituting this value for b into the first equation yields

$$10m + 41.5 = 50$$

$$m = 0.85$$

Therefore the line of best fit is $C(x) = 0.85x + 8.3$.

There are many uses to which the firm can put this information. Note that the linear model $C(x)$ determines that the firm's fixed cost is $C(0) = 8.3$ thousand dollars per month whereas the empirically computed data reflected a fixed cost of $c(0) = 7.5$ thousand dollars. This might be an indication that the method of computing fixed cost from empirical data is incorrect or it may be reasoned that initial variable costs, the cost of making only a few units, is high, which would cause the model to be skewed as x approaches zero. In any case, this difficulty must be explained somehow.

Another point worthy of note is the discrepancy between the empirical data and the mathematical model at $x = 3$, that is, $c(3) = 11$ whereas $C(3) = 0.85 \cdot 3 + 8.3 = 10.85$. Thus the predicted cost is lower than the actual cost to produce 300 items. This might indicate that the firm is not operating efficiently when it is producing 300 items per month.

EXERCISE 9-8

1. Use the method of least squares to find the line of regression of the points $(0,-1)$, $(1,0)$, $(2,1)$, $(3,3)$, $(4,2)$.
2. Use the method of least squares to find the line of regression of the points $(0,1)$, $(1,3)$, $(2,5)$, $(3,4)$.
3. Use the method of least squares to find the line of regression of the points $(-2,2)$, $(-1,0)$, $(0,2)$, $(1,1)$, $(3,1)$.
4. A manufacturing firm determines that the data in the following table reflects the cost $c(x)$ of producing x units per month, where $c(x)$ is measured in thousands of dollars and x in hundreds of units. Use the method of least squares to determine the regression line of the data points.

x	10	11	12	13
$c(x)$	2	3	3	6

5. A biological researcher determines that correlation between the number of grams of vitamin C taken per day and the percentage of the population

taking that amount of vitamin C who contacted a cold in December is given by the following data. Use the method of least squares to determine the line of regression of the data points.

Number of grams x of vitamin C taken per day (hundreds of grams)	1	2	3	4	5
Percentage p of the population which contacted a cold in December	21	20	18	16	15

10

Trigonometric Functions

10-1 ANGLES

The study of circles, triangles, and angles dates back to the dawn of civilization. The trigonometric functions, which are defined in terms of these concepts, allow us to create models in such diverse disciplines as astronomy, ecology, and music.

The study of trigonometry must commence with the notion of an *angle*. An angle is defined by two intersecting line segments. One segment is called the *initial side* and the other is called the *terminal side*. The point of intersection of the line segments is called the *vertex*. In order to measure the angle, the angle is pictured as having been generated by rotating the terminal side from the initial side to its present position.

If the terminal side was rotated in a counterclockwise position, the angle is said to be positive, otherwise negative.

If the vertex is at the origin of a rectangular coordinate system and the initial side lies on the positively directed x-axis, then the angle is said to be in *standard position*.

The standard unit used to measure an angle θ is the *degree*, which is defined as $\frac{1}{360}$ of a circle. Thus one degree, written 1°, is the measure of an angle generated by a line segment rotating $\frac{1}{360}$ of a complete rotation. One complete rotation is 360°. One half of a complete rotation is 180°. A *right angle*, an angle of 90°, is therefore one quarter of a complete rotation.

One minute, written 1′, is $\frac{1}{60}$ of a degree and one second, written 1″, is $\frac{1}{60}$ of a minute.

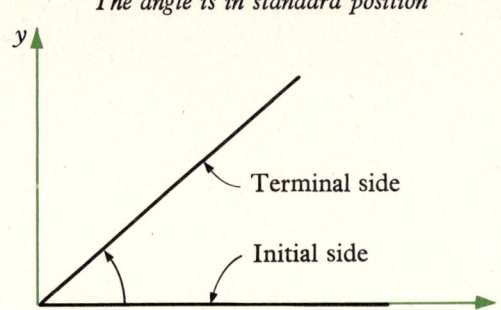

The angle is in standard position

FIGURE 10-1

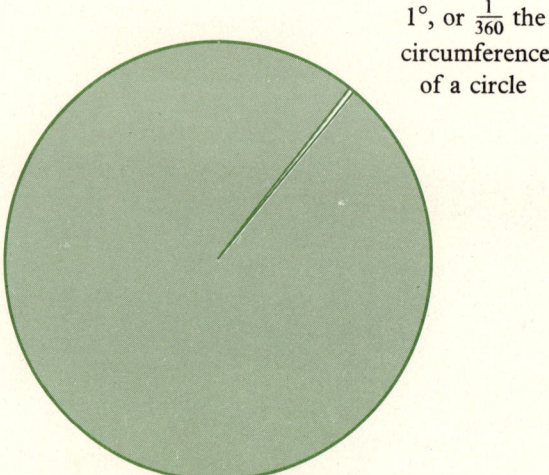

$1°$, or $\frac{1}{360}$ the circumference of a circle

FIGURE 10-2

While the degree system of measure is perhaps the most common system of angular measurement, it has serious drawbacks in many branches of science. The *radian* measure system is of essential importance in calculus.

Definition Place the vertex of the angle to be measured at the center of a circle. Let r be the radius of the circle and let s be the length of the arc of the circle subtended by the sides of the angle. Then the radian measure x of the angle is defined by $x = s/r$. See Fig. 10-3.

The definition is independent of the size of the circle selected because, while the radius and the subtended arc of a larger circle would both be larger, the ratio would remain the same.

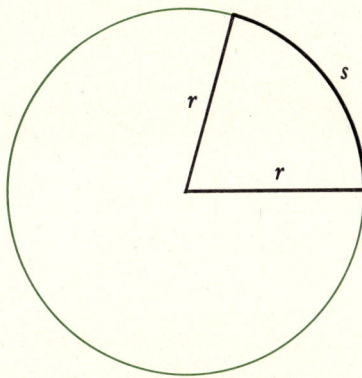

An angle of radian measure $\frac{s}{r}$

FIGURE 10-3

The correspondence between the degree system and the radian system can be described by the fact that a straight angle is 180° or one half of a complete rotation in the degree system while the arc subtended by a straight angle is one half the circumference of a circle, that is, $\frac{1}{2}(2\pi r)$, so that a straight angle is π radians. (Recall that the circumference of a circle is $2\pi r$ units, where r is the radius of the circle.) Thus the radian measure of a complete rotation, an angle of 360°, is $2\pi r/r = 2\pi$ radians. Therefore, in measuring angles

$$2\pi \text{ radians} = 360°$$

$$\pi \text{ radians} = 180°$$

$$\pi/2 \text{ radians} = 90°$$

$$\pi \cdot x \text{ radians} = (180 \cdot x)°$$

Thus, to convert from degree measure to radian measure we use the formula

$$x° = \frac{\pi x}{180} \text{ radians}$$

and to convert from radian measure to degree measure we use the formula

$$x \text{ radians} = \left(\frac{180x}{\pi}\right)°$$

The following table gives the correspondence between degree and radian measure of some important angles.

Angles measured in	
degrees	radians
0	0
30	$\pi/6$
45	$\pi/4$
60	$\pi/3$
90	$\pi/2$
135	$3\pi/4$
180	π
270	$3\pi/2$
360	2π

TABLE 10-1

EXAMPLE 10-1

One radian is $(180 \cdot 1/\pi)° = (180/\pi)° \cong 57°17'45''$, that is, 57 degrees, 17 minutes, and 45 seconds.

EXAMPLE 10-2

To convert 450° to radians we write

$$450° = \frac{\pi 450}{180} = \frac{5\pi}{2} \text{ radians}$$

EXAMPLE 10-3

To convert $5\pi/12$ radians to degrees we write

$$\frac{5\pi}{12} \text{ radians} = \left(\frac{180 \cdot 5\pi/12}{\pi}\right)° = \left(\frac{150}{2}\right)° = 75°$$

EXERCISE 10-1

1. Convert to radian measure.
   - (a) 20°
   - (b) 40°
   - (c) 135°
   - (d) 210°
   - (e) 300°
   - (f) 400°
   - (g) 155°
   - (h) 800°

2. Convert to radian measure.
   - (a) −12°
   - (b) 52°30′
   - (c) 57°17′45″
   - (d) 1′

3. Convert to degree measure.
   - (a) 3π
   - (b) $\pi/180$
   - (c) $\pi/18$
   - (d) $\pi/5$
   - (e) $2\pi/5$
   - (f) $3\pi/4$
   - (g) $7\pi/6$
   - (h) $7\pi/12$

4. Convert to degree measure.
   (a) $-\pi$
   (b) $-5\pi/6$
   (c) 2
   (d) 2.5

5. The area of a circle is 100 ft². Find the length of its arc subtended by an angle of 45° and the area of the corresponding sector.

6. The area of a circle is 400 cm². Find the length of its arc subtended by an angle of $\pi/12$ radians and the area of the corresponding sector.

10-2 DEFINITION OF THE TRIGONOMETRIC FUNCTIONS

The classical definition of the functions sine θ and cosine θ, which we abbreviate $\sin \theta$ and $\cos \theta$, can be given in a direct fashion by assuming that the angle θ is drawn at the center of a unit circle, that is, a circle with radius one, as in Fig. 10-2. Such an angle is said to be in *standard position*.

Using the definition generally attributed to the ancient Babylonians, we consider the right triangle pictured in Fig. 10-4 and can define $\sin \theta$ and $\cos \theta$ by the following,

$$\sin \theta = \frac{\text{length of side opposite angle } \theta}{\text{length of hypotenuse}}$$

$$\cos \theta = \frac{\text{length of side adjacent to angle } \theta}{\text{length of hypotenuse}}$$

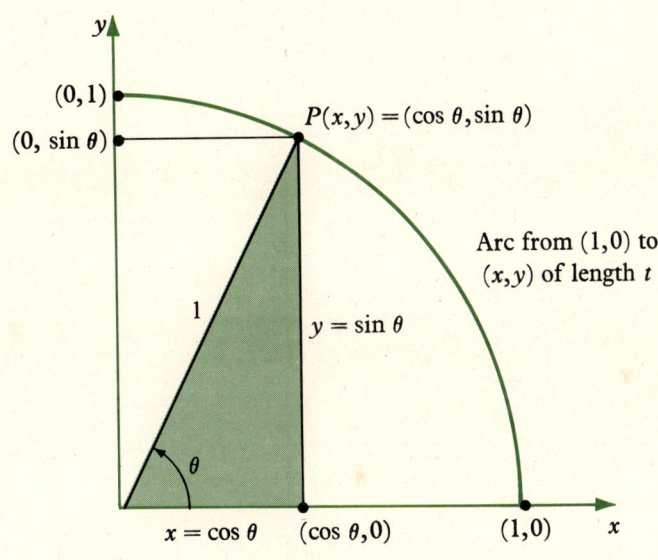

FIGURE 10-4

Note that the length of the hypotenuse of the triangle is one unit, since the hypotenuse is a radius of the unit circle. If (x,y) is the point P where the hypotenuse intersects the circle, then we can formulate these definitions in the following way:

$$\sin \theta = \frac{y \text{ units}}{1 \text{ unit}} = y$$

$$\cos \theta = \frac{x \text{ units}}{1 \text{ unit}} = x$$

A major obstacle in defining $\sin \theta$ and $\cos \theta$ in terms of lengths of sides of a triangle is that these functions are then only defined for angles between 0 and $\pi/2$ radians. However, if we realize that the point P has coordinates $(\cos \theta, \sin \theta)$, then we can assign values to any angle θ, even if θ is less than 0 (measured clockwise) or greater than $\pi/2$, by assigning to $\cos \theta$ and $\sin \theta$ the first and second coordinates, respectively, of the point of intersection of the unit circle and the terminal side of the angle (see Fig. 10-5).

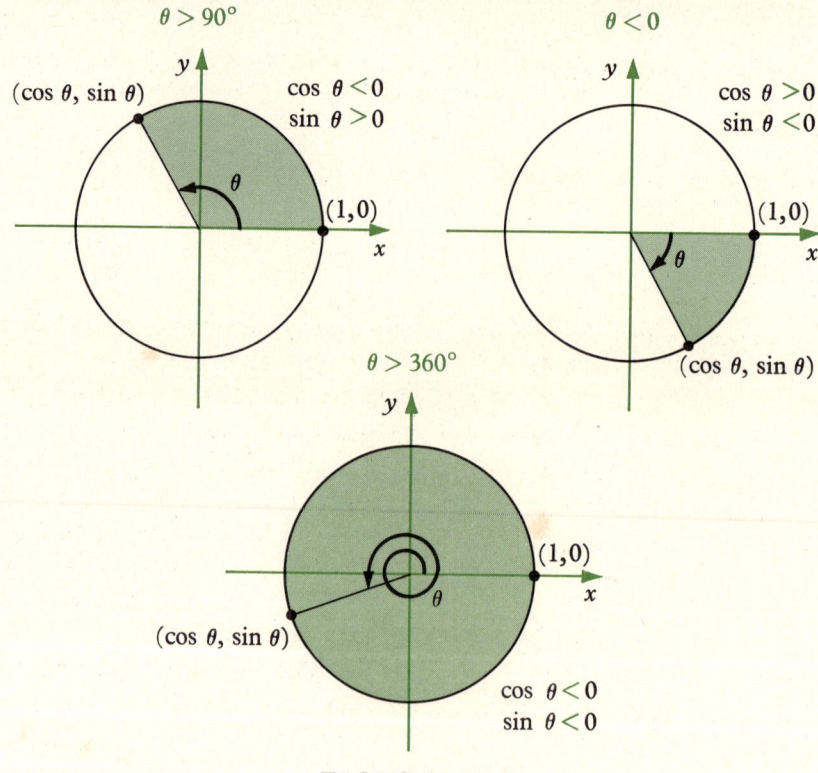

FIGURE 10-5

Definition If x is any real number, then $\sin x$ is defined to be the second coordinate of the point of intersection of the unit circle and the terminal side of an angle of x radians in standard position. Similarly, $\cos x$ is the first coordinate of this point.

The unit circle was chosen for convenience. A more general definition would utilize any circle, and then the functional value of the sine and cosine would be the respective coordinates divided by the radius of the circle. A simple argument using similar triangles shows that the two definitions are equivalent.

One can see that this definition of the trigonometric functions agrees with the classical "right-triangle" definition presented earlier in this section. It will sometimes be advantageous to use the classical definition to calculate trigonometric functions.

From the definition it is clear that the sine and cosine functions are defined for all real numbers. Thus the domain of each function is the set of all real numbers.

Since the coordinates of the points on the unit circle are less than or equal to one, the range of the sine and cosine functions is $[-1,1]$.

The other four trigonometric functions are defined in terms of $\sin x$ and $\cos x$ by the following:

$$\tan x = \frac{\sin x}{\cos x} \qquad \sec x = \frac{1}{\cos x}$$

$$\cot x = \frac{\cos x}{\sin x} \qquad \csc x = \frac{1}{\sin x}$$

where "tan" is the abbreviation for tangent, "cot" is the abbreviation for cotangent, "sec" for secant, and "csc" for cosecant.

The tangent and the secant are not defined for all real numbers x such that $\cos x = 0$, that is, for all x such that the related angle has its terminal side on the y-axis. The set of all such x is $A = \{x \mid x = (2k+1)\pi/2\}$. For example, $\pi/2$, $3\pi/2$, $5\pi/2$ are elements of A. Therefore the domain of the tangent and the secant is all real numbers except those in A.

Similarly, the domain of the cotangent and cosecant functions are not defined on the set $B = \{x \mid x = k\pi\}$ and the domain of these functions is the set of all real numbers except those in B.

It might be beneficial at this juncture to emphasize that the domain of the trigonometric functions are sets of *real numbers* and not angles. For example, we compute $\sin x$ for the *real number x*. In certain instances, one discusses the trigonometric function for the angle x, but what is meant is that

The quadrantal angle $\frac{\pi}{2}$

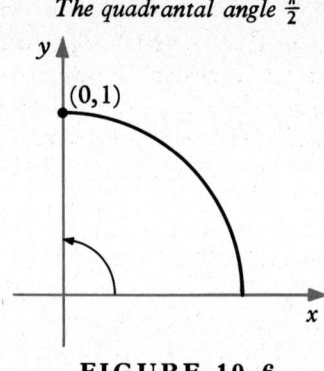

FIGURE 10-6

angle whose radian measure is x. Often we will encounter situations where $\sin x$ is computed and no mention of angles is given.

A *quadrantal* angle is one whose terminal side lies on one of the axes. We now find the trigonometric functions of some quadrantal angles and of some additional special angles.

EXAMPLE 10-4

If we let $x = 0$, then the terminal side of the angle whose measure is 0 radians intersects the unit circle at the point (1,0). Hence $\sin 0 = 0$ and $\cos 0 = 1$. From the definition of the other four trigonometric functions we have $\tan 0 = \sin 0/\cos 0 = 0/1 = 0$, $\cot 0$ and $\csc 0$ are undefined, and $\sec 0 = 1$.

EXAMPLE 10-5

If we let $x = \pi/2$, as in Ex. 10-4, we note that the terminal side of the angle intersects the unit circle at the point (0,1). Hence $\sin \pi/2 = 1$, $\cos \pi/2 = 0$, $\cot \pi/2 = 0$, $\csc \pi/2 = 1$, and $\tan \pi/2$ and $\sec \pi/2$ are undefined.

The quadrantal angle π

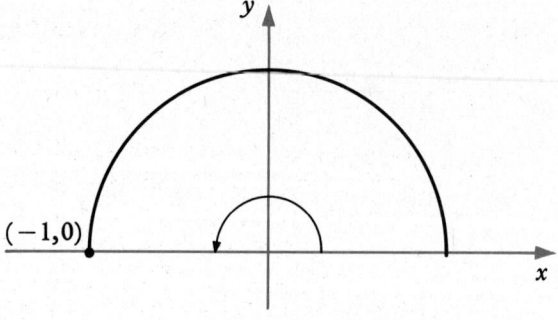

FIGURE 10-7

EXAMPLE 10-6

As in Ex. 10-4 and 10-5, to compute the trigonometric functions of π and $3\pi/2$ we consider the points $(-1,0)$ and $(0,-1)$ respectively. Thus we find that $\cos \pi = -1$, $\sin \pi = \tan \pi = 0$, $\sec \pi = -1$, and $\cot \pi$ and $\csc \pi$ are undefined. Similarly, $\sin 3\pi/2 = -1$, $\cos 3\pi/2 = \cot 3\pi/2 = 0$, $\csc 3\pi/2 = -1$, and $\tan 3\pi/2$ and $\sec 3\pi/2$ are undefined.

EXAMPLE 10-7

To compute the trigonometric functional values of $x = \pi/4$, we will use the classical definition of the trigonometric functions and Fig. 10-8. The right triangle in Fig. 10-8 is an isosceles triangle, i.e., the two legs are equal in length. Since we chose the length of the two legs to be 1 unit, then the length of the hypotenuse, by the Pythagorean Theorem, is $\sqrt{1^2 + 1^2} = \sqrt{2}$. Consequently,

$$\sin \frac{\pi}{4} = \cos \frac{\pi}{4} = \frac{1}{\sqrt{2}} = \frac{\sqrt{2}}{2}$$

$$\tan \frac{\pi}{4} = \cot \frac{\pi}{4} = 1$$

$$\sec \frac{\pi}{4} = \csc \frac{\pi}{4} = \sqrt{2}$$

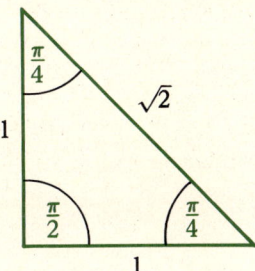

FIGURE 10-8

EXAMPLE 10-8

To calculate the trigonometric functional values of $x = \pi/6$ and $x = \pi/3$ we shall also utilize the classical definition of $\sin x$ and $\cos x$ and Fig. 10-9. The triangle in Fig. 10-9 is an equilaterial triangle, i.e., all the sides are equal in length, whose sides are of length 2 and consequently each angle is 60° or $\pi/3$ radians. By dropping the perpendicular bisector from one vertex, we bisect the angle into two equal angles, each $\pi/6$ radians. and one side is bisected. By the Pythagorean Theorem, the height of the triangle is $\sqrt{3}$ since $1^2+(\sqrt{3})^2 = 2^2$. Therefore

$$\sin\frac{\pi}{6} = \cos\frac{\pi}{3} = \frac{1}{2}$$

$$\sin\frac{\pi}{3} = \cos\frac{\pi}{6} = \frac{\sqrt{3}}{2}$$

$$\tan\frac{\pi}{6} = \cot\frac{\pi}{3} = \frac{1}{\sqrt{3}} = \frac{\sqrt{3}}{3}$$

$$\tan\frac{\pi}{3} = \cot\frac{\pi}{6} = \sqrt{3}$$

$$\sec\frac{\pi}{6} = \csc\frac{\pi}{3} = \frac{2}{\sqrt{3}} = \frac{2\sqrt{3}}{3}$$

$$\sec\frac{\pi}{3} = \csc\frac{\pi}{6} = 2$$

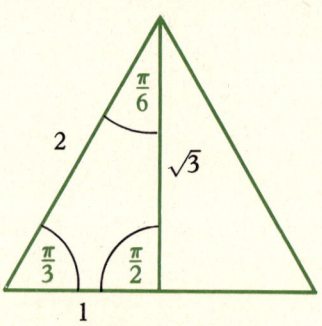

FIGURE 10-9

The procedure for finding the trigonometric functional values of numbers not in the interval $[0, \pi/2]$ is demonstrated in the next example. Usually a familiar angle must be located and then the point of intersection of the terminal side and the unit circle can be found. In addition, Table B-3 in the Appendix is a table of the trigonometric functional values for numbers in the interval $[0, \pi/2]$.

EXAMPLE 10-9

We will find the trigonometric functional values of $x = -3\pi/4$. The angle whose measure is $-3\pi/4$ has its terminal side in the third quadrant. Note that the angle that this terminal side makes with the negatively directly

x-axis is $\pi/4$ radians. Thus the point of intersection of this terminal side and the unit circle is $(-\sqrt{2}/2, -\sqrt{2}/2)$. (See Fig. 10-10 and Ex. 10-7.) Therefore

$$\sin\frac{-3\pi}{4} = \cos\frac{-3\pi}{4} = \frac{-\sqrt{2}}{2}$$

$$\tan\frac{-3\pi}{4} = \cot\frac{-3\pi}{4} = 1$$

$$\sec\frac{-3\pi}{4} = \csc\frac{-3\pi}{4} = -\sqrt{2}$$

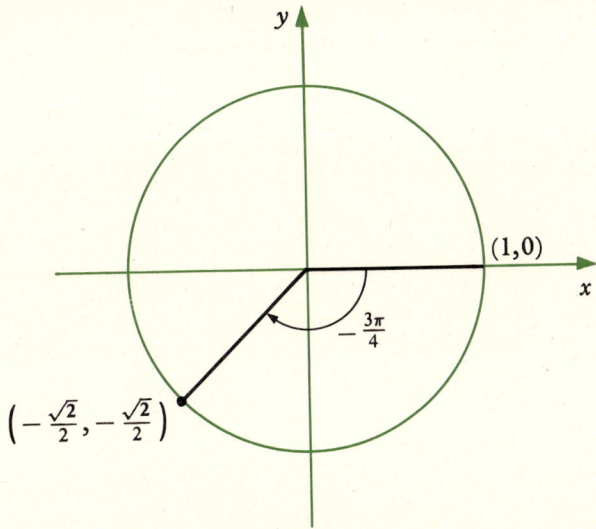

FIGURE 10-10

In Sec. 10-4 we present some identities which will give us another method to evaluate the trigonometric functional values of numbers not in $[0, \pi/2]$.

In the previous examples we have calculated the trigonometric functional values of some familiar angles. In Table B-3 in the Appendix a table of the trigonometric functional values is given. Angular measure is given in both degrees and radians. The trigonometric functions of numbers from 0 to $\pi/4$ (about 0.7854) radians can be obtained from the table by reading down. Trigonometric functions of numbers from $\pi/4$ to $\pi/2$ (about 1.5708) radians can be obtained by locating the number in the far right-hand column and reading up.

CHAP. 10   Trigonometric Functions

EXAMPLE 10-10

From Table B-3, we find that $\sin 0.524 = 0.0523$, $\sin 0.2705 = 0.2672$, $\sin 0.6109 = 0.5736$, $\sin 0.7854 = 0.7071$. Also $\cos 0.0873 = 0.9962$, $\tan 0.0873 = 0.0875$, and $\cot 0.0873 = 11.4301$.

Recall that for numbers in $[\pi/4, \pi/2]$ one must read up the table. Thus $\sin 1.4835 = 0.9962$, $\tan 1.4835 = 11.4301$, $\cos 1.0385 = 0.5075$, and $\cot 0.8639 = 0.8541$.

EXERCISE 10-2

1. Fill in the table (u stands for undefined).

x	$\sin x$	$\cos x$	$\tan x$	$\cot x$	$\sec x$	$\csc x$
0	0	1	0	u	1	u
$\pi/2$	1	0	u	0	u	
π	0	-1	0			
$3\pi/2$	-1	0				
2π						
$-\pi/2$						
$-\pi$						
$-3\pi/2$						
$5\pi/2$						

2. Use an argument utilizing similar triangles to show that the definition of the sine and cosine functions does not depend upon the radius of the circle chosen.

3. Calculate the sine, cosine, and tangent of the given real numbers.
    (a) $\pi/4$
    (b) $-\pi/4$
    (c) $3\pi/4$
    (d) $-3\pi/4$
    (e) $\pi/3$
    (f) $-\pi/3$
    (g) $2\pi/3$
    (h) $-2\pi/3$
    (i) $\pi/6$
    (j) $-\pi/6$
    (k) $5\pi/6$
    (l) $-5\pi/6$

4. Use Table B-3 to find the sine, cosine, tangent, and cotangent of the given numbers.
    (a) 0.0349
    (b) 0.1833
    (c) 0.2967
    (d) 0.4189
    (e) 0.7069
    (f) 0.8988
    (g) 1.1694
    (h) 1.5010
    (i) 0.2910
    (j) 1.3800

10–3 GRAPHS OF THE TRIGONOMETRIC FUNCTIONS

If two numbers differ by 2π, then when we draw the corresponding angles in standard position the terminal sides are the same because one has simply gone through one more revolution. It follows that the trigonometric functions must be the same for the two numbers. Such a function which repeats its values as the domain increases is called *periodic*.

Definition If there is a positive number p such that $f(x+p)=f(x)$, for all x, and if no smaller positive number has this property, then $f(x)$ is said to be *periodic* with period p.

Thus the functions $\sin x$ and $\cos x$ have period 2π, as well as $\sec x$ and $\csc x$. The functions $\tan x$ and $\cot x$ have period π.

Thus to obtain a sketch of the graph of $\sin x$ one must ascertain its behavior in the interval $[0, 2\pi]$. A quick inspection should convince the reader of the validity of the following table.

As x increases from	$\sin x$ varies from
0 to $\pi/2$	0 to 1
$\pi/2$ to π	1 to 0
π to $3\pi/2$	0 to -1
$3\pi/2$ to 2π	-1 to 0

Armed with this information concerning $\sin x$, a good sketch of the graph can be given by plotting a few points $(x, \sin x)$.

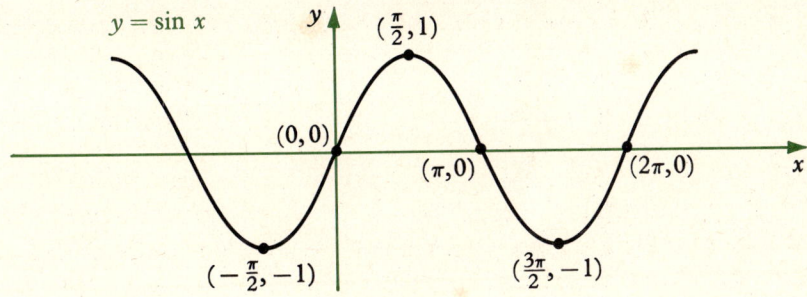

FIGURE 10-11

Similar arguments will yield good sketches of $\cos x$ and $\tan x$. The following tables will prove helpful.

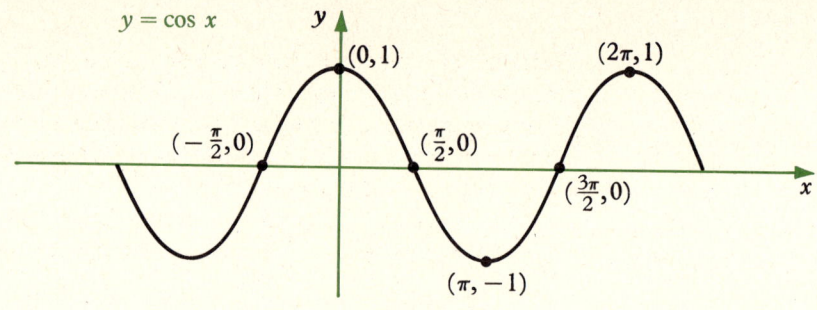

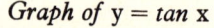

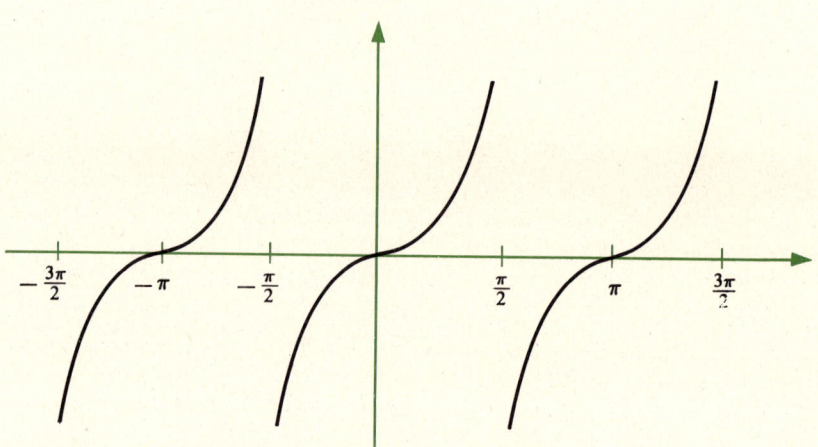

FIGURE 10-13

As x increases from	$\cos x$ varies from
0 to $\pi/2$	1 to 0
$\pi/2$ to π	0 to -1
π to $3\pi/2$	-1 to 0
$3\pi/2$ to 2π	0 to 1

As x increases from	$\tan x$ varies from
$-\pi/2$ to 0	$-\infty$ to 0
0 to $\pi/2$	0 to ∞
$\pi/2$ to π	$-\infty$ to 0
π to $3\pi/2$	0 to ∞

EXAMPLE 10-11

Consider the function $f(x) = \sin 3x$. Note that $f(0) = \sin 0 = 0$ and $f(\pi/6) = \sin[3(\pi/6)] = \sin \pi/2 = 1$. Thus as x increases from 0 to $\pi/6$, $f(x)$ varies from 0 to 1; and as x increases from $\pi/6$ to $\pi/3$, $f(x)$ varies from 1 to $f(\pi/3) = \sin \pi = 0$. Similarly, as x increases from $\pi/3$ to $\pi/2$, $f(x)$ varies from 0 to -1; and as x increases from $\pi/2$ to $2\pi/3$, $f(x)$ varies from -1 to 0. Then $f(x)$ repeats this cycle. Thus $f(x)$ is periodic of period $2\pi/3$. (See Fig. 10-14). The shape of the graph is similar to that of $\sin x$.

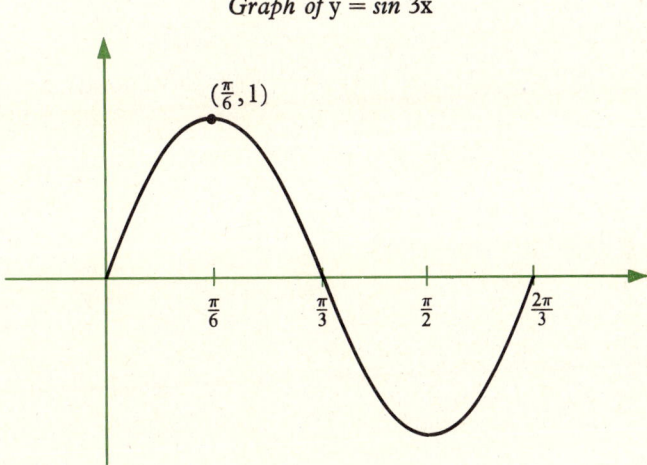

FIGURE 10-14

EXAMPLE 10-12

Consider the function $f(x) = 2\cos 3x$. As in Ex. 10-11, this function will have a period of $2\pi/3$. The shape of its graph is similar to $\cos x$ except that the maximum value that this function attains is $f(0) = 2\cos 0 = 2$ and its minimum value is $f(\pi/3) = 2\cos 3\pi/3 = 2\cos \pi = -2$. Hence it varies from -2 to 2. Intuitively, if you "squeeze" and "stretch" the graph of $\cos x$ you get the graph of $2\cos 3x$.

In general, the period of $\sin kx$ is $2\pi/k$ because,

$$\sin(x + 2\pi) = \sin x$$

$$\sin(kx + 2\pi) = \sin kx$$

$$\sin k(x + 2\pi/k) = \sin kx$$

Hence if x is increased by $2\pi/k$, the value of $\sin kx$ is the same. A similar argument shows that the period of $\cos kx$ is also $2\pi/k$.

EXERCISE 10-3

1. Determine the period of the functions.
   - (a) $y = \cos 2x$
   - (b) $y = \sin(x/2)$
   - (c) $y = \sin(2x+1)$
   - (d) $y = \cos 5x$
   - (e) $y = \tan 2\pi x$

2. Find the function of the form $y = a \sin bx$, for real numbers a and b, which satisfies the given conditions.
   - (a) Period π, range $[-2,2]$
   - (b) Period 4π, range $[1/2, 1/2]$
   - (c) Period $3\pi/2$, range $[-3,3]$

3. Graph the functions.
   - (a) $y = \sin 2x$
   - (b) $y = \cos 3x$
   - (c) $y = \tan 4x$
   - (d) $y = \sin \frac{1}{2}x$
   - (e) $y = \cos \frac{1}{3}x$

4. Graph the functions.
   - (a) $y = 2\sin(x+\pi)$
   - (b) $y = 3\cos(x-\pi)$
   - (c) $y = 2\sin(x+\pi/2)$
   - (d) $y = \sin^2 x$

10-4 IDENTITIES

In this section we study relationships, or identities, between the six trigonometric functions. The first is a direct consequence of the Pythagorean Theorem. Consider the point $(\cos x, \sin x)$ on the unit circle which is plotted in Fig. 10-15. In the shaded triangle, the legs have length $\sin x$ and $\cos x$ while the hypotenuse has length 1. Thus

$$(1) \qquad \sin^2 x + \cos^2 x = 1$$

which is true for all real numbers x. Note that $\sin^2 x$ means $(\sin x)^2$. It is more convenient to delete the parentheses. Note also that $\sin x^2$ is not equal to $\sin^2 x$. This can be seen by letting $x = \pi$ so that $\sin^2 \pi = (\sin \pi)^2 = 0^2 = 0$, but $(\pi)^2$ is about 9.86 and not equal to $n\pi$ for any integer n, and thus $\sin \pi^2 \neq 0$. Thus $\sin(\pi)^2 \neq \sin^2 \pi$.

By dividing equation (1) by $\cos^2 x$ we get

$$\frac{\sin^2 x}{\cos^2 x} + \frac{\cos^2 x}{\cos^2 x} = \frac{1}{\cos^2 x}$$

Recalling that $\sin x / \cos x = \tan x$ and $1/\cos x = \sec x$, the last equation becomes,

$$(2) \qquad \tan^2 x + 1 = \sec^2 x$$

EXAMPLE 10-11

Consider the function $f(x) = \sin 3x$. Note that $f(0) = \sin 0 = 0$ and $f(\pi/6) = \sin [3(\pi/6)] = \sin \pi/2 = 1$. Thus as x increases from 0 to $\pi/6$, $f(x)$ varies from 0 to 1; and as x increases from $\pi/6$ to $\pi/3$, $f(x)$ varies from 1 to $f(\pi/3) = \sin \pi = 0$. Similarly, as x increases from $\pi/3$ to $\pi/2$, $f(x)$ varies from 0 to -1; and as x increases from $\pi/2$ to $2\pi/3$, $f(x)$ varies from -1 to 0. Then $f(x)$ repeats this cycle. Thus $f(x)$ is periodic of period $2\pi/3$. (See Fig. 10-14). The shape of the graph is similar to that of $\sin x$.

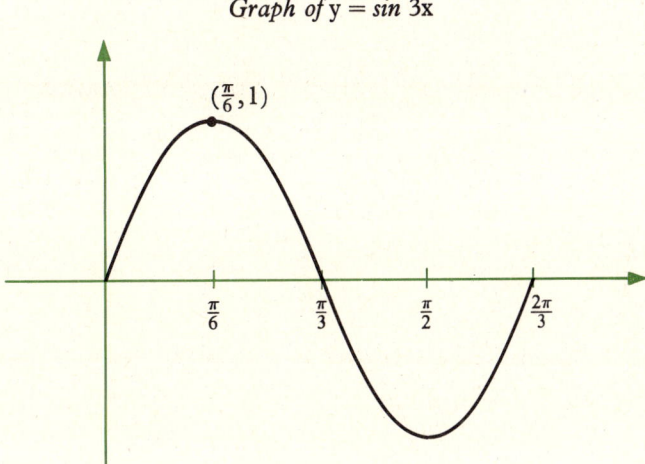

Graph of $y = \sin 3x$

FIGURE 10-14

EXAMPLE 10-12

Consider the function $f(x) = 2 \cos 3x$. As in Ex. 10-11, this function will have a period of $2\pi/3$. The shape of its graph is similar to $\cos x$ except that the maximum value that this function attains is $f(0) = 2 \cos 0 = 2$ and its minimum value is $f(\pi/3) = 2 \cos 3\pi/3 = 2 \cos \pi = -2$. Hence it varies from -2 to 2. Intuitively, if you "squeeze" and "stretch" the graph of $\cos x$ you get the graph of $2 \cos 3x$.

In general, the period of $\sin kx$ is $2\pi/k$ because,

$$\sin(x + 2\pi) = \sin x$$

$$\sin(kx + 2\pi) = \sin kx$$

$$\sin k(x + 2\pi/k) = \sin kx$$

Hence if x is increased by $2\pi/k$, the value of $\sin kx$ is the same. A similar argument shows that the period of $\cos kx$ is also $2\pi/k$.

EXERCISE 10-3

1. Determine the period of the functions.
   (a) $y = \cos 2x$
   (b) $y = \sin(x/2)$
   (c) $y = \sin(2x+1)$
   (d) $y = \cos 5x$
   (e) $y = \tan 2\pi x$

2. Find the function of the form $y = a \sin bx$, for real numbers a and b, which satisfies the given conditions.
   (a) Period π, range $[-2,2]$
   (b) Period 4π, range $[1/2, 1/2]$
   (c) Period $3\pi/2$, range $[-3,3]$

3. Graph the functions.
   (a) $y = \sin 2x$
   (b) $y = \cos 3x$
   (c) $y = \tan 4x$
   (d) $y = \sin \tfrac{1}{2}x$
   (e) $y = \cos \tfrac{1}{3}x$

4. Graph the functions.
   (a) $y = 2\sin(x+\pi)$
   (b) $y = 3\cos(x-\pi)$
   (c) $y = 2\sin(x+\pi/2)$
   (d) $y = \sin^2 x$

10-4 IDENTITIES

In this section we study relationships, or identities, between the six trigonometric functions. The first is a direct consequence of the Pythagorean Theorem. Consider the point $(\cos x, \sin x)$ on the unit circle which is plotted in Fig. 10-15. In the shaded triangle, the legs have length $\sin x$ and $\cos x$ while the hypotenuse has length 1. Thus

(1) $$\sin^2 x + \cos^2 x = 1$$

which is true for all real numbers x. Note that $\sin^2 x$ means $(\sin x)^2$. It is more convenient to delete the parentheses. Note also that $\sin x^2$ is not equal to $\sin^2 x$. This can be seen by letting $x = \pi$ so that $\sin^2 \pi = (\sin \pi)^2 = 0^2 = 0$, but $(\pi)^2$ is about 9.86 and not equal to $n\pi$ for any integer n, and thus $\sin \pi^2 \neq 0$. Thus $\sin(\pi)^2 \neq \sin^2 \pi$.

By dividing equation (1) by $\cos^2 x$ we get

$$\frac{\sin^2 x}{\cos^2 x} + \frac{\cos^2 x}{\cos^2 x} = \frac{1}{\cos^2 x}$$

Recalling that $\sin x/\cos x = \tan x$ and $1/\cos x = \sec x$, the last equation becomes,

(2) $$\tan^2 x + 1 = \sec^2 x$$

By the Pythagorean Theorem,
$$\sin^2 x + \cos^2 x = 1$$

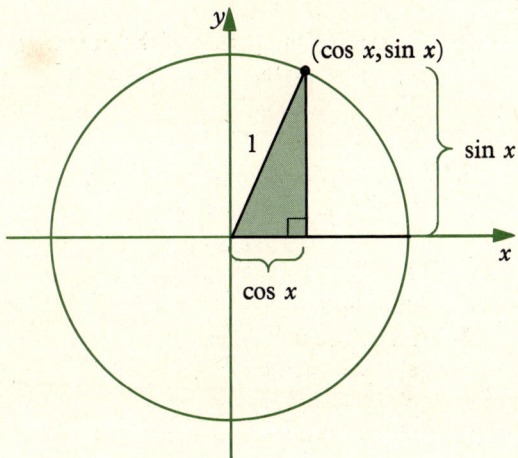

FIGURE 10-15

Dividing equation (1) by $\sin^2 x$ yields

$$\frac{\sin^2 x}{\sin^2 x} + \frac{\cos^2 x}{\sin^2 x} = \frac{1}{\sin^2 x}$$

and since $\cos x/\sin x = \cot x$ and $1/\sin x = \csc x$, we have

$$1 + \cot^2 x = \csc^2 x$$

We have derived three identities. There is almost an endless list of such formulae relating the trigonometric functions. We present a partial list and prove a few more of these identities.

Theorem 10-1 If x is a real number then

(1) $\sin^2 x + \cos^2 x = 1$
(2) $1 + \tan^2 x = \sec^2 x$
(3) $1 + \cot^2 x = \csc^2 x$
(4) $\sin(-x) = -\sin x$
(5) $\cos(-x) = \cos x$

(6) $\sin\left(\dfrac{\pi}{2} - x\right) = \cos x$

(7) $\cos\left(\dfrac{\pi}{2} - x\right) = \sin x$

(8) $\sin\left(\dfrac{\pi}{2} + x\right) = \cos x$

$$(9) \quad \cos\left(\frac{\pi}{2} + x\right) = -\sin x$$
$$(10) \quad \sin(\pi + x) = -\sin x$$
$$(11) \quad \cos(\pi + x) = -\cos x$$
$$(12) \quad \sin(\pi - x) = \sin x$$
$$(13) \quad \cos(\pi - x) = -\cos x$$

PROOF We will prove (4), (5), (6), and (7) and leave the remaining proofs for the reader. Before we commence with the proof we will briefly discuss the signs of the sine and cosine in the four quadrants. From their definitions one sees that $\sin x$ is positive in the first and second quadrants, negative in the third and fourth, while the cosine is positive in the first and fourth, negative in the second and third. See Fig. 10-16.

As in Fig. 10-17, we will assume that the terminal side of the angle in question is in the first quadrant. Similar arguments can be used to prove the results for the other quadrants. Consider the angles x and $-x$ in standard position. Let the point A be the point $(\cos x, \sin x)$, B the point $(\cos(-x), \sin(-x))$, and O the origin $(0,0)$; then the point C is the point $(\cos x, 0)$, which is the intersection of the x-axis and the line segment from A to B. The triangles OAC and OBC are congruent since the angles are equal and they share the side OC. Hence the length of the side AC is equal to the side BC. But the length of AC is $\sin x$ and the length of BC is $-\sin(-x)$. Hence $\sin x = -\sin(-x)$, which proves

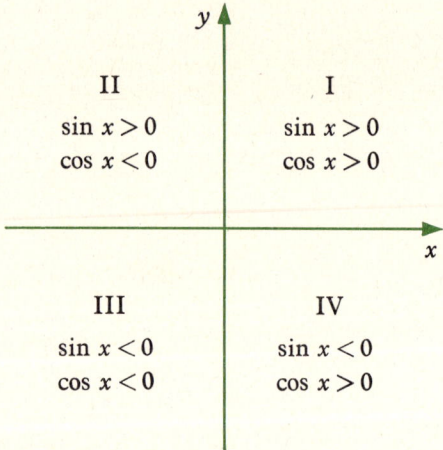

FIGURE 10-16

Signs of the sine and cosine

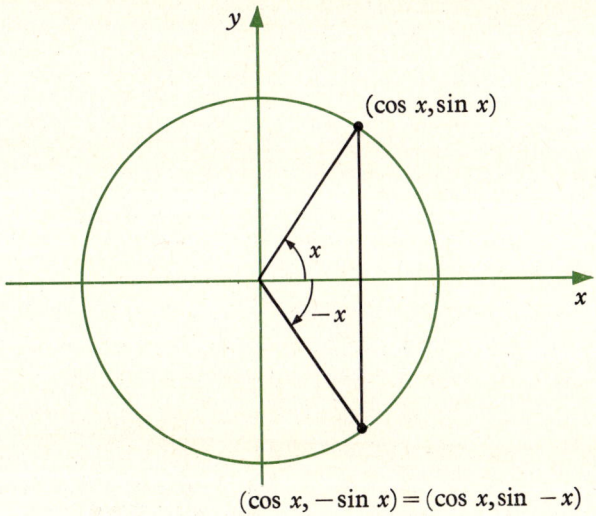

FIGURE 10-17

(4). To prove (5) we note that the length of OC is the first coordinate of A and of B. Hence $\cos x = \cos(-x)$.

To prove (6) and (7) we utilize the classical definition of the sine and cosine functions. Consider Fig. 10-18. In triangle ABC, angle A is a right angle and hence $\pi/2$ radians, and angle B is x radians. Thus angle C is $(\pi/2)-x$ radians since the sum of the three angles, $(\pi/2)+x+((\pi/2)-x)$ must equal π radians. From the definition of $\cos x$, we have

$$\cos x = \frac{\text{length of } AB}{\text{length of } BC}$$

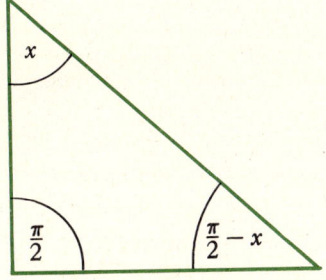

FIGURE 10-18

From the definition of $\sin((\pi/2)-x)$ we have

$$\sin\left(\frac{\pi}{2}-x\right) = \frac{\text{length of } AB}{\text{length of } BC}$$

Hence $\sin((\pi/2)-x) = \cos x$ proving (6). Similarly (7) is proved by noting that

$$\sin x = \frac{\text{length of } AC}{\text{length of } BC}$$

$$= \cos\left(\frac{\pi}{2}-x\right)$$

EXAMPLE 10-13

The identities in the above theorem enable us to calculate trigonometric functional values of angles whose terminal side is not in the first quadrant. For example

$$\sin\frac{3\pi}{4} = \sin\left(\frac{\pi}{2}+\frac{\pi}{4}\right)$$

$$= \cos\frac{\pi}{4} \qquad \textit{Identity (8)}$$

$$= \frac{\sqrt{2}}{2}$$

$$\cos\frac{7\pi}{6} = \cos\left(\pi+\frac{\pi}{6}\right)$$

$$= -\cos\frac{\pi}{6} \qquad \textit{Identity (11)}$$

$$= -\frac{\sqrt{3}}{2}$$

$$\sin\frac{11\pi}{6} = \sin\left(\pi+\frac{5\pi}{6}\right)$$

$$= -\sin\frac{5\pi}{6} \qquad \textit{Identity (10)}$$

$$= -\sin\left(\pi-\frac{\pi}{6}\right)$$

$$= -\sin\frac{\pi}{6} \qquad \textit{Identity (12)}$$

$$= -\frac{1}{2}$$

EXAMPLE 10-14

If $\tan x = -4/3$ and $\pi/2 < x < \pi$, then we can calculate the other trigo-

nometric functions. From identity (2) we have
$$\sec^2 x = 1 + \tan^2 x$$
$$= 1 + \left(-\frac{4}{3}\right)^2$$
$$= 1 + \frac{16}{9}$$
$$= \frac{25}{9}$$

Hence
$$\sec x = \frac{5}{3} \quad \text{or} \quad -\frac{5}{3}$$

Since $\pi/2 < x < \pi$, $\sec x$ is negative so that
$$\sec x = -\frac{5}{3}$$

By definition of $\sec x$, we have
$$\cos x = \frac{1}{\sec x} = -\frac{3}{5}$$

Since $\tan x = \sin x/\cos x$, we have
$$\sin x = \tan x \cdot \cos x$$
$$= -\frac{4}{3} \cdot -\frac{3}{5}$$
$$= \frac{4}{5}$$

From the definition of $\cot x$ and $\csc x$ we have
$$\cot x = \frac{1}{\tan x}$$
$$= \frac{1}{(-4/3)}$$
$$= -\frac{3}{4}$$
$$\csc x = \frac{1}{\sin x}$$
$$= \frac{1}{(4/5)}$$
$$= \frac{5}{4}$$

CHAP. 10   Trigonometric Functions

EXERCISE 10-4

1. Prove identities (8) through (13).
2. Prove the following identities for suitable values of x.
   - (a) $\sec(-x) = \sec x$
   - (b) $\csc(-x) = -\csc x$
   - (c) $\sec\left(\dfrac{\pi}{2} - x\right) = \csc x$
   - (d) $\csc\left(\dfrac{\pi}{2} + x\right) = \sec x$
   - (e) $\sin^4 x - \cos^4 x = \sin^2 x - \cos^2 x$
   - (f) $\cot^4 x + \cot^2 x = \csc^4 x - \csc^2 x$
   - (g) $\dfrac{1}{1 + \tan x} = \dfrac{\cot x}{1 + \cot x}$

3. Use identities (1) through (13) to find the sine and cosine of the given numbers.
   - (a) $3\pi/4$
   - (b) $5\pi/4$
   - (c) $7\pi/4$
   - (d) $5\pi/6$
   - (e) $7\pi/6$
   - (f) $11\pi/6$
   - (g) $2\pi/3$
   - (h) $4\pi/3$
   - (i) $-\pi/4$
   - (j) $-3\pi/4$
   - (k) $-5\pi/6$
   - (l) $-4\pi/3$

4. Show that $\cos^2 \pi/3 \neq \cos(\pi/3)^2$.
5. Find the five other trigonometric functional values if $\sin x = -3/5$ and $x \in [\pi/2, 3\pi/2]$.
6. Find the five other trigonometric functional values if $\tan x = 4/3$ and $x \in [\pi, 2\pi]$.

10-5 DERIVATIVES OF THE TRIGONOMETRIC FUNCTIONS

To rigorously derive the derivative of $\sin x$ one must use the fact that

$$\lim_{h \to 0} \frac{\sin h}{h} = 1$$

as well as the identity

$$\sin(A + B) = \sin A \cos B + \sin B \cos A$$

In the exercises, we derive the first of these results and then outline the proof of the fact that

$$D_x \sin x = \cos x$$

We record this important result in the next theorem.

Theorem 10-2 *If $f(x) = \sin x$, then $f'(x) = \cos x$.*

For the present, we content ourselves with presenting an intuitive, geometrical argument intended to convince you of the validity of Theorem 10-2.

SEC. 10-5   Derivatives of the Trigonometric Functions

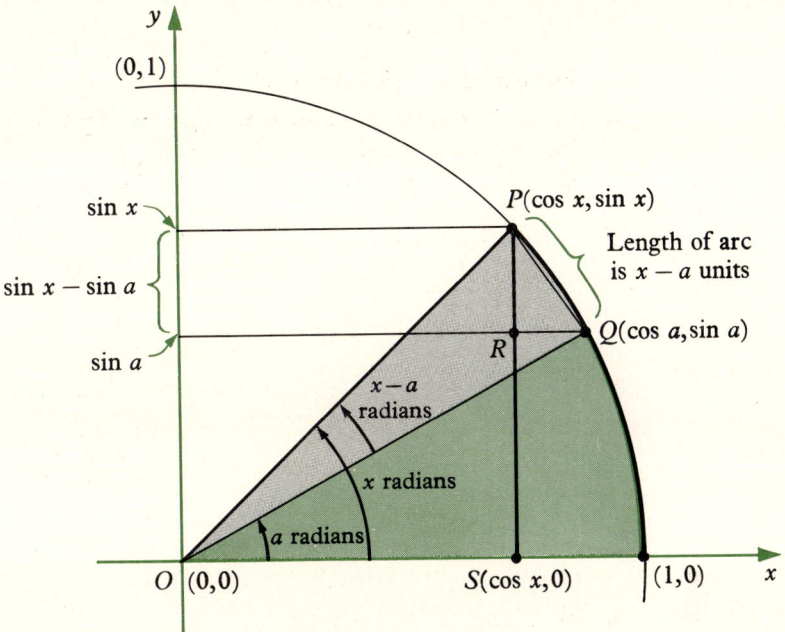

FIGURE 10-19

The calculation of the derivative of $\sin x$ will involve the difference quotient

$$\frac{\sin x - \sin a}{x - a}$$

where a is the measurement in radians of a fixed angle. If x is the measurement of an angle whose size is close to a, then $x - a$ is the measure in radians of a small angle. Consider Fig. 10-19.

The length of arc from the point $(1,0)$ to $Q(\cos a, \sin a)$ is a units and the length of arc from $(1,0)$ to $P(\cos x, \sin x)$ is x units. Thus the length of arc from Q to P is $x - a$ units. The triangle PQR is a right triangle that is similar to triangle PSO; hence the ratios of corresponding sides are equal. If we let $\overline{PQ}$ be the length of the hypotenuse of triangle PQR, we therefore have

$$\frac{\sin x - \sin a}{\overline{PQ}} = \frac{\cos x}{1} = \cos x$$

If x is very close to a, note that $\overline{PQ}$ is very close to the length of arc from P to Q, namely $x - a$ units. Hence $\cos x$ is very close to

$$\frac{\sin x - \sin a}{x - a}$$

and we say that they are approximately equal and write

$$\frac{\sin x - \sin a}{x - a} \cong \cos x$$

If we take the limit as x approaches a, it seems reasonable to predict that if $f(x) = \sin x$, then

$$f'(a) = \lim_{x \to a} \frac{\sin x - \sin a}{x - a} = \lim_{x \to a} \cos x = \cos a$$

because $\cos x$ is a continuous function.

If we apply the Chain Rule, Theorem 4-4, to Theorem 10-2 we get

$$D_x \sin u = (\cos u) D_x u$$

where u is some function of x.

EXAMPLE 10-15

If $f(x) = \sin x^2$, then to find $f'(x)$ we let $u = x^2$ in the Chain Rule and get

$$f'(x) = D_x \sin x^2 = \cos x^2 (D_x x^2) = 2x \cos x^2$$

EXAMPLE 10-16

From the Chain Rule we get

$$D_x \sin(x^3 - 5x) = [\cos(x^3 - 5x)](3x^2 - 5)$$

and

$$D_x \sin e^x = (\cos e^x) e^x$$

EXAMPLE 10-17

To find the derivative of $\sin^2 x$, we again use the Chain Rule. However, we use the form of the Chain Rule given by $D_x u^n = n u^{n-1} D_x u$. Thus

$$D_x \sin^2 x = 2(\sin x) D_x \sin x = 2 \sin x \cdot \cos x$$

EXAMPLE 10-18

We can use the Chain Rule twice to find

$$D_x (1 - \sin^2 x)^{1/2} = \frac{1}{2}(1 - \sin^2 x)^{-1/2} D_x (-\sin^2 x)$$

$$= \frac{1}{2}(1 - \sin^2 x)^{-1/2}(-2 \sin x \, D_x \sin x)$$

$$= -(1 - \sin^2 x)^{-1/2}(\sin x \cdot \cos x)$$

$$= \frac{-\sin x \cos x}{(1 - \sin^2 x)^{1/2}}$$

Our next task is to find the derivative of $\cos x$. A major difficulty is that in order to derive this formula, we must be able to evaluate another difficult limit, which is

$$\lim_{x \to 0} \frac{\cos x - 1}{x} = 0$$

Consider, however, the following discussion. We know that

$$\sin^2 x + \cos^2 x = 1$$

for every x. If $\cos x > 0$, then $\cos x = (1 - \sin^2 x)^{1/2}$. Thus $D_x \cos x = D_x (1 - \sin^2 x)^{1/2}$. The derivative on the right was found in Ex. 10-18 by utilizing the Chain Rule twice. Hence

$$D_x \cos x = D_x (1 - \sin^2 x)^{1/2} = \frac{-\sin x \cos x}{(1 - \sin^2 x)^{1/2}}$$

Since we are assuming that $\cos x > 0$ and thus $\cos x = (1 - \sin^2 x)^{1/2}$, we have

$$D_x \cos x = -\sin x$$

One can similarly show that this formula is valid for $\cos x \leq 0$. We therefor have the following theorem.

Incorporating Theorem 10-3 into the Chain Rule yields the formula

$$D_x \cos u = -\sin u \, D_x u$$

where u is some function of x.

EXAMPLE 10-19

Using the Chain Rule we calculate the following derivatives.

$$D_x \cos x^2 = -\sin x^2 (2x)$$

$$D_x \cos(x^3 + x) = -\sin(x^3 + x)(3x^2 + 1)$$

$$D_x \cos(\ln x) = -\sin(\ln x)\left(\frac{1}{x}\right)$$

$$D_x \cos^3 x = 3 \cos^2 x (-\sin x)$$

$$= -3 \cos^2 x \sin x$$

$$D_x \cos^4 x^3 = 4 \cos^3 x^3 D_x \cos x^3$$

$$= 4 \cos^3 x^3 [-\sin x^3 (3x^2)]$$

$$= -12 x^2 \cos^3 x^3 \sin x^3$$

One can use Theorems 10-2 and 10-3 to calculate the derivatives of the remaining trigonometric functions. In the following example, we illustrate this procedure.

EXAMPLE 10-20

Since $\tan x = \sin x / \cos x$, we have

$$D_x \tan x = D_x \frac{\sin x}{\cos x}$$

$$= \frac{\cos x \, D_x \sin x - \sin x \, D_x \cos x}{\cos^2 x}$$

$$= \frac{\cos x \cos x - \sin x (-\sin x)}{\cos^2 x}$$

$$= \frac{\cos^2 x + \sin^2 x}{\cos^2 x}$$

$$= \frac{1}{\cos^2 x}$$

$$= \sec^2 x$$

Therefore $D_x \tan x = \sec^2 x$.

In a similar way we can verify the following formulas.

$$D_x \cot x = -\csc^2 x$$

$$D_x \sec x = \sec x \tan x$$

$$D_x \csc x = -\csc x \cot x$$

Incorporating the above formulae into the Chain Rule yields

$$D_x \tan u = \sec^2 u \, D_x u$$

$$D_x \cot u = -\csc^2 u \, D_x u$$

$$D_x \sec u = \sec u \tan u \, D_x u$$

$$D_x \csc u = -\csc u \cot u \, D_x u$$

where u is a differentiable function of x.

EXAMPLE 10-21

Using the Chain Rule we calculate the following derivatives

$$D_x \tan x^2 = \sec^2 x^2 \, D_x x^2 = 2x \sec^2 x^2$$

$$D_x \cot(2x^2+1) = -4x \csc^2(2x^2+1)$$

$$D_x \sec^3 x = 3 \sec^2 x \, D_x \sec x$$

$$= 3 \sec^3 x \tan x$$

$$D_x \csc^3 x^2 = 3 \csc^2 x^2 \, D_x \csc x^2$$

$$= 3 \csc^2 x^2 (-\csc x^2 \cot x^2) D_x x^2$$

$$= -6x \csc^3 x^2 \cot x^2$$

EXERCISE 10-5

1. Find the derivatives of the functions.
   - (a) $y = \sin 2x$
   - (b) $y = \sin 5x$
   - (c) $y = \sin x^3$
   - (d) $y = \sin 4x^2$
   - (e) $y = \sin 2e^x$
   - (f) $y = \sin(x + e^x)$
   - (g) $y = \sin(x^2 + \ln x)$
   - (h) $y = \cos(x^2 + x)$
   - (i) $y = \cos(2x+1)^2$
   - (j) $y = \cos e^{2x}$
   - (k) $y = \cos e^{x^2}$
   - (l) $y = \sin^3 x$
   - (m) $y = \cos^4 x$
   - (n) $y = \sqrt{\cos x^2}$
   - (o) $y = \sqrt{\sin x^2 + \cos^2 x}$
   - (p) $y = \tan x^2$
   - (q) $y = \tan^2 x$
   - (r) $y = \sin(\cos x)$
   - (s) $y = \tan(\sin x)$
   - (t) $y = \dfrac{\tan x}{\sin x}$

2. Derive the differentiation formulas by expressing $\cot x$, $\sec x$, and $\csc x$ in terms of $\sin x$ and $\cos x$.
   - (a) $D_x \cot x = -\csc^2 x$
   - (b) $D_x \sec x = \sec x \tan x$
   - (c) $D_x \csc x = -\csc x \cot x$

3. Find the derivatives of the functions.
   - (a) $y = \cot^2 x$
   - (b) $y = \sec x^2$
   - (c) $y = \csc(x^2 + x)$
   - (d) $y = \ln(\cot x)$
   - (e) $y = e^{\sec x}$
   - (f) $y = \sec^2 x$
   - (g) $y = \cot(\sec x)$
   - (h) $y = e^x \cot x^2$
   - (i) $y = \csc^2(x + x^2)$
   - (j) $y = \sin(\csc x)$

The next two exercises outline the proof that

$$\lim_{h \to 0} \frac{\sin h}{h} = 1$$

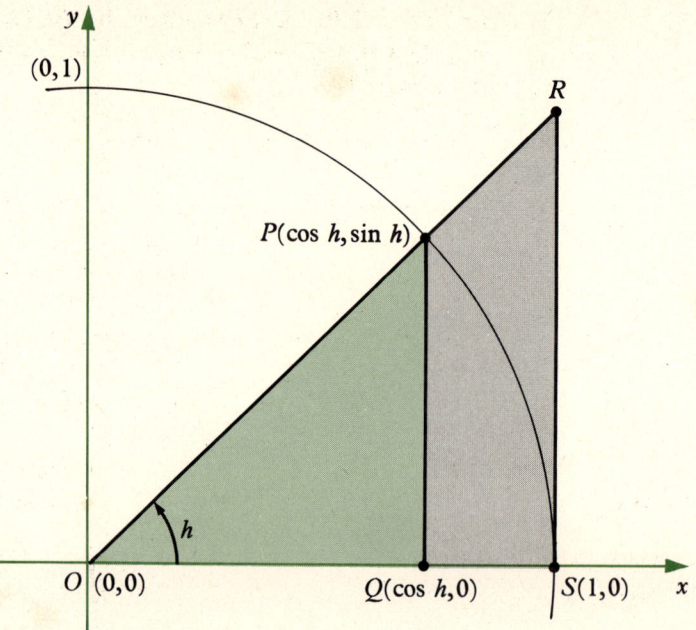

4. Construct a unit circle, as in the figure above, with an angle of h radians in standard position and the two triangles OPQ and ORS. Let A_1 be the area of triangle OPQ, A_2 the area of the sector OPS, and A_3 the area of the triangle ORS. Convince yourself that if $h \in (0, \pi/2)$, then $A_1 < A_2 < A_3$. Prove the following statements.

(a) $A_1 = \frac{1}{2}(\sin h)(\cos h)$.

(b) Use the fact that the two triangles OPQ and ORS are similar to prove that
$$A_3 = \frac{1}{2} \cdot 1 \cdot \frac{\sin h}{\cos h} = \frac{\sin h}{2 \cos h}$$

(c) Since the area of a sector is one-half the radius times the arc length, show that $A_2 = h/2$.

(d) Show that
$$\cos h < \frac{h}{\sin h} < \frac{1}{\cos h}$$

(e) Show that
$$\frac{1}{\cos h} > \frac{\sin h}{h} > \cos h$$

(f) Show that
$$\lim_{h \to 0^+} \frac{\sin h}{h} = 1$$

What is left to prove?

5. Let $h \in (-\pi/2, 0)$. Let $h = -k$ so that $k \in (0, \pi/2)$. From our definition of $\sin x$, show that $\sin(-x) = -\sin x$. Use this fact and the above substitution for h to show that

$$\lim_{h \to 0^-} \frac{\sin h}{h} = 1$$

6. To prove that $D_x \sin x = \cos x$, we need to show that

$$\lim_{h \to 0} \frac{\cos h - 1}{h} = 0$$

Verify each step in the following string of equalities.

$$\lim_{h \to 0} \frac{\cos h - 1}{h} = \lim_{h \to 0} \frac{\cos h - 1}{h} \cdot \frac{\cos h + 1}{\cos h + 1}$$

$$= \lim_{h \to 0} \frac{\cos^2 h - 1}{h(\cos h + 1)}$$

$$= \lim_{h \to 0} \frac{-\sin^2 h}{h(\cos h + 1)}$$

$$= \lim_{h \to 0} \frac{-\sin h}{h} \cdot \frac{\sin h}{\cos h + 1}$$

$$= 0$$

10–6 INTEGRATION OF THE TRIGONOMETRIC FUNCTIONS

The differentiation formulas of the preceding section can be rewritten as integration formulas.

$$\int \sin x \, dx = -\cos x + C$$

$$\int \cos x \, dx = \sin x + C$$

$$\int \sec^2 x \, dx = \tan x + C$$

$$\int \csc^2 x \, dx = -\cot x + C$$

$$\int \sec x \tan x \, dx = \sec x + C$$

$$\int \csc x \cot x \, dx = -\csc x + C$$

The antiderivatives of $\tan x$, $\cot x$, $\sec x$, and $\csc x$ are not as easily obtained. For their calculation we utilize the techniques of integration presented in Chapter 5.

To calculate $\int \tan x \, dx$ we use the substitution $u = \cos x$, $du = -\sin x \, dx$ and obtain

$$\int \tan x \, dx = \int \frac{\sin x}{\cos x} \, dx$$

$$= \int \frac{-du}{u}$$

$$= -\ln|u| + C$$

$$= -\ln|\cos x| + C$$

$$= \ln|\cos x|^{-1} + C$$

$$= \ln\left|\frac{1}{\cos x}\right| + C$$

$$= \ln|\sec x| + C$$

To calculate $\int \cot x \, dx$ we use the substitution $u = \sin x$, $du = \cos x \, dx$ and obtain

$$\int \cot x \, dx = \int \frac{\cos x}{\sin x} \, dx$$

$$= \int \frac{du}{u}$$

$$= \ln|u| + C$$

$$= \ln|\sin x| + C$$

To calculate $\int \sec x \, dx$, we use the algebraic ploy of multiplying and dividing the integrand, $\sec x$, by the quantity $\sec x + \tan x$. We then let $u = \sec x + \tan x$, $du = (\sec x \tan x + \sec^2 x) \, dx = \sec x (\sec x + \tan x) \, dx$. Thus

$$\int \sec x \, dx = \int \frac{\sec x (\sec x + \tan x)}{\sec x + \tan x} \, dx$$

$$= \int \frac{du}{u}$$

$$= \ln|u| + C$$

$$= \ln|\sec x + \tan x| + C$$

We use a similar ploy to calculate $\int \csc x \, dx$ and let $u = \csc x + \cot x$,

$$du = (-\csc^2 x - \csc x \cot x)\,dx = -\csc x(\csc x + \cot x)\,dx \text{ and hence}$$

$$\int \csc x \, dx = \int \frac{\csc x(\csc x + \cot x)}{\csc x + \cot x}\,dx$$

$$= \int -\frac{du}{u}$$

$$= -\ln|u| + C$$

$$= -\ln|\csc x + \cot x| + C$$

We have derived the following four formulas

$$\int \tan x \, dx = \ln|\sec x| + C$$

$$\int \cot x \, dx = \ln|\sin x| + C$$

$$\int \sec x \, dx = \ln|\sec x + \tan x| + C$$

$$\int \csc x \, dx = -\ln|\csc x + \cot x| + C$$

EXAMPLE 10-22

To integrate

$$\int x \sec^2 x^2 \, dx$$

we make the substitution $u = x^2$, $du = 2x\,dx$, $\frac{1}{2}du = x\,dx$, so that

$$\int x \sec^2 x^2 \, dx = \int \tfrac{1}{2} \sec^2 u \, du$$

$$= \tfrac{1}{2} \tan u + C$$

$$= \tfrac{1}{2} \tan x^2 + C$$

EXAMPLE 10-23

To integrate

$$\int \sin^6 x \cos x \, dx$$

we use the substitution $u = \sin x$, $du = \cos x\,dx$, so that

$$\int \sin^6 x \cos x \, dx = \int u^6 \, du$$

$$= (1/7) u^7 + C$$

$$= (1/7) \sin^7 x + C$$

EXAMPLE 10-24

To integrate

$$\int \sin^5 x \, dx$$

we write $\sin^5 x = (\sin^2 x)^2 \sin x = (1-\cos^2 x)^2 \sin x$ and make the substitution $u = \cos x$, $du = -\sin x \, dx$, $-du = \sin x \, dx$, so that

$$\int \sin^5 x \, dx = \int (1-\cos^2 x)^2 \sin x \, dx$$

$$= \int (1-u^2)^2 (-du)$$

$$= -\int (1-2u^2+u^4) \, du$$

$$= -(u-(2/3)u^3+(1/5)u^5) + C$$

$$= -\cos x + (2/3)\cos^3 x - (1/5)\cos^5 x + C$$

EXERCISE 10-6

1. Evaluate the integrals.
   (a) $\int (2\sin 3x + 5\cos 2x) \, dx$
   (b) $\int (2\tan 4x + 3\sec 2x) \, dx$
   (c) $\int \sin(4x+5) \, dx$
   (d) $\int \sec 5x \tan 5x \, dx$

2. Evaluate the integrals.
   (a) $\int x \cos x^2 \, dx$
   (b) $\int x \sin(2x^2+1) \, dx$
   (c) $\int x^2 \sec x^3 \, dx$
   (d) $\int x \csc x^2 \, x^2 \cot x^2 \, dx$

3. Evaluate the integrals.
   (a) $\int \cos^6 x \sin x \, dx$
   (b) $\int \sec^3 x \tan x \, dx$

(c) $\int \sec^2 x \tan^2 x \, dx$

(d) $\int \csc^2 x \cot^4 x \, dx$

(e) $\int \sqrt{3 \sin^3 x} \cos x \, dx$

(f) $\int \dfrac{\sin 2t}{\cos^3 2t} \, dt$

4. Find the area of the region bounded by one arc of $y = \sin x$ and the x-axis.
5. Find the area of the region bounded by one arc of $y = 2 \sin 3x$ and the x-axis.
6. Find the area of the region in the first quadrant bounded by $y = \sin \pi x$ and $y = 2x$.

10-7 TECHNIQUES OF INTEGRATION IV: TRIGONOMETRIC SUBSTITUTION

We now augment our discussion of techniques of integration in Chapter 5 by the technique of trigonometric substitution. An algebraic expression of degree two can often be expressed in terms of the three trigonometric Pythagorean identities,

(1) $\qquad 1 - \sin^2 \theta = \cos^2 \theta$

(2) $\qquad 1 + \tan^2 \theta = \sec^2 \theta$

(3) $\qquad \sec^2 \theta - 1 = \tan^2 \theta$

For an expression of the form $a^2 - x^2$, identity (1) will be used. For an expression of the form $a^2 + x^2$, identity (2) will be used and identity (3) will be used for an expression of the form $x^2 - a^2$. Let us first demonstrate the method utilized in this type of substitution.

EXAMPLE 10-25

Consider the expressions (i) $\sqrt{9 - x^2}$, (ii) $\sqrt{9 + x^2}$, and (iii) $\sqrt{4x^2 - 9}$. For (i), if we let $x = 3 \sin \theta$, then $9 - x^2 = 9 - 9 \sin^2 \theta = 9(1 - \sin^2 \theta) = 9 \cos^2 \theta$ from identity (1). Thus

$$\sqrt{9 - x^2} = \sqrt{9 \cos^2 \theta} = 3 \cos \theta \qquad \text{if } -\pi/2 < \theta < \pi/2$$

For (ii) if we let $x = 3 \tan \theta$, then $9 + x^2 = 9 + 9 \tan^2 \theta = 9(1 + \tan^2 \theta) = 9 \sec^2 \theta$ from identity (2). Thus

$$\sqrt{9 + x^2} = \sqrt{9 \sec^2 \theta} = 3 \sec \theta \qquad \text{if } -\pi/2 < \theta < \pi/2$$

For (iii), if we let $x = (3/2) \sec \theta$ then $x^2 = (9/4) \sec^2 \theta$ and $4x^2 - 9 = 4(9/4) \sec^2 \theta - 9 = 9 \sec^2 \theta - 9 = 9(\sec^2 \theta - 1) = 9 \tan^2 \theta$. Thus, if $x > 0$, we write

$$\sqrt{4x^2-9} = \sqrt{9\tan^2\theta} = 3\tan\theta \quad \text{if } 0 < \theta < \pi/2$$

Note that in each case the substitution had the effect of eliminating the square root sign.

In the next two examples we use the technique of integration by trigonometric substitution. Note that the first integral could be solved by a direct substitution. It is included here to demonstrate the technique.

EXAMPLE 10-26
Consider

$$\int x\sqrt{9-x^2}\, dx$$

As in Ex. 10-25, we let $x = 3\sin\theta$, $-\pi/2 < \theta < \pi/2$, so that $x^2 = 9\sin^2\theta$, $9 - x^2 = 9 - 9\sin^2\theta = 9(1-\sin^2\theta) = 9\cos^2\theta$ and so $\sqrt{9-x^2} = \sqrt{9\cos^2\theta} = 3\cos\theta$. Also $dx = 3\cos\theta\, d\theta$. These substitutions yield

$$\int x\sqrt{9-x^2}\, dx = \int 3\sin\theta(3\cos\theta)3\cos\theta\, d\theta$$

$$= \int 27\sin\theta\cos^2\theta\, d\theta$$

Now make the substitution $u = \cos\theta$, $du = -\sin\theta\, d\theta$ and get

$$\int \sin\theta\cos^2\theta\, d\theta = \int -u^2\, du$$

$$= -\frac{u^3}{3} + C$$

$$= -\frac{\cos^3\theta}{3} + C$$

Thus

$$\int x\sqrt{9-x^2}\, dx = 27\int \sin\theta\cos^2\theta\, d\theta$$

$$= -9\cos^3\theta + C$$

$$= -9\left(\frac{\sqrt{9-x^2}}{3}\right)^3 + C$$

$$= -\left(\frac{1}{3}\right)(9-x^2)^{3/2} + C$$

EXAMPLE 10-27
Consider
$$\int \frac{1}{\sqrt{x^2-4}}\,dx \qquad x > 4$$

Recognize that $\sqrt{x^2-4}$ is of third type in Ex. 10-18, so we make the substitution $x = 2\sec\theta$, $0 < \theta < \pi/2$, $x^2 = 4\sec^2\theta$ and $x^2 - 4 = 4\sec^2\theta - 4 = 4(\sec^2\theta - 1) = 4\tan^2\theta$. Hence $\sqrt{x^2-4} = \sqrt{4\tan^2\theta} = 2\tan\theta$ and $dx = 2\sec\theta\tan\theta\,d\theta$. We now make these substitutions and get

$$\int \frac{1}{\sqrt{x^2-4}}\,dx = \int \frac{2\sec\theta\tan\theta\,d\theta}{2\tan\theta}$$

$$= \int \sec\theta\,d\theta$$

$$= \ln(\sec\theta + \tan\theta) + C$$

From the above substitutions $\sec\theta = x/2$ and $\tan^2\theta = (x^2-4)/4$, so $\tan\theta = \sqrt{x^2-4}/2$. Thus

$$\int \frac{1}{\sqrt{x^2-4}}\,dx = \ln\left(\frac{x}{2} + \frac{\sqrt{x^2-4}}{2}\right) + C$$

EXERCISE 10-7

1. Evaluate the integrals.

   (a) $\int \sqrt{9+x^2}\,dx$

   (b) $\int \sqrt{16+x^2}\,dx$

   (c) $\int \sqrt{1+9x^2}\,dx$

   (d) $\int \sqrt{x^2-9}\,dx$

   (e) $\int \sqrt{4x^2-9}\,dx$

   (f) $\int x\sqrt{16-x^2}\,dx$

   (g) $\int x\sqrt{9-4x^2}\,dx$

   (h) $\int \frac{dx}{\sqrt{9x^2-16}}$

   (i) $\int \frac{dx}{\sqrt{1+4x^2}}$

2. Evaluate the definite integrals.

   (a) $\int_0^2 \frac{dx}{\sqrt{x^2+4}}$

   (b) $\int_0^3 x\sqrt{16-x^2}\,dx$

   (c) $\int_2^3 \frac{dx}{\sqrt{4x^2-9}}$

10-8 INVERSE TRIGONOMETRIC FUNCTIONS

The concept of *inverse* function was first mentioned in Chapter 1. In Chapter 7 we noted that $y = \ln x$ and $y = e^x$ are inverses of each other. In Chapter 1, it was mentioned that $y = x^2$ does not have an inverse function because it is not one-to-one. However, if the domain of $y = x^2$ is restricted to $[0, \infty)$, then $y = \sqrt{x}$ is its inverse.

The same situation takes place with $y = \sin x$ and $y = \cos x$. Neither is one-to-one. Indeed, each is periodic and repeats its values every period of 2π. Hence, in order to define an inverse function, we must restrict the domain of each function.

From the graph of the sine function one sees that there are many ways of restricting the domain so that the sine has an inverse. For example, one could choose the restricted domain to be $[-\pi/2, \pi/2]$, $[\pi/2, 3\pi/2]$, or $[3\pi/2, 5\pi/2]$. In each interval, the sine takes on its full range of values, i.e., it takes on all the real numbers in $[-1,1]$, and the function is one-to-one in each interval. It is customary to choose the interval such that $|x|$ is the smallest number corresponding to a given value of the function. That is, $0 \in [-1,1]$ and $\sin 0 = 0$, and certainly 0 has the smallest absolute value of all x such that $\sin x = 0$. Hence 0 should be in our restricted domain. Hence we choose $[-\pi/2, \pi/2]$. The inverse sine function will therefore have the range $[-\pi/2, \pi/2]$. Its domain is $[-1,1]$.

Definition The inverse sine function, written $y = \sin^{-1} x$, with the range $[-\pi/2, \pi/2]$, is defined by

$$y = \sin^{-1} x \text{ if and only if } x = \sin y$$

where $-\pi/2 \leq y \leq \pi/2$

The function $y = \sin^{-1} x$ is often called the "arcsine function" and written $y = \arcsin x$. Do not confuse $\sin^{-1} x$ with $(\sin x)^{-1} = 1/\sin x$.

Another way of thinking of $\sin^{-1} x$ is "$\sin^{-1} x$ is that number whose sine is x". Hence, since $\sin 0 = 0$, $\sin^{-1} 0 = 0$; and since $\sin \pi/6 = 1/2$, $\sin^{-1} 1/2 = \pi/6$. Note that $\sin^{-1} x$ must be in $[-\pi/2, \pi/2]$.

EXAMPLE 10-28

$\sin^{-1} \sqrt{3}/2 = \pi/3$ because $\sin \pi/3 = \sqrt{3}/2$ and $\pi/3 \in [-\pi/2, \pi/2]$. Note also that $\sin 2\pi/3 = \sqrt{3}/2$ but $\sin^{-1} \sqrt{3}/2 \neq 2\pi/3$ because $2\pi/3 \notin [-\pi/2, \pi/2]$. Similarly, $\sin^{-1}(-1/2) = -\pi/6$ because $\sin(-\pi/6) = -1/2$ and $-\pi/6 \in [-\pi/2, \pi/2]$. Note that $\sin(7\pi/6) = -1/2$, but $7\pi/6 \notin [-\pi/2, \pi/2]$, so that $\sin^{-1}(-1/2)$ is *not* $7\pi/6$.

EXAMPLE 10-29

Evaluate $\cos(\sin^{-1} x)$ for $x \in [-1, 1]$, in terms of x. We first write $y = \sin^{-1} x$ so that $\sin y = x$ and $y \in [-\pi/2, \pi/2]$. Thus $\cos(\sin^{-1} x) = \cos y$, and hence we must find $\cos y$ in terms of x. Now $\cos y = \pm\sqrt{1 - \sin^2 y} = \pm\sqrt{1 - x^2}$. Since $y \in [-\pi/2, \pi/2]$, we have $\cos y \geq 0$ and hence $\cos y = \sqrt{1 - x^2}$.

To obtain a table of values for $\sin^{-1} x$, we can simply take Table 10-1 in Sec. 10-1 and switch the columns and change the column headings. In Table 10-2 we also have added a few values.

x	$\sin^{-1} x$
-1	$-\pi/2$
$-\sqrt{3}/2$	$-\pi/3$
$-1/\sqrt{2}$	$-\pi/4$
$-1/2$	$-\pi/6$
0	0
$1/2$	$\pi/6$
$1/\sqrt{2}$	$\pi/4$
$\sqrt{3}/2$	$\pi/3$
1	$\pi/2$

TABLE 10-2

In Chapter 7 we discovered that the graph of the inverse f^{-1} of the function f is the mirror image of f in the line $y = x$. To obtain the graph of $y = \sin^{-1} x$, we first graph $y = \sin x$ for $x \in [-\pi/2, \pi/2]$ and then take the reflection of $y = \sin x$ in the line $y = x$. Note that the domain of $\sin^{-1} x$ is $[-1, 1]$. We do this in Figure 10-20.

We now turn our attention to deriving the derivative of the $\sin^{-1}$ function. We write $y = \sin^{-1} x$ so that $y \in [-\pi/2, \pi/2]$ and $\sin y = x$. Differentiating the last equation implicitly yields

$$D_x \sin y = D_x x$$

$$\cos y \, D_x y = 1$$

$$D_x y = \frac{1}{\cos y}$$

$$D_x y = \frac{1}{\cos(\sin^{-1} x)}$$

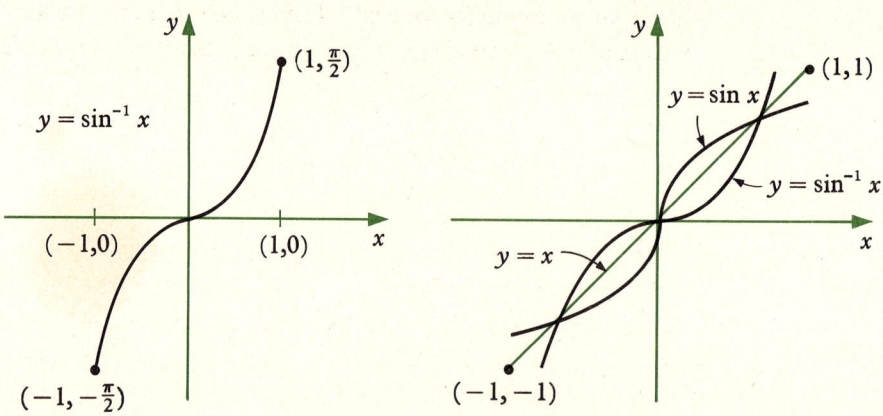

FIGURE 10-20

From Ex. 10-29 we have $\cos(\sin^{-1} x) = \sqrt{1-x^2}$ and hence we have

$$D_x \sin^{-1} x = \frac{1}{\sqrt{1-x^2}}$$

Incorporating this formula into the Chain Rule yields

$$D_x \sin^{-1} u = \frac{D_x u}{\sqrt{1-u^2}}$$

EXAMPLE 10-30

To calculate $D_x \sin^{-1}(1-x^2)$, we utilize the above formula and get

$$D_x \sin^{-1}(1-x^2) = \frac{-2x}{\sqrt{1-(1-x^2)^2}}$$

$$= \frac{-2x}{\sqrt{2x^2 - x^4}}$$

$$= \frac{-2}{\sqrt{2-x^2}}$$

EXAMPLE 10-31

To calculate $D_x \sin^{-1} e^x$ we utilize the above formula and get

$$D_x \sin^{-1} e^x = \frac{e^x}{\sqrt{1-e^{2x}}}$$

Just as we had to restrict the domain of the sine function in order to define its inverse, so too must we restrict the domain of the tangent function in order to define its inverse. Using the same criteria we used for the sine function, the interval we choose for the tangent is $(-\pi/2, \pi/2)$.

Definition The inverse tangent function, written $y = \tan^{-1} x$, with range $(-\pi/2, \pi/2)$, is defined by

$$y = \tan^{-1} x \text{ if and only if } x = \tan y$$

where $-\pi/2 < y < \pi/2$

The function $y = \tan^{-1} x$ is often called the "arctangent function" and written $y = \arctan x$. Another way of thinking of $y = \tan^{-1} x$ is "that number y whose tangent is x".

EXAMPLE 10-32

We have $\tan^{-1} x = \pi/4$ because $\tan \pi/4 = 1$ and $\pi/4 \in (-\pi/2, \pi/2)$. Also, $\tan^{-1}(-\sqrt{3}) = -\pi/3$ because $\tan(-\pi/3) = -\sqrt{3}$ and $-\pi/3 \in (-\pi/2, \pi/2)$.

The graph of $y = \tan^{-1} x$ can be obtained by reflecting the graph of the tangent function in the line $y = x$. See Fig. 10-21.

To calculate $D_x \tan^{-1} x$ we write $y = \tan^{-1} x$ where $y \in (-\pi/2, \pi/2)$. Then $\tan y = x$. By differentiating the latter equation implicitly we get

$$D_x \tan y = D_x x$$

$$\sec^2 y \, D_x y = 1$$

$$D_x y = \frac{1}{\sec^2 y}$$

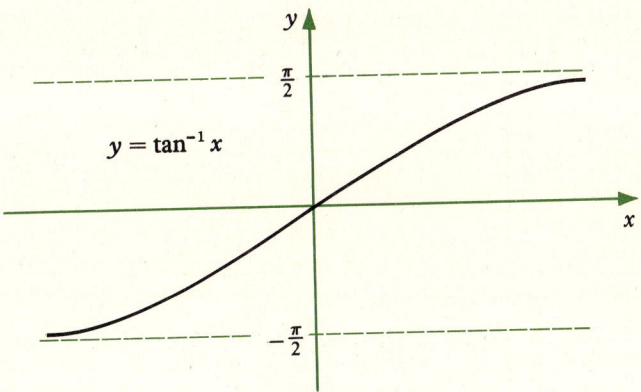

FIGURE 10-21

But $\sec^2 y = 1 + \tan^2 y$ for all y and since $\tan y = x$ we have

$$D_x \tan^{-1} x = \frac{1}{1+x^2}$$

Incorporating this formula into the Chain Rule yields

$$D_x \tan^{-1} u = \frac{D_x u}{1+u^2}$$

EXAMPLE 10-33

To calculate $D_x \tan 5x$ we utilize the above formula and get

$$D_x \tan 5x = \frac{5}{1+25x^2}$$

EXAMPLE 10-34

To calculate $D_x \tan e^{2x}$ we utilize the above formula and get

$$D_x \tan e^{2x} = \frac{2e^{2x}}{1+e^{4x}}$$

Since $\csc x = 1/\sin x$ for $\sin x \neq 0$, these functions have the same period. We can thus define the inverse cosecant as we did the inverse sine and the inverse tangent.

Definition The inverse cosecant function, written $y = \csc^{-1} x$, with range $[-\pi/2, \pi/2]$, is defined by

$$y = \csc^{-1} x \text{ if and only if } \csc y = x$$

where $-\pi/2 \leq y \leq \pi/2$

The function $y = \csc^{-1} x$ is often called the "arccosecant function" and written $y = \operatorname{arccsc} x$.

The graph of $y = \csc^{-1} x$ can be obtained by reflecting the graph of $y = \csc x$ in the line $y = x$. See Fig. 10-22. The definitions of the inverse functions of the cosine, secant, and cotangent functions can be defined similarly. The formulation of their definitions and the constructions of their graphs are left for the exercises.

We conclude this section by restating the differentiation formulas involving $\sin^{-1} x$ and $\tan^{-1} x$ as antidifferentiation or integral formulas together with some examples of their application.

$$\int \frac{1}{\sqrt{1-x^2}} \, dx = \sin^{-1} x + C$$

$$\int \frac{1}{1+x^2} \, dx = \tan^{-1} x + C$$

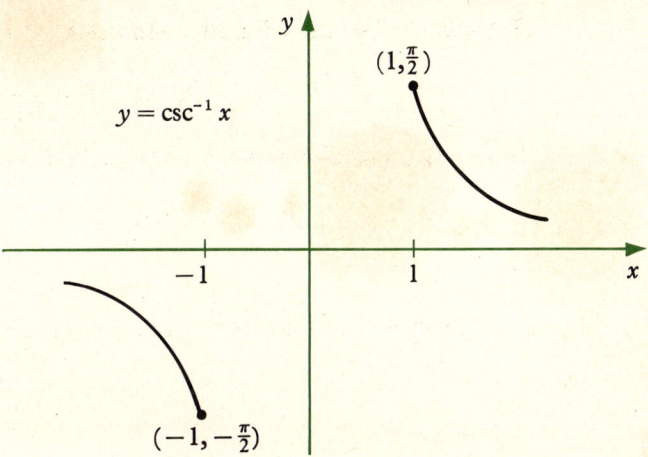

FIGURE 10-22

EXAMPLE 10-35

To calculate $\int x\,dx/\sqrt{1-x^4}$ we make the substitution $u = x^2$, $du = 2x\,dx$, $\tfrac{1}{2}du = x\,dx$ and hence

$$\int \frac{x\,dx}{\sqrt{1-x^4}} = \int \frac{\tfrac{1}{2}du}{\sqrt{1-u^2}}$$
$$= \tfrac{1}{2}\sin^{-1} u + C$$
$$= \tfrac{1}{2}\sin^{-1} x^2 + C$$

EXAMPLE 10-36

To calculate $\int dx/(4+9x^2)$ we write $4+9x^2 = 4(1+9x^2/4)$ and make the substitution $u = 3x/2$ so that $(2/3)du = dx$ and $u^2 = (9/4)x^2$. Making these substitutions yields

$$\int \frac{dx}{4+9x^2} = \int \frac{(2/3)\,du}{4(1+u^2)}$$
$$= \frac{1}{6}\int \frac{du}{1+u^2}$$
$$= (1/6)\tan^{-1} u + C$$
$$= (1/6)\tan^{-1}(3x/2) + C$$

EXAMPLE 10-37

To integrate

$$\int \frac{\sqrt{9-x^2}}{x^2}\,dx$$

we make the substitution, as in Example 10-26

$$x = 3 \sin \theta \quad \text{for } -\pi/2 < \theta < \pi/2$$
$$dx = 3 \cos \theta \, d\theta$$

and $\sqrt{9-x^2} = \sqrt{9-9\sin^2\theta} = \sqrt{9(1-\sin^2\theta)} = 3\cos\theta$

Hence

$$\int \frac{\sqrt{9-x^2}}{x^2} dx = \int \frac{3\cos\theta}{9\sin^2\theta}(3\cos\theta \, d\theta)$$
$$= \int \frac{\cos^2\theta}{\sin^2\theta} d\theta$$
$$= \int \cot^2\theta \, d\theta$$

To evaluate this trigonometric integral, we use the identity $\cot^2\theta = \csc^2\theta - 1$, and thus

$$\int \cot^2\theta \, d\theta = \int (\csc^2\theta - 1) \, d\theta$$
$$= -\cot\theta - \theta + C$$

Since $x = 3\sin\theta$ and $-\pi/2 < \theta < \pi/2$, we have $\theta = \sin^{-1}(x/3)$. Also, $\cot^2\theta = \csc^2\theta - 1 = (1/\sin^2\theta) - 1 = 1/(x^2/9) - 1 = (9 - x^2)/x^2$. Hence

$$\int \frac{\sqrt{9-x^2}}{x^2} dx = \frac{\sqrt{x^2-9}}{x} - \sin^{-1}\left(\frac{x}{3}\right) + C$$

EXERCISE 10-8

1. Evaluate the number described.

    (a) $\sin^{-1}\dfrac{1}{2}$

    (b) $\sin^{-1}\left(-\dfrac{1}{2}\right)$

    (c) $3\sin^{-1}0$

    (d) $4\sin^{-1}\dfrac{\sqrt{2}}{2}$

    (e) $\sin^{-1}(\cos 0)$

    (f) $2\tan^{-1}\left(\tan\dfrac{\pi}{4}\right)$

    (g) $\sec\left(\tan^{-1}\dfrac{1}{\sqrt{3}}\right)$

    (h) $\sin(\sin^{-1}0.1)$

2. Graph the functions.
   (a) $y = 2\sin^{-1} x$
   (b) $y = 5\tan^{-1} x$
   (c) $y = \sin^{-1} 2x$
   (d) $y = 2\tan^{-1} 3x$

3. Find $D_x y$.
   (a) $y = 2\sin^{-1} 3x$
   (b) $y = \sin^{-1} x^2$
   (c) $y = \sin^{-1}(x^2 + 5x)$
   (d) $y = 4\tan^{-1}(2x^3)$
   (e) $y = \tan^{-1} \dfrac{1}{x}$
   (f) $y = x\sin^{-1} 2x$
   (g) $y = \sin^{-1}(\tan x)$
   (h) $y = \ln \sin^{-1} x$

4. Evaluate the integrals.
   (a) $\displaystyle\int \dfrac{dx}{\sqrt{1-4x^2}}$
   (b) $\displaystyle\int \dfrac{x\,dx}{1+x^4}$
   (c) $\displaystyle\int \dfrac{x\,dx}{\sqrt{1-x^4}}$
   (d) $\displaystyle\int \dfrac{dx}{1+4x^2}$
   (e) $\displaystyle\int_0^{1/2} \dfrac{dx}{\sqrt{1-x^2}}$
   (f) $\displaystyle\int_0^1 \dfrac{dx}{1+x^2}$

10-9 APPLICATIONS

In this section we present some applications of the trigonometric functions to maxima and minima problems as well as to differential equations.

EXAMPLE 10-38

A lighthouse stands 0.3 miles off a long straight shore. The beam of light from the lighthouse has a constant speed of $\frac{1}{5}$ rpm. Suppose a car, travelling on the shore, is keeping in the beam of light for at least a little while. How fast is the car travelling when it is 0.4 miles from the point A in Fig. 10-23?

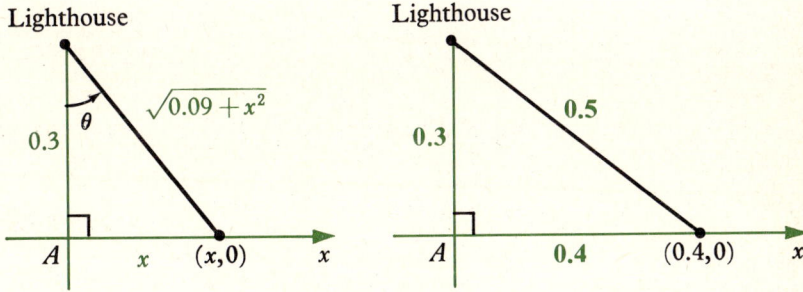

FIGURE 10-23

If we set up our coordinate axes as in Fig. 10-23, then the car and the beam are at point $(x,0)$ and our problem is to find dx/dt at $x = 0.4$. From the diagram we see that

$$\tan \theta = \frac{x}{0.3} \quad \text{or} \quad x = 0.3 \tan \theta$$

while, when $x = 0.4$, we have

$$\sec \theta = \frac{1}{\cos \theta} = \frac{1}{0.3/0.5} = \frac{5}{3}$$

From the Chain Rule we get

$$\frac{dx}{dt} = 0.3 \frac{d}{dt} \tan \theta$$

$$= 0.3 \frac{d}{d\theta} \tan \theta \cdot \frac{d\theta}{dt}$$

$$= 0.3 (\sec^2 \theta) \left(\frac{2\pi}{5} \right)$$

$$= \frac{0.6\pi}{5} (\sec^2 \theta)$$

Hence

$$\left. \frac{dx}{dt} \right|_{x=0.4} = \frac{0.6\pi}{5} \left(\frac{5}{3} \right)^2 = \frac{\pi}{3} \cong 1 \text{ mi per min}$$

where $x = 0.4$. Thus the car is travelling approximately 60 mph.

The trigonometric functions are used to describe wavelike patterns and oscillatory behavior. The next few examples illustrate such methods.

EXAMPLE 10-39

Consider the pendulum depicted in Fig. 10-24. As the pendulum oscillates back and forth, its motion is dependent upon $\sin x$, as can be understood by imagining a pen attached to the pendulum in such a way that it is always in contact with the writing surface below. If a roll of paper is passed along the surface and if we set up the axes as in Fig. 10-25, the shape of the curve traced by the motion of the pendulum would be *sinusoidal*, i.e., it will have the general shape of $y = \sin x$.

Such a function has an equation of the form

$$A \sin (Bt + C)$$

for constants A, B, and C. The constant A stretches (or contracts) the function vertically, B stretches the function horizontally, and C shifts the

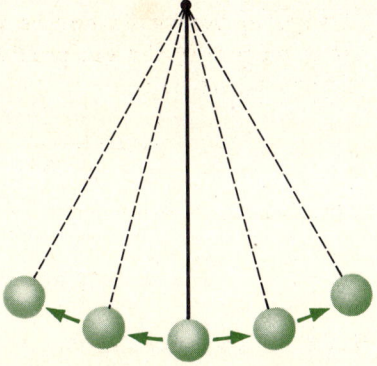

FIGURE 10-24

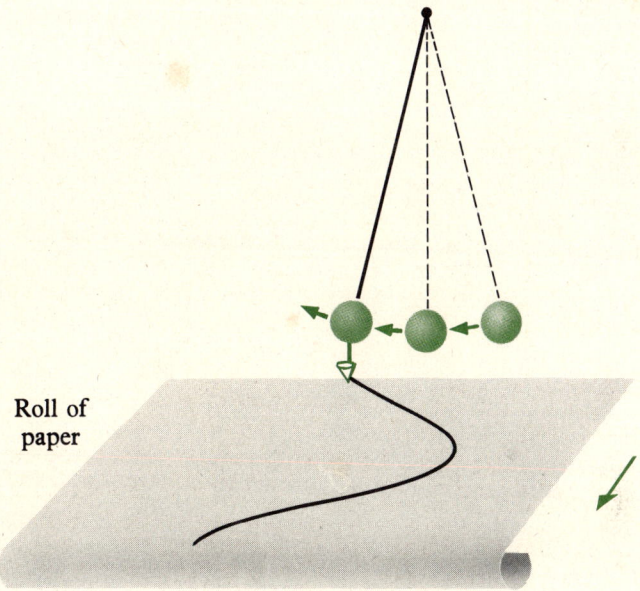

Roll of paper

FIGURE 10-25

whole graph horizontally. These concepts are amplified more in the exercises.

The motion described in Ex. 10-39 is called *simple harmonic motion*, and the theory of harmonic motion can be applied to physical phenomena including sound and ecology. In physics, a *wave* is defined to be a continuous train of disturbances or vibrations. One can think of an ordinary wave in water caused by a pebble dropped into the water. If a stretched string is

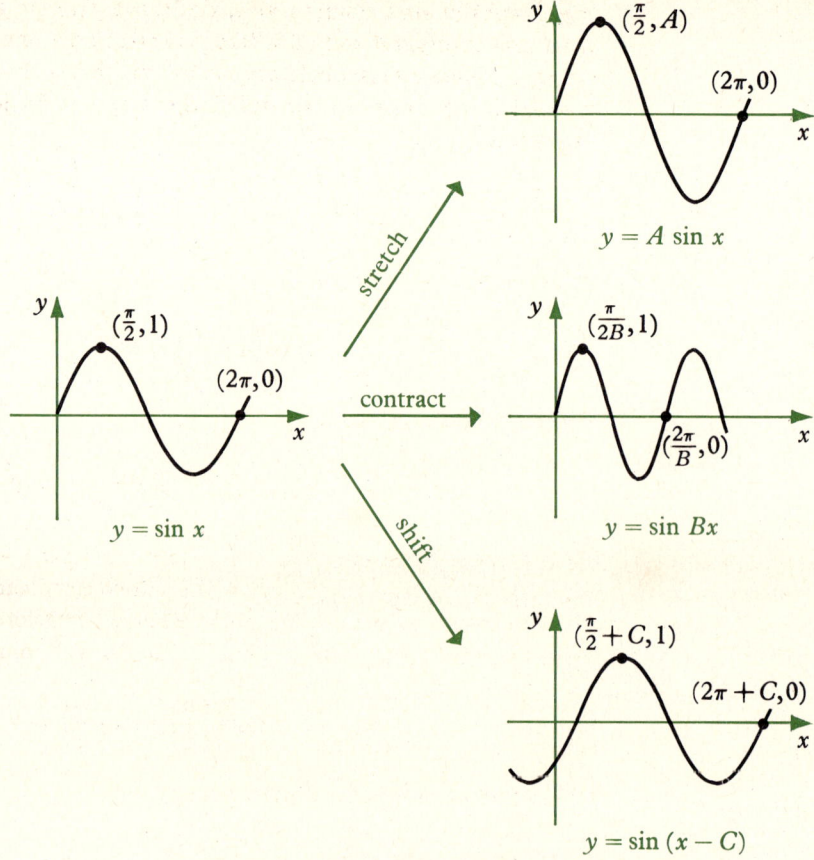

FIGURE 10-26

forced to vibrate, its motion will be of a simple harmonic nature, and the disturbance in the air caused by this vibration is heard by our ear and called *sound*. Thus we see that the phenomenon of sound is decsribed mathematically by use of the trigonometric functions.

EXAMPLE 10-40

In many physical phenomena exhibiting oscillatory behavior, the most accurate model usually will depict a function that is the *sum* of the wavelike components mentioned in the previous example. Thus the sound of a guitar is the resultant sum of the sound of each string, each vibrating at its own frequency. Suppose that the voltage representing a certain sound is given by

$$f(t) = 2\sin\left(t + \frac{\pi}{4}\right) - \sin 2t$$

The graphs of $\sin t$, $2\sin(t+\pi/4)$, $\sin 2t$, and $f(t)$ are given in Fig. 10-27. A natural question is to find where $f(t)$ takes on a minimum or maximum voltage. Since $f(t)$ is oscillatory, we assume that such an extremum will be reached many times, so we will find the extremum for t in a restricted interval, say $(0,\pi/2)$. Now

$$f'(t) = 2\cos\left(t + \frac{\pi}{4}\right) - 2\cos 2t$$

If we let $f'(t) = 0$, we get

$$\cos\left(t + \frac{\pi}{4}\right) = \cos 2t$$

Since $t \in (0,\pi/2)$, we have

$$t + \frac{\pi}{4} \in \left(\frac{\pi}{4}, \frac{3\pi}{4}\right) \quad \text{and} \quad 2t \in (0,\pi)$$

In $(0,\pi)$, $\cos t$ is one-to-one, i.e., if r is any number in $(-1,1,)$ which together with $\{1,-1\}$ is the range of $\cos t$, then there exists only one value of t in $(0,\pi)$ such that $\cos t = r$. The above equation therefore yields $t + \pi/4 = 2t$, which implies that $t = \pi/4$. Thus $f(t)$ has an extremum value of

$$f\left(\frac{\pi}{4}\right) = 2\sin\left(\frac{\pi}{2}\right) - \sin\left(\frac{\pi}{2}\right) = 1$$

We see that this value is a minimum since

$$f''(t) = -2\sin\left(t + \frac{\pi}{4}\right) + 4\sin 2t \quad \text{and} \quad f''\left(\frac{\pi}{4}\right) = 2$$

and the second derivative test implies the result.

One characteristic property of the trigonometric functions is that they provide solutions to the differential equation

$$y'' = -a^2 y$$

for the constant a. One can immediately calculate to see that $y = A\sin ax$ and $y = A\cos ax$ are solutions to the above equation. In fact, one can prove, although it is fairly difficult, that every solution to the above equation is of the form

$$y = A\sin ax + B\cos ax$$

Thus the field of differential equations is dependent upon the trigonometric functions.

CHAP. 10   Trigonometric Functions

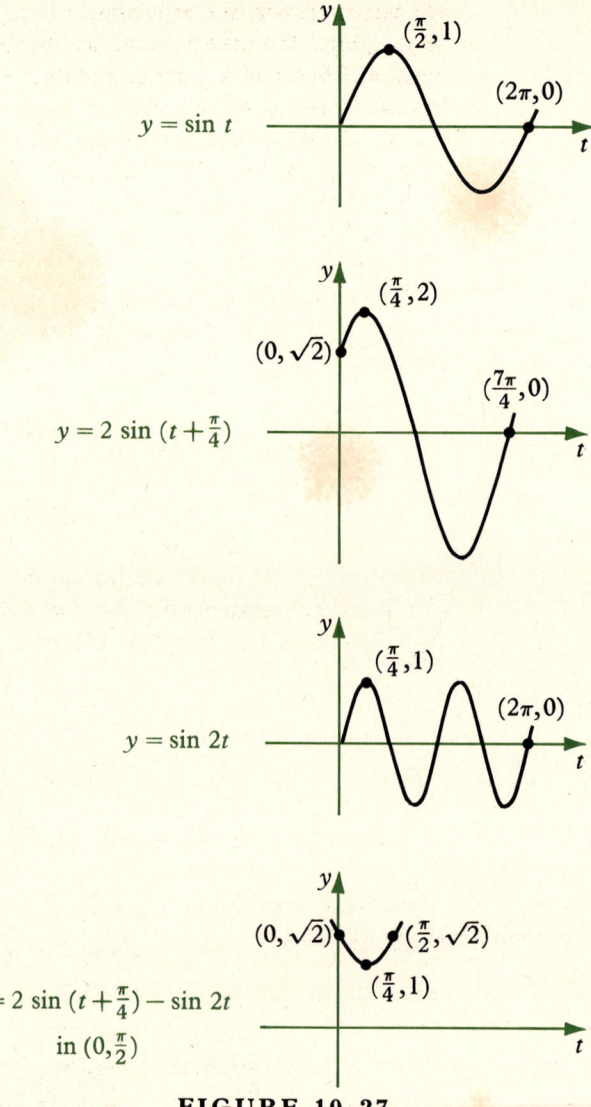

FIGURE 10-27

EXAMPLE 10-41

Let us once again consider the ecology model of the predator-prey population fluctuation first mentioned in Chap. 4. We expanded that model by introducing the work of Lotka and Volterra in Chap. 8, where we discussed methods of population fluctuation of an idealized situation of one species occupying a territory without interaction from other species. We now have at hand tools to handle the more complex model of two species occupying

The graphs of $\sin t$, $2\sin(t+\pi/4)$, $\sin 2t$, and $f(t)$ are given in Fig. 10-27. A natural question is to find where $f(t)$ takes on a minimum or maximum voltage. Since $f(t)$ is oscillatory, we assume that such an extremum will be reached many times, so we will find the extremum for t in a restricted interval, say $(0,\pi/2)$. Now

$$f'(t) = 2\cos\left(t + \frac{\pi}{4}\right) - 2\cos 2t$$

If we let $f'(t) = 0$, we get

$$\cos\left(t + \frac{\pi}{4}\right) = \cos 2t$$

Since $t \in (0,\pi/2)$, we have

$$t + \frac{\pi}{4} \in \left(\frac{\pi}{4}, \frac{3\pi}{4}\right) \quad \text{and} \quad 2t \in (0,\pi)$$

In $(0,\pi)$, $\cos t$ is one-to-one, i.e., if r is any number in $(-1,1,)$ which together with $\{1,-1\}$ is the range of $\cos t$, then there exists only one value of t in $(0,\pi)$ such that $\cos t = r$. The above equation therefore yields $t + \pi/4 = 2t$, which implies that $t = \pi/4$. Thus $f(t)$ has an extremum value of

$$f\left(\frac{\pi}{4}\right) = 2\sin\left(\frac{\pi}{2}\right) - \sin\left(\frac{\pi}{2}\right) = 1$$

We see that this value is a minimum since

$$f''(t) = -2\sin\left(t + \frac{\pi}{4}\right) + 4\sin 2t \quad \text{and} \quad f''\left(\frac{\pi}{4}\right) = 2$$

and the second derivative test implies the result.

One characteristic property of the trigonometric functions is that they provide solutions to the differential equation

$$y'' = -a^2 y$$

for the constant a. One can immediately calculate to see that $y = A\sin ax$ and $y = A\cos ax$ are solutions to the above equation. In fact, one can prove, although it is fairly difficult, that every solution to the above equation is of the form

$$y = A\sin ax + B\cos ax$$

Thus the field of differential equations is dependent upon the trigonometric functions.

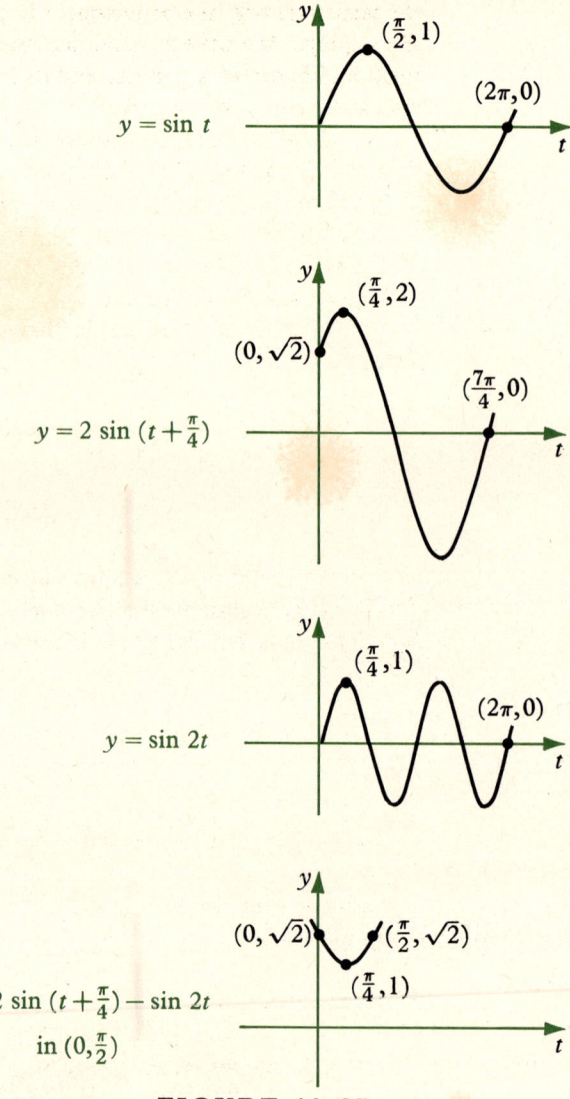

FIGURE 10-27

EXAMPLE 10-41

Let us once again consider the ecology model of the predator-prey population fluctuation first mentioned in Chap. 4. We expanded that model by introducing the work of Lotka and Volterra in Chap. 8, where we discussed methods of population fluctuation of an idealized situation of one species occupying a territory without interaction from other species. We now have at hand tools to handle the more complex model of two species occupying the same territory, one population F being predator and the other population H the prey. We have in mind here the biological association between, say, foxes and hares, or a parasite and its host, or perhaps the spread of a fatal disease by means of bacteria.

In addition to the discussion in Chap. 8 in which we assumed that the rate of growth $D_t F(t)$ of the population F depended upon F, we further assume that the change of the size of F is proportional to H also. If there is an abundance of prey (H is large) at time t, then F will increase since its food supply is large. If H is small, the threat of starvation impedes the increase of F, and soon F will decrease.

The simplest model is that originally described by Lotka and Volterra, namely,

$$D_t F = (a - bH) F = \frac{dF}{dt}$$

$$D_t H = (-c + gF) H = \frac{dH}{dt}$$

where a, b, c, and g are positive constants. We'd like to solve this system of differential equations for F and H. We note that if we divide the two equations, we can formally (in the same sense as in Chap. 5) write

$$\frac{dF}{dH} = \frac{(a - bH) F}{(-c + gF) H}$$

and so

$$c \frac{dF}{F} - g\, dF + a \frac{dH}{H} - b\, dH = 0$$

Upon integrating we get

$$c \cdot \ln F - g \cdot F + a \cdot \ln H - b \cdot H = k$$

for some constant k.

The above equation yields a family of curves, one for each choice of a constant k. The initial conditions for F and H determine the constant. Three curves of the family are given in Fig. 10-28 for choices of $a = 1$, $b = 0.1$, $c = 0.5$ and $g = 0.02$.

As one can readily see from the diagrams, when the curves are very small, the equation can be approximated by an ellipse. For such cases, it is possible to derive formulas for F and H.

Let us investigate behavioral patterns described by the inner curve. Consider the point A on the curve, that is, $F \cong 8$ and $H \cong 20$. As F increases, H stays approximately constant until the predators kill too many prey, then F will continue to increase, but the rate of increase will

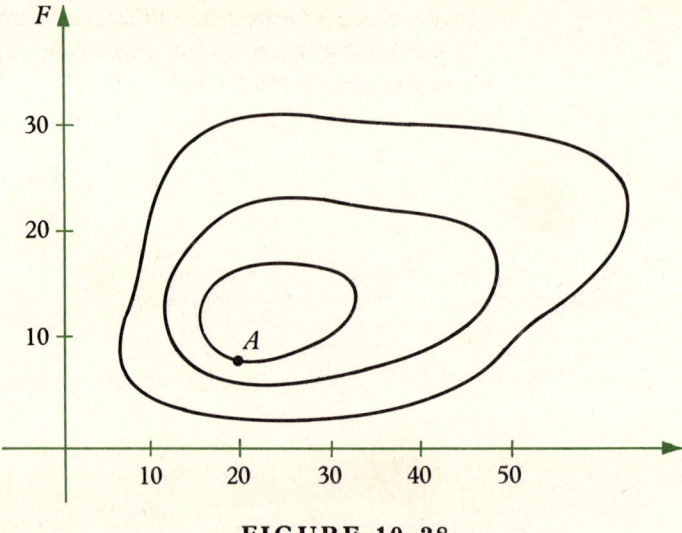

FIGURE 10-28

slow. Then H will reflect this shift and begin to increase slowly until F reaches its maximum. As F then decreases, H increases more rapidly till it reaches its maximum.

Given the above discussion it seems reasonable to assume that the functions F and H themselves can be expressed in terms of the trigonometric functions. In fact, for small values of k, we have F and H of the form

$$F(t) = A_1 + B_1 \cos(C_1 t + D_1)$$
$$H(t) = A_2 + B_2 \sin(C_2 t + D_2)$$

Only trigonometric functions can adequately describe the fluctuations and oscillation of such population behavior.

EXERCISE 10-9

1. A lighthouse is 11 miles off a long straight shore. The light beam rotates at a constant rate of 2 rpm. How fast is the beam moving along the shore as it passes a point 5 miles from the point on the shore closest to the lighthouse?

2. Suppose a point A is moving counterclockwise at the rate of 2 rpm around a unit circle whose center is at the origin. How fast is the distance between A and the point $(3,0)$ increasing when the line from the origin to A makes an angle of $\pi/4$ radians with the x-axis, i.e., when A is at the point $(\frac{\sqrt{2}}{2}, \frac{\sqrt{2}}{2})$?

3. What are the dimensions of the isosceles triangle of smallest area which circumscribes a circle of radius one?

4. Suppose a man who lives in a lighthouse 4 miles off shore wants to travel to a spot 3 miles farther up along the shore. If he can walk twice as fast as he can row, what is his fastest route?

5. Find the extrema in the interval $(0, \pi/2)$ of the function
$$y = \sin\left(2t + \frac{\pi}{4}\right) - 2\sin\left(t + \frac{3\pi}{8}\right)$$

6. Find the extrema in the interval $(0, \pi/2)$ of the function
$$y = 2\sin\left(3t + \frac{\pi}{6}\right) - 3\sin\left(2t + \frac{\pi}{3}\right)$$

7. Show that $y = A \cos ax$ and $y = A \sin ax$ are solutions to the differential equation $y'' = -a^2 y$.

8. Show that $y = A \cos ax + B \sin ax$ is a solution to the differential equation $y'' = -a^2 y$.

9. Show that $y = Ae^{ax}$ is not a solution, as you may have suspected, to the differential equation $y'' = -a^2 y$.

Appendix A

Basic Integration Formulas

1. $\int x^r \, dx = \dfrac{x^{r+1}}{r+1}, \; r \neq -1$ *(power rule)*

2. $\int \sin x \, dx = -\cos x$

3. $\int \cos x \, dx = \sin x$

4. $\int \sec^2 x \, dx = \tan x$

5. $\int \csc^2 x \, dx = -\cot x$

6. $\int \sec x \tan x \, dx = \sec x$

7. $\int \csc x \cot x \, dx = -\csc x$

8. $\int e^x \, dx = e^x$

9. $\int \dfrac{1}{x} \, dx = \ln|x|$

APPENDIX A   Basic Integration Formulas

10. $\displaystyle\int a^x\, dx = \frac{a^x}{\ln a},\ a>0,\ a\ne 1$

11. $\displaystyle\int \frac{1}{\sqrt{a^2+x^2}}\, dx = \ln\left|x+\sqrt{x^2+a^2}\right|$

12. $\displaystyle\int \frac{1}{\sqrt{x^2-a^2}}\, dx = \ln\left|x+\sqrt{x^2-a^2}\right|$

13. $\displaystyle\int \frac{1}{a^2-x^2}\, dx = \frac{1}{2a}\ln\left|\frac{a+x}{a-x}\right|,\ x^2<a^2$

14. $\displaystyle\int \frac{1}{x^2-a^2}\, dx = -\frac{1}{2a}\ln\left|\frac{x+a}{x-a}\right|,\ a^2<x^2$

15. $\displaystyle\int \frac{1}{x\sqrt{a^2-x^2}}\, dx = -\frac{1}{a}\ln\left|\frac{a+\sqrt{a^2-x^2}}{x}\right|,\ 0<x<a$

16. $\displaystyle\int x^n \ln x\, dx = x^{n+1}\left[\frac{\ln x}{n+1}-\frac{1}{(n+1)^2}\right]$

17. $\displaystyle\int \frac{1}{x\sqrt{a^2+x^2}}\, dx = -\frac{1}{a}\ln\left|\frac{a+\sqrt{a^2+x^2}}{x}\right|$

18. $\displaystyle\int \ln|x|\, dx = x(\ln|x|-1)$

19. $\displaystyle\int \frac{x}{ax+b}\, dx = \frac{x}{a}-\frac{b}{a^2}\ln|ax+b|$

20. $\displaystyle\int \frac{x}{(ax+b)^2}\, dx = \frac{b}{a^2(ax+b)}+\frac{1}{a^2}\ln|ax+b|$

21. $\displaystyle\int \frac{1}{x(ax+b)}\, dx = \frac{1}{b}\ln\left|\frac{x}{ax+b}\right|$

22. $\displaystyle\int \frac{1}{x(ax+b)^2}\, dx = \frac{1}{b(ax+b)}+\frac{1}{b^2}\ln\left|\frac{x}{ax+b}\right|$

23. $\displaystyle\int \sqrt{x^2\pm a^2}\, dx = \tfrac{1}{2}x\sqrt{x^2\pm a^2}+\frac{a^2}{2}\ln\left|x+\sqrt{x^2\pm a^2}\right|$

24. $\displaystyle\int \tan x\, dx = \ln|\sec x|$

25. $\displaystyle\int \cot x\, dx = \ln|\sin x|$

APPENDIX A Basic Integration Formulas

26. $\int \sec x \, dx = \ln|\sec x + \tan x|$

27. $\int \csc x \, dx = \ln|\csc x - \cot x|$

28. $\int x^n \ln x \, dx = x^{n+1} \left[\dfrac{\ln x}{n+1} - \dfrac{1}{(n+1)^2} \right], \ n \neq -1$

29. $\int x^n e^{ax} \, dx = \dfrac{1}{a} x^n e^{ax} - \dfrac{n}{a} \int x^{n-1} e^{ax} \, dx$

30. $\int e^{ax} \sin bx \, dx = \dfrac{e^{ax}}{a^2 + b^2} (a \sin bx - b \cos bx)$

31. $\int e^{ax} \cos bx \, dx = \dfrac{e^{ax}}{a^2 + b^2} (b \sin bx + a \cos bx)$

32. $\int \dfrac{1}{\sqrt{a^2 - x^2}} \, dx = \sin^{-1}(x/a), \ a > 0$

33. $\int \dfrac{1}{a^2 + x^2} \, dx = \tan^{-1}(x/a), \ a > 0$

34. $\int \sqrt{a^2 - x^2} \, dx = \dfrac{x}{2} \sqrt{a^2 - x^2} + \dfrac{a^2}{2} \sin^{-1} \dfrac{x}{a}$

35. $\int x^2 \sqrt{a^2 - x^2} \, dx = -\tfrac{1}{4} x (a^2 - x^2)^{3/2} + \tfrac{1}{8} a^2 x \sqrt{a^2 - x^2} + \tfrac{1}{8} a^4 \sin^{-1} \dfrac{x}{a}$

36. $\int \dfrac{dx}{ax^2 + bx + c} = \dfrac{2}{\sqrt{4ac - b^2}} \tan^{-1} \left(\dfrac{2ax + b}{\sqrt{4ac - b^2}} \right) \quad (b^2 < 4ac)$

37. $\int \dfrac{x \, dx}{ax^2 + bx + c} = \dfrac{1}{2a} \ln|ax^2 + bx + c|$

$\quad - \dfrac{b}{a \sqrt{4ac - b^2}} \tan^{-1} \left(\dfrac{2ax + b}{\sqrt{4ac - b^2}} \right) \quad (b^2 < 4ac)$

Appendix B

Tables of Natural Logarithms and the Exponential Function

TABLE B-1 NATURAL LOGARITHMS

Numbers 0.00 to 1.99 (Base e = 2.718281)

N		0	1	2	3	4	5	6	7	8	9
0.0			5.395	6.088	6.493	6.781	7.004	7.187	7.341	7.474	7.592
0.1		7.697	7.793	7.880	7.960	8.034	8.103	8.167	8.228	8.285	8.339
0.2	Take tabular value −10	8.391	8.439	8.486	8.530	8.573	8.614	8.653	8.691	8.727	8.762
0.3		8.796	8.829	8.861	8.891	8.921	8.950	8.978	9.006	9.032	9.058
0.4		9.084	9.108	9.132	9.156	9.179	9.201	9.223	9.245	9.266	9.287
0.5		9.307	9.327	9.346	9.365	9.384	9.402	9.420	9.438	9.455	9.472
0.6		9.489	9.506	9.522	9.538	9.554	9.569	9.584	9.600	9.614	9.629
0.7		9.643	9.658	9.671	9.685	9.699	9.712	9.726	9.739	9.752	9.764
0.8		9.777	9.789	9.802	9.814	9.826	9.837	9.849	9.861	9.872	9.883
0.9		9.895	9.906	9.917	9.927	9.938	9.949	9.959	9.970	9.980	9.990
1.0	0.0	0000	0995	1980	2956	3922	4879	5827	6766	7696	8618
1.1		9531	*0436	*1333	*2222	*3103	*3976	*4842	*5700	*6551	*7395
1.2	0.1	8232	9062	9885	*0701	*1511	*2314	*3111	*3902	*4686	*5464
1.3	0.2	6236	7003	7763	8518	9267	*0010	*0748	*1481	*2208	*2930
1.4	0.3	3647	4359	5066	5767	6464	7156	7844	8526	9204	9878
1.5	0.4	0547	1211	1871	2527	3178	3825	4469	5103	5742	6373
1.6		7000	7623	8243	8858	9470	*0078	*0682	*1282	*1879	*2473
1.7	0.5	3063	3649	4232	4812	5389	5962	6531	7098	7661	8222
1.8		8779	9333	9884	*0432	*0977	*1519	*2058	*2594	*3127	*3658
1.9	0.6	4185	4710	5233	5752	6269	6783	7294	7803	8310	8813

log. 0.10 = 7.69741 49070 − 10

TABLE B–1 (Continued)

Numbers 2.00 to 5.99

N		0	1	2	3	4	5	6	7	8	9
2.0	0.6	9315	9813	*0310	*0804	*1295	*1784	*2271	*2755	*3237	*3716
2.1	0.7	4194	4669	5142	5612	6081	6547	7011	7473	7932	8390
2.2		8846	9299	9751	*0200	*0648	*1093	*1536	*1978	*2418	*2855
2.3	0.8	3291	3725	4157	4587	5015	5442	5866	6289	6710	7129
2.4		7547	7963	8377	8789	9200	9609	*0016	*0422	*0826	*1228
2.5	0.9	1629	2028	2426	2822	3216	3609	4001	4391	4779	5166
2.6		5551	5935	6317	6698	7078	7456	7833	8208	8582	8954
2.7		9325	9695	*0063	*0430	*0796	*1160	*1523	*1885	*2245	*2604
2.8	1.0	2962	3318	3674	4028	4380	4732	5082	5431	5779	6126
2.9		6471	6815	7158	7500	7841	8181	8519	8856	9192	9527
3.0		9861	*0194	*0526	*0856	*1186	*1514	*1841	*2168	*2493	*2817
3.1	1.1	3140	3462	3783	4103	4422	4740	5057	5373	5688	6002
3.2		6315	6627	6938	7248	7557	7865	8173	8479	8784	9089
3.3		9392	9695	9996	*0297	*0597	*0896	*1194	*1491	*1788	*2083
3.4	1.2	2378	2671	2964	3256	4547	3837	4127	4415	4703	4990
3.5		5276	5562	5846	6130	6413	6695	6976	7257	7536	7815
3.6		8093	8371	8647	8923	9198	9473	9746	*0019	*0291	*0563
3.7	1.3	0833	1103	1372	1641	1909	2176	2442	2708	2972	3237
3.8		3500	3763	4025	4286	4547	4807	5067	5325	5584	5841
3.9		6098	6354	6609	6864	7118	7372	7624	7877	8128	8379
4.0		8629	8879	9128	9377	9624	9872	*0118	*0364	*0610	*0854
4.1	1.4	1099	1342	1585	1828	2070	2311	2552	2792	3031	3270
4.2		3508	3746	3984	4220	4456	4692	4927	5161	5395	5629
4.3		5862	6094	6326	6557	6787	7018	7247	7476	7705	7933
4.4		8160	8387	8614	8840	9065	9200	9515	9739	9962	*0185
4.5	1.5	0408	0630	0851	1072	1293	1513	1732	1951	2170	2388
4.6		2606	2823	3039	3256	3471	3687	3902	4116	4330	4543
4.7		4756	4969	5181	5393	5604	5814	6025	6235	6444	6653
4.8		6862	7070	7277	7485	7691	7898	8104	8309	8515	8719
4.9		8924	9127	9331	9534	9737	9939	*0141	*0342	*0543	*0744
5.0	1.6	0944	1144	1343	1542	1741	1939	2137	2334	2531	2728
5.1		2924	3120	3315	3511	3705	3900	4094	4287	4481	4673
5.2		4866	5058	5250	5441	5632	5823	6013	6203	6393	6582
5.3		6771	6959	7147	7335	7523	7710	7896	8083	8269	8455
5.4		8640	8825	9010	9194	9378	9562	9745	9928	*0111	*0293
5.5	1.7	0475	0656	0838	1019	1199	1380	1560	1740	1919	2098
5.6		2277	2455	2633	2811	2988	3166	3342	3519	3695	3871
5.7		4047	4222	4397	4572	4746	4920	5094	5267	5440	5613
5.8		5786	5958	6130	6302	6473	6644	6815	6985	7156	7326
5.9		7495	7665	7834	8002	8171	8339	8507	8675	8842	9009

TABLE B-1 (Continued)

Numbers 6.00 to 10.09

N		0	1	2	3	4	5	6	7	8	9
6.0	1.7	9176	9342	9509	9675	9840	*0006	*0171	*0336	*0500	*0665
6.1	1.8	0829	0993	1156	1319	1482	1645	1808	1970	2132	2294
6.2		2455	2616	2777	2938	3098	3258	3418	3578	3737	3896
6.3		4055	4214	4372	4530	4688	4845	5003	5160	5317	5473
6.4		5630	5786	5942	6097	6253	6408	6563	6718	6872	7026
6.5		7180	7334	7487	7641	7794	7947	8099	8251	8403	8555
6.6		8707	8858	9010	9160	9311	9462	9612	9762	9912	*0061
6.7	1.9	0211	0360	0509	0658	0806	0954	1102	1250	1398	1545
6.8		1692	1839	1986	2132	2279	2425	2571	2716	2862	3007
6.9		3152	3297	3442	3586	3730	3874	4018	4162	4305	4448
7.0		4591	4734	4876	5019	5161	5303	5445	5586	5727	5869
7.1		6009	6150	6291	6431	6571	6711	6851	6991	7130	7269
7.2		7408	7547	7685	7824	7962	8100	8238	8376	8513	8650
7.3		8787	8924	9061	9198	9334	9470	9606	9742	9877	*0013
7.4	2.0	0148	0283	0418	0553	0687	0821	0956	1089	1223	1357
7.5		1490	1624	1757	1890	2022	2155	2287	2419	2551	2683
7.6		2815	2946	3078	3209	3340	3471	3601	3732	3862	3992
7.7		4122	4252	4381	4511	4640	4769	4898	5027	5156	5284
7.8		5412	5540	5668	5796	5924	6051	6179	6306	6433	6560
7.9		6686	6813	6939	7065	7191	7317	7443	7568	7694	7819
8.0		7944	8069	8194	8318	8443	8567	8691	8815	8939	9063
8.1		9186	9310	9433	9556	9679	9802	9924	*0047	*0169	*0291
8.2	2.1	0413	0535	0657	0779	0900	1021	1142	1263	1384	1505
8.3		1626	1746	1866	1986	2106	2226	2346	2465	2585	2704
8.4		2823	2942	3061	3180	3298	3417	3535	3653	3771	3889
8.5		4007	4124	4242	4359	4476	4593	4710	4827	4943	5060
8.6		5176	5292	5409	5524	5640	5756	5871	5987	6102	6217
8.7		6332	6447	6562	6677	6791	6905	7020	7134	7248	7361
8.8		7475	7589	7702	7816	7929	8042	8155	8267	8380	8493
8.9		8605	8717	8830	8942	9054	9165	9277	9389	9500	9611
9.0		9722	9834	9944	*0055	*0166	*0276	*0387	*0497	*0607	*0717
9.1	2.2	0827	0937	1047	1157	1266	1375	1485	1594	1703	1812
9.2		1920	2029	2138	2246	2354	2462	2570	2678	2786	2894
9.3		3001	3109	3216	3324	3431	3538	3645	3751	3858	3965
9.4		4071	4177	4284	4390	4496	4601	4707	4813	4918	5024
9.5		5129	5234	5339	5444	5549	5654	5759	5863	5968	6072
9.6		6176	6280	6384	6488	6592	6696	6799	6903	7006	7109
9.7		7213	7316	7419	7521	7624	7727	7829	7932	8034	8136
9.8		8238	8340	8442	8544	8646	8747	8849	8950	9051	9152
9.9		9253	9354	9455	9556	9657	9757	9858	9958	*0058	*0158
10.0	2.3	0259	0358	0458	0558	0658	0757	0857	0956	1055	1154

TABLE B-1 (Continued)

Numbers 10 to 99

N	0	1	2	3	4	5	6	7	8	9
1	2.30259	39790	48491	56495	63906	70805	77259	83321	89037	94444
2	99573	*04452	*09104	*13549	*17805	*21888	*25810	*29584	*33220	*36730
3	3.40120	43399	46574	49651	52636	55535	58352	61092	63759	66356
4	68888	71357	73767	76120	78419	80666	82864	85015	87120	89182
5	91202	93183	95124	97029	98898	*00733	*02535	*04305	*06044	*07754
6	4.09434	11087	12713	14313	15888	17439	18965	20469	21951	23411
7	24850	26268	27667	29046	30407	31749	33073	34381	35671	36945
8	38203	39445	40672	41884	43082	44265	45435	46591	47734	48864
9	49981	51086	52179	53260	54329	55388	56435	57471	58497	59512

$\log 10 = 2.30258\ 50930$

TABLE B-2 EXPONENTIAL FUNCTION

x	e^x	e^{-x}	x	e^x	e^{-x}
0.00	1.000	1.000	0.25	1.284	0.779
0.01	1.010	0.990	0.26	1.297	0.771
0.02	1.020	0.980	0.27	1.310	0.763
0.03	1.031	0.970	0.28	1.323	0.756
0.04	1.041	0.960	0.29	1.336	0.748
0.05	1.051	0.951	0.30	1.350	0.741
0.06	1.062	0.942	0.31	1.363	0.733
0.07	1.073	0.932	0.32	1.377	0.726
0.08	1.083	0.923	0.33	1.391	0.719
0.09	1.094	0.914	0.34	1.405	0.712
0.10	1.105	0.905	0.35	1.419	0.705
0.11	1.116	0.896	0.36	1.433	0.698
0.12	1.128	0.887	0.37	1.478	0.691
0.13	1.139	0.878	0.38	1.462	0.684
0.14	1.150	0.869	0.39	1.477	0.677
0.15	1.162	0.861	0.40	1.492	0.670
0.16	1.174	0.852	0.41	1.507	0.664
0.17	1.185	0.844	0.42	1.522	0.657
0.18	1.197	0.835	0.43	1.537	0.651
0.19	1.209	0.827	0.44	1.553	0.644
0.20	1.221	0.819	0.45	1.568	0.638
0.21	1.234	0.811	0.46	1.584	0.631
0.22	1.246	0.802	0.47	1.600	0.625
0.23	1.259	0.795	0.48	1.616	0.619
0.24	1.271	0.787	0.49	1.632	0.613

TABLE B-2 (Continued)

x	e^x	e^{-x}	x	e^x	e^{-x}
0.50	1.649	0.607	0.90	2.460	0.407
0.51	1.665	0.601	0.91	2.484	0.403
0.52	1.682	0.595	0.92	2.509	0.399
0.53	1.699	0.589	0.93	2.535	0.395
0.54	1.716	0.583	0.94	2.560	0.391
0.55	1.733	0.577	0.95	2.586	0.387
0.56	1.751	0.571	0.96	2.612	0.383
0.57	1.768	0.566	0.97	2.638	0.379
0.58	1.786	0.560	0.98	2.665	0.375
0.59	1.804	0.554	0.99	2.691	0.372
0.60	1.822	0.549	1.0	2.718	0.368
0.61	1.840	0.543	1.1	3.004	0.333
0.62	1.859	0.538	1.2	3.320	0.301
0.63	1.878	0.533	1.3	3.669	0.273
0.64	1.897	0.527	1.4	4.055	0.247
0.65	1.916	0.522	1.5	4.482	0.223
0.66	1.935	0.517	1.6	4.953	0.202
0.67	1.954	0.512	1.7	5.474	0.183
0.68	1.974	0.507	1.8	6.050	0.165
0.69	1.994	0.502	1.9	6.686	0.150
0.70	2.014	0.497	2.0	7.389	0.135
0.71	2.034	0.492	2.1	8.166	0.122
0.72	2.054	0.487	2.2	9.025	0.111
0.73	2.075	0.482	2.3	9.974	0.100
0.74	2.096	0.477	2.4	11.023	0.091
0.75	2.117	0.472	2.5	12.182	0.082
0.76	2.138	0.468	2.6	13.464	0.074
0.77	2.160	0.463	2.7	14.880	0.067
0.78	2.181	0.458	2.8	16.445	0.061
0.79	2.203	0.453	2.9	18.174	0.055
0.80	2.226	0.449	3.0	20.086	0.050
0.81	2.248	0.445	3.1	22.198	0.045
0.82	2.271	0.440	3.2	24.533	0.041
0.83	2.293	0.436	3.3	27.113	0.037
0.84	2.316	0.432	3.4	29.964	0.033
0.85	2.340	0.427	3.5	33.115	0.030
0.86	2.363	0.423	3.6	36.598	0.027
0.87	2.387	0.419	3.7	40.447	0.025
0.88	2.411	0.415	3.8	44.701	0.022
0.89	2.435	0.411	3.9	49.402	0.020

TABLE B-3 TRIGONOMETRIC FUNCTIONS $\sin\theta$, $\cos\theta$, $\tan\theta$, AND $\cot\theta$

Angle θ		$\sin\theta$	$\cos\theta$	$\tan\theta$	$\cot\theta$		
Radians	Degrees						
0.0000	0.0	1.0000	1.0000	0.0000	—	90.0	1.5708
0.0087	0.5	0.0087	1.0000	0.0087	114.5887	89.5	1.5621
0.0175	1.0	0.0175	0.9998	0.0175	57.2900	89.0	1.5533
0.0262	1.5	0.0262	0.9997	0.0262	38.1885	88.5	1.5446
0.0349	2.0	0.0349	0.9994	0.0349	28.6363	88.0	1.5359
0.0436	2.5	0.0436	0.9990	0.0437	22.9038	87.5	1.5272
0.0524	3.0	0.0523	0.9986	0.0524	19.0811	87.0	1.5184
0.0611	3.5	0.0610	0.9981	0.0612	16.3499	86.5	1.5097
0.0698	4.0	0.0698	0.9976	0.0699	14.3007	86.0	1.5010
0.0785	4.5	0.0785	0.9969	0.0787	12.7062	85.5	1.4923
0.0873	5.0	0.0872	0.9962	0.0875	11.4301	85.0	1.4835
0.0960	5.5	0.0958	0.9954	0.0963	10.3854	84.5	1.4748
0.1047	6.0	0.1045	0.9945	0.1051	9.5144	84.0	1.4661
0.1134	6.5	0.1132	0.9936	0.1139	8.7769	83.5	1.4574
0.1222	7.0	0.1219	0.9925	0.1228	8.1443	83.0	1.4486
0.1309	7.5	0.1305	0.9914	0.1317	7.5958	82.5	1.4399
0.1396	8.0	0.1392	0.9903	0.1405	7.1154	82.0	1.4312
0.1484	8.5	0.1478	0.9890	0.1495	6.6912	81.5	1.4224
0.1571	9.0	0.1564	0.9877	0.1584	6.3138	81.0	1.4137
0.1658	9.5	0.1650	0.9863	0.1673	5.9758	80.5	1.4050
0.1745	10.0	0.1736	0.9848	0.1763	5.6713	80.0	1.3963
0.1833	10.5	0.1822	0.9833	0.1853	5.3955	79.5	1.3875
0.1920	11.0	0.1908	0.9816	0.1944	5.1446	79.0	1.3788
0.2007	11.5	0.1994	0.9799	0.2035	4.9152	78.5	1.3701
0.2094	12.0	0.2079	0.9781	0.2126	4.7046	78.0	1.3614
0.2182	12.5	0.2164	0.9763	0.2217	4.5107	77.5	1.3526
0.2269	13.0	0.2250	0.9744	0.2309	4.3315	77.0	1.3439
0.2356	13.5	0.2334	0.9724	0.2401	4.1653	76.5	1.3352
0.2443	14.0	0.2419	0.9703	0.2493	4.0108	76.0	1.3265
0.2531	14.5	0.2504	0.9681	0.2586	3.8667	75.5	1.3177
0.2618	15.0	0.2588	0.9659	0.2679	3.7321	75.0	1.3090
0.2705	15.5	0.2672	0.9636	0.2773	3.6059	74.5	1.3003
0.2793	16.0	0.2756	0.9613	0.2867	3.4874	74.0	1.2915
0.2880	16.5	0.2840	0.9588	0.2962	3.3759	73.5	1.2828
0.2967	17.0	0.2924	0.9563	0.3057	3.2709	73.0	1.2741
0.3054	17.5	0.3007	0.9537	0.3153	3.1716	72.5	1.2654
		$\cos\theta$	$\sin\theta$	$\cot\theta$	$\tan\theta$	Degrees	Radians
						Angle θ	

TABLE B-3 (Continued)

Angle θ		sin θ	cos θ	tan θ	cot θ		
Radians	Degrees						
0.3142	18.0	0.3090	0.9511	0.3249	3.0777	72.0	1.2566
0.3229	18.5	0.3173	0.9483	0.3346	2.9887	71.5	1.2479
0.3316	19.0	0.3256	0.9455	0.3443	2.9042	71.0	1.2392
0.3403	19.5	0.3338	0.9426	0.3541	2.8239	70.5	1.2305
0.3491	20.0	0.3420	0.9397	0.3640	2.7475	70.0	1.2217
0.3578	20.5	0.3502	0.9367	0.3739	2.6746	69.5	1.2130
0.3665	21.0	0.3584	0.9336	0.3839	2.6051	69.0	1.2043
0.3752	21.5	0.3665	0.9304	0.3939	2.5386	68.5	1.1956
0.3840	22.0	0.3746	0.9272	0.4040	2.4751	68.0	1.1868
0.3927	22.5	0.3827	0.9239	0.4142	2.4142	67.5	1.1781
0.4014	23.0	0.3907	0.9205	0.4245	2.3559	67.0	1.1694
0.4102	23.5	0.3987	0.9171	0.4348	2.2998	66.5	1.1606
0.4189	24.0	0.4067	0.9135	0.4452	2.2460	66.0	1.1519
0.4276	24.5	0.4147	0.9100	0.4557	2.1943	65.5	1.1432
0.4363	25.0	0.4226	0.9063	0.4663	2.1445	65.0	1.1345
0.4451	25.5	0.4305	0.9026	0.4770	2.0965	64.5	1.1257
0.4538	26.0	0.4384	0.8988	0.4877	2.0503	64.0	1.1170
0.4625	26.5	0.4462	0.8949	0.4986	2.0057	63.5	1.1083
0.4712	27.0	0.4540	0.8910	0.5095	1.9626	63.0	1.0996
0.4800	27.5	0.4617	0.8870	0.5206	1.9210	62.5	1.0908
0.4887	28.0	0.4695	0.8829	0.5317	1.8807	62.0	1.0821
0.4974	28.5	0.4772	0.8788	0.5430	1.8418	61.5	1.0734
0.5061	29.0	0.4848	0.8746	0.5543	1.8040	61.0	1.0647
0.5149	29.5	0.4924	0.8704	0.5658	1.7675	60.5	1.0559
0.5236	30.0	0.5000	0.8660	0.5774	1.7321	60.0	1.0472
0.5323	30.5	0.5075	0.8616	0.5890	1.6977	59.5	1.0385
0.5411	31.0	0.5150	0.8572	0.6009	1.6643	59.0	1.0297
0.5498	31.5	0.5225	0.8526	0.6128	1.6319	58.5	1.0210
0.5585	32.0	0.5299	0.8480	0.6249	1.6003	58.0	1.0123
0.5672	32.5	0.5373	0.8434	0.6371	1.5697	57.5	1.0036
0.5760	33.0	0.5446	0.8387	0.6494	1.5399	57.0	0.9948
0.5847	33.5	0.5519	0.8339	0.6619	1.5108	56.5	0.9861
0.5934	34.0	0.5592	0.8290	0.6745	1.4826	56.0	0.9774
0.6021	34.5	0.5664	0.8241	0.6873	1.4550	55.5	0.9687
0.6109	35.0	0.5736	0.8192	0.7002	1.4281	55.0	0.9599
		cos θ	sin θ	cot θ	tan θ	Degrees	Radians
						Angle θ	

TABLE B-3 (Continued)

Angle θ		sin θ	cos θ	tan θ	cot θ		
Radians	Degrees						
0.6196	35.5	0.5807	0.8141	0.7133	1.4019	54.5	0.9512
0.6283	36.0	0.5878	0.8090	0.7265	1.3764	54.0	0.9425
0.6370	36.5	0.5948	0.8039	0.7400	1.3514	53.5	0.9338
0.6458	37.0	0.6018	0.7986	0.7536	1.3270	53.0	0.9250
0.6545	37.5	0.6088	0.7934	0.7673	1.3032	52.5	0.9163
0.6632	38.0	0.6157	0.7880	0.7813	1.2799	52.0	0.9076
0.6720	38.5	0.6225	0.7826	0.7954	1.2572	51.5	0.8988
0.6807	39.0	0.6239	0.7771	0.8098	1.2349	51.0	0.8901
0.6894	39.5	0.6361	0.7716	0.8243	1.2131	50.5	0.8814
0.6981	40.0	0.6428	0.7660	0.8391	1.1918	50.0	0.8727
0.7069	40.5	0.6494	0.7604	0.8541	1.1708	49.5	0.8639
0.7156	41.0	0.6561	0.7547	0.8693	1.1504	49.0	0.8552
0.7243	41.5	0.6626	0.7490	0.8847	1.1303	48.5	0.8465
0.7330	42.0	0.6691	0.7431	0.9004	1.1106	48.0	0.8378
0.7418	42.5	0.6756	0.7373	0.9163	1.0913	47.5	0.8290
0.7505	43.0	0.6820	0.7314	0.9325	1.0724	47.0	0.8203
0.7592	43.5	0.6884	0.7254	0.9490	1.0538	46.5	0.8116
0.7679	44.0	0.6947	0.7193	0.9657	1.0355	46.0	0.8029
0.7767	44.5	0.7009	0.7133	0.9827	1.0176	45.5	0.7941
0.7854	45.0	0.7071	0.7071	1.0000	1.0000	45.0	0.7854
		cos θ	sin θ	cot θ	tan θ	Degrees	Radians
						Angle θ	

Appendix C

Mathematical Induction

The principle of mathematical induction states that if a set is a subset of the natural numbers N and if the set obeys two properties, then the set must be N. These two properties characterize the natural numbers.

C-1 PRINCIPLE OF MATHEMATICAL INDUCTION

Suppose a set S is a subset of the natural numbers N, and S obeys the following properties:
(i) $1 \in S$
(ii) if $n \in S$ then $n+1 \in S$.
Then $S = N$.

We shall prove the following two theorems by utilizing the Principle of Mathematical Induction.

Theorem C-1 *For every natural number n, we have*
$$1 + 2 + 3 + \cdots + n = \tfrac{1}{2}n(n+1)$$

PROOF Let S be the set of natural numbers for which the statement is true. One can directly verify that $1 \in S$ since $1 = \tfrac{1}{2} \cdot 1 \cdot (1+1)$. Hence the first property in the Principle of Mathematical Induction is satisfied. To prove that the second property is also satisfied, we suppose $n \in S$. We must now show that $n+1 \in S$. Since $n \in S$, we have
$$1 + 2 + 3 + \cdots + n = \tfrac{1}{2}n(n+1)$$

If we add $n+1$ to each side of the equation, we have

$$1 + 2 + 3 + \cdots + n + (n+1) = \tfrac{1}{2}n(n+1) + n + 1$$
$$= \tfrac{1}{2}(n+1)(n+2)$$

which shows that $n+1 \in S$.

Theorem C-2 *For every natural number n, we have*

$$1^2 + 2^2 + 3^2 + \cdots + n^2 = \frac{n(n+1)(2n+1)}{6}$$

PROOF Let S be the set of natural numbers for which the statement is true. One can directly verify that $1 \in S$ since

$$1^2 = 1 = \frac{1(1+1)(2 \cdot 1 + 1)}{6}$$

Hence the first property in the Principle of Mathematical Induction is satisfied. To prove that the second property is also satisfied, we suppose $n \in S$. We must now show that $n+1 \in S$. Since $n \in S$, we have

$$1^2 + 2^2 + 3^2 + \cdots + n^2 = \frac{n(n+1)(2n+1)}{6}$$

If we add $(n+1)^2$ to each side of the equation, we have

$$1^2 + 2^2 + 3^2 + \cdots + n^2 + (n+1)^2 = \frac{n(n+1)(2n+1)}{6} + (n+1)^2$$
$$= \frac{(n+1)}{6}[n(2n+1) + 6(n+1)]$$
$$= \frac{(n+1)}{6}(2n^2 + 7n + 6)$$
$$= \frac{(n+1)}{6}(n+2)(2n+3)$$

which shows that $n+1 \in S$.

Appendix D

A Brief Review of Algebra

D-1 EXPONENTS

Exponents or powers play an important role in all of mathematics and especially calculus. In this section we review the basic formulas governing the use of exponents.

If n is a positive integer, then a^n means $a \cdot a \cdots a$ n times. That is, $2^3 = 2 \cdot 2 \cdot 2 = 8$ and $3^4 = 3 \cdot 3 \cdot 3 \cdot 3 = 81$. If $a > 0$ and n is a positive integer, then $a^{1/n}$ is that number b such that $b^n = a$. The number $a^{1/n} = \sqrt[n]{a}$ is called the (positive) nth root of a. For example, $4^{1/2} = \sqrt{4} = 2$ and $8^{1/3} = \sqrt[3]{8} = 2$.

The basic properties of exponents are presented below.

1. $a^0 = 1$
2. $a^{-1} = \dfrac{1}{a}$
3. $a^{-n} = \dfrac{1}{a^n}$
4. $(a^n)^m = a^{nm}$
5. $a^n a^m = a^{n+m}$
6. $a^n b^n = (ab)^n$

We present a few illustrations of these properties.

$$
\begin{aligned}
(2^3)^2 3^6 &= 2^6 \cdot 3^6 &&\text{Property 4}\\
&= (2 \cdot 3)^6 &&\text{Property 6}\\
&= 6^6
\end{aligned}
$$

APPENDIX D   A Brief Review of Algebra

$$(10^{-3} \cdot 10^2)^3 = (10^{-3+2})^3 \quad \text{Property 5}$$
$$= (10^{-1})^3$$
$$= 10^{(-1)3} \quad \text{Property 4}$$
$$= 10^{-3}$$
$$= \frac{1}{10^3} \quad \text{Property 3}$$

$$\frac{3^{-3}}{3^{-2}} = 3^{-3}3^2 \quad \text{Property 3}$$
$$= 3^{-3+2} \quad \text{Property 5}$$
$$= 3^{-1}$$
$$= \tfrac{1}{3} \quad \text{Property 2}$$

EXERCISE D-1

Simplify the expressions.
(1) $(2^2)(2^3)$
(2) $2^2 \, 5^2$
(3) $(2^2)^3 \, 2^{-3}$
(4) $4^{-1/2}$
(5) $9^{-3/2}$
(6) $8^{4/3}$
(7) $8^{-1/3}$
(8) $(16^{1/2})^3 \, 4^{-2}$
(9) $(10^{-2})^4 \, 3^{-8} \, 2^{-8}$
(10) $(3^{-2})^3 \, 3^2 \, 3^4$
(11) $(3^2 \, 8^4)^0$
(12) $2^0 \, 3^{-2} \, 6^0$
(13) $a^2 b^{-3} a^{-3} b^2$
(14) $a^3 b^3 (ab)^{-3}$
(15) $(a^{1/3})^{3/2} b^{1/2}$
(16) $(a+b)^2 (a+b)^{-3}$

D-2 FACTORING QUADRATIC EXPRESSIONS

To multiply two linear expressions of the form $x+a$ and $x+b$ we use the formula
$$(x+a)(x+b) = x^2 + (a+b)x + ab$$
This same formula helps to factor a quadratic expression of the form $x^2 + cx + d$.

EXAMPLE D-1

To factor the quadratic expression
$$x^2 + 4x + 3$$
into two linear factors, $x+a$ and $x+b$, we set
$$a + b = 4$$
$$ab = 3$$
Hence we look for numbers a and b whose sum is 4 and whose product is 3.

This is a trial and error method and one sees that in this case, we can let $a = 1$ and $b = 3$. Hence

$$x^2 + 4x + 3 = (x+1)(x+3)$$

EXAMPLE D-2

To factor the quadratic expression

$$x^2 - 5x + 6$$

we set

$$a + b = -5$$
$$ab = 6$$

Since $ab = 6$, one might select $a = 1$ and $b = 6$ but then $a+b \neq -5$. Since $a+b = -5$, one might select $a = 1$ and $b = -6$, but then $ab \neq 6$. The correct choice is $a = -2$ and $b = -3$. Thus

$$x^2 - 5x + 6 = (x + (-2))(x + (-3))$$
$$= (x-2)(x-3)$$

In the above examples the coefficient of x^2 has always been 1. When the coefficient of x^2 is not 1 we use the more general formula

$$abx^2 + (ad+bc)x + cd = (ax+c)(bx+d)$$

EXAMPLE D-3

To factor

$$6x^2 + 7x + 2$$

we set

$$ab = 6$$
$$ad + bc = 7$$
$$cd = 2$$

We look for numbers a and b whose product is 6. We could choose 1 and 6, 6 and 1, 2 and 3, or 3 and 2, or their negatives. We also need $cd = 2$ so we could choose 2 and 1, or 1 and 2, or their negatives. These possibilities are tested until we find a combination such that $ad + cb = 7$. If we choose $a = 6$, $b = 1$, $c = 1$, and $d = 2$ we get $ad + bc = 6 \cdot 2 + 1 \cdot 1 = 13 \neq 7$, so this is not the correct choice. The correct choice is $a = 2$, $b = 3$, $c = 1$, and $d = 2$ since then $ad + bc = 2 \cdot 2 + 3 \cdot 1 = 7$. (Note also we could have chosen $a = 3$, $b = 2$, $c = 2$, and $d = 1$ and we would get the same two linear factors below). Hence we write

$$6x^2 + 7x + 2 = (2x+1)(3x+2)$$

EXAMPLE D-4

To factor

$$2x^2 + 7x - 4$$

we set
$$ab = 2$$
$$ad + bc = 7$$
$$cd = -4$$

We might arrange the work as follows: since $ab = 2$, select $a = 2$ and $b = 1$ and write
$$(2x+\square)(x+\square)$$
since $cd = -4$, select one of the pairs 1 and -4, -1 and 4, 2 and -2, or -2 and 2 for c and d, substitute them into the space above, and multiply the linear factors. If we choose $c = 2$, $d = 2$, we get
$$(2x+2)(x-2) = 2x^2 - 2x - 4$$
which is not what is desired. The correct choice is $c = -1$, $d = 4$ because
$$(2x-1)(x+4) = x^2 + 7x - 4$$

EXAMPLE D-5

To factor
$$4x^2 - 9$$
we set
$$ab = 4$$
$$ad + bc = 0$$
$$cd = -9$$

The choices for a and b are 4 and 1, 1 and 4, 2 and 2, and their negatives. For c and d we could select 1 and -9, -1 and 9, 3 and -3, or -3 and 3. The correct choices are $a = 2$, $b = 2$, $c = 3$, $d = -3$. Hence
$$4x^2 - 9 = (2x+3)(2x-3)$$

EXERCISE D-2

Factor the following.
(1) $x^2 + 3x + 2$
(2) $x^2 + 4x + 4$
(3) $x^2 + 6x + 5$
(4) $x^2 + 8x + 12$
(5) $x^2 - x - 2$
(6) $x^2 - 4x + 3$
(7) $x^2 - 8x + 15$
(8) $x^2 - 1$
(9) $x^2 - 9$
(10) $x^2 - 16$
(11) $2x^2 + 3x + 1$
(12) $2x^2 + 5x + 2$
(13) $3x^2 + 4x + 1$
(14) $3x^2 + 5x - 2$
(15) $4x^2 + 5x + 1$
(16) $4x^2 - 1$
(17) $6x^2 - 11x - 10$
(18) $6x^2 + 19x + 10$
(19) $-x^2 - x + 2$
(20) $9x^2 - 4$

D-3 MORE ON FACTORING

We can often factor more than once.

EXAMPLE D-6

To factor the cubic expression
$$x^3 + 3x^2 + 2x$$
we can first factor an x from each term and get
$$x^3 + 3x^2 + 2x = x(x^2 + 3x + 2)$$
and now we factor the quadratic expression so that
$$x^3 + 3x^2 + 2x = x(x+2)(x+1)$$
To factor the expression
$$x^4 + x^3 - 2x^2$$
we first factor x^2 from each term and then factor the quadratic expression. Thus
$$x^4 + x^3 - 2x^2 = x^2(x^2 + x - 2)$$
$$= x^2(x+2)(x-1)$$

We can often simplify a more complicated expression by factoring out of the expression the appropriate quantity.

EXAMPLE D-7

To simplify the expression
$$x(x+2)^2 + x^2(x+2)^3$$
we factor the quantity $x(x+2)^2$ from each term which yields
$$x(x+2)^2 + x^2(x+2)^3 = x(x+2)^2[1 + x(x+2)]$$
$$= x(x+2)^2[x^2 + 2x + 1]$$
$$= x(x+2)^2(x+1)^2$$
since $(x+1)^2 = x^2 + 2x + 1$.

EXERCISE D-3

Factor the following.

(1) $x^3 + 3x^2 + 2x$
(2) $x^3 + 6x^2 + 5x$
(3) $x^4 - x^3 - 2x^2$
(4) $x^6 - x^4$
(5) $x^3 - 9x$
(6) $2x^3 + 3x^2 + x$
(7) $2x^3 + 5x^2 + 2x$
(8) $x^4 - 7x^3 + 12x^2$
(9) $x^4 - 16x^2$
(10) $x^4 - 3x^3 - 10x^2$
(11) $3x^4 + 4x^3 + x^2$
(12) $4x^4 + 5x^3 - 2x^2$
(13) $4x^2 + 12x + 8$
(14) $3x^6 - 3x^4$
(15) $6x^3 + 19x^2 + 10x$
(16) $18x^3 - 8x$

D-4 COMPLETING THE SQUARE

It is often advantageous to express a quadratic expression in the form
$$(x+a)^2 + b$$
This process, called *completing the square*, depends upon the formula
$$(x+a)^2 = x^2 + 2ax + a^2$$
In the expression on the right-hand side of the above equation, note that the constant term a^2 is the square of one-half the coefficient of x.

EXAMPLE D-8

To express the quadratic expression
$$x^2 + 4x + 1$$
in the form
$$(x+a)^2 + b$$
we first consider $x^2 + 4x$, then one-half the coefficient of x is $\frac{1}{2} \cdot 4 = 2$ and hence $a = 2$ and $a^2 = 4$. Thus we add 4 to $x^2 + 4x$ in order to complete the square, i.e., $x^2 + 4x + 4 = (x+2)^2$. If we add 4 to the original quadratic expression, $x^2 + 4x + 1$, we must also subtract 4 so as not to alter the expression. Thus

$$x^2 + 4x + 1 = x^2 + 4x + 4 - 4 + 1$$
$$= (x+2)^2 - 4 + 1$$
$$= (x+2)^2 - 3$$

EXAMPLE D-9

To complete the square of the quadratic expression
$$x^2 + 6x - 3$$
we take one-half the coefficient of x, i.e., $\frac{1}{2} \cdot 6 = 3$, so $a = 3$ and $a^2 = 9$. Hence we add and subtract 9 and get

$$x^2 + 6x - 3 = x^2 + 6x + 9 - 9 - 3$$
$$= (x+3)^2 - 9 - 3$$
$$= (x+3)^2 - 12$$

If the coefficient of x^2 is not 1, we first factor it from the x^2 and x terms and operate inside the parentheses. This method is given in the next example.

EXAMPLE D-10

To complete the square of
$$3x^2 + 9x - 5$$

we first consider
$$3x^2 + 9x = 3(x^2+3x)$$
The coefficient of x in the parentheses is 3 so we set $a = \frac{1}{2} \cdot 3 = \frac{3}{2}$ and so $a^2 = \frac{9}{4}$. We have
$$\begin{aligned}
3x^2 + 9x - 5 &= 3(x^2+3x) - 5 \\
&= 3(x^2+3x+\tfrac{9}{4}-\tfrac{9}{4}) - 5 \\
&= 3(x^2+3x+\tfrac{9}{4}) - 3 \cdot \tfrac{9}{4} - 5 \\
&= 3(x+\tfrac{3}{2})^2 - \tfrac{27}{4} - 5 \\
&= 3(x+\tfrac{3}{2})^2 - \tfrac{47}{4}
\end{aligned}$$

EXERCISE D-4

Complete the square.

(1) $x^2 + 2x + 2$
(2) $x^2 + 2x - 3$
(3) $x^2 + 4x + 6$
(4) $x^2 + 6x - 1$
(5) $x^2 - 2x - 1$
(6) $x^2 - 8x + 1$
(7) $x^2 + 5x - 1$
(8) $x^2 - 7x + 2$
(9) $2x^2 + 4x + 1$
(10) $2x^2 + 6x - 1$
(11) $2x^2 + 7x + 4$
(12) $3x^2 + 6x + 1$
(13) $3x^2 + 5x - 2$
(14) $4x^2 + 8x + 2$
(15) $4x^2 - 7x - 1$
(16) $5 - 2x + 2x^2$

D-5 QUADRATIC EQUATIONS

Some examples of quadratic equations are

$$x^2 + 2x - 3 = 0 \qquad x^2 - x + 6 = 0$$
$$x^2 = 6x \qquad 3x^2 = 2x - 1$$

A quadratic equation in one variable is a polynomial equation in which the highest power of the variable is two. The standard form of a quadratic equation is

$$ax^2 + bx + c = 0$$

We will discuss two ways of solving quadratic equations, that is, of finding those real numbers which are solutions of the equation. The first method is by factoring and the second is by the quadratic formula. In general, solution by factoring is quicker than using the quadratic formula, but it doesn't always work.

To solve a quadratic equation by factoring, one writes the equation in standard form, then factors the quadratic polynomial on the left-hand side of the equation. The property of real numbers which states that $ab = 0$ implies that $a = 0$ or $b = 0$ is then applied.

EXAMPLE D-11

To solve the quadratic equation
$$x^2 + 2x - 3 = 0$$
by factoring we write
$$x^2 + 2x - 3 = (x-1)(x+3) = 0$$
and then set each factor equal to zero and solve for x.
$$x - 1 = 0$$
$$x + 3 = 0$$
$$x = 1, -3$$

Hence the solutions are 1 and -3. One checks to see that they are indeed solutions by substituting them back into the original equation. Thus
$$(1)^2 + 2(1) - 3 = 0$$
$$(-3)^2 + 2(-3) - 3 = 0$$

EXAMPLE D-12

To solve the equation
$$5x^2 + 7x - 6 = 0$$
we factor the polynomial and set each factor equal to zero.
$$5x^2 + 7x - 6 = (5x-3)(x+2) = 0$$
$$5x - 3 = 0$$
$$x + 2 = 0$$
$$x = \tfrac{3}{5}, -2$$

EXAMPLE D-13

To solve the equation
$$3x^2 = 4 - 11x$$
we first put the equation in standard form.
$$3x^2 + 11x - 4 = 0$$
Then we factor the polynomial and set each factor equal to zero.
$$3x^2 + 11x - 4 = (3x-1)(x+4) = 0$$
$$3x - 1 = 0$$
$$x + 4 = 0$$
$$x = \tfrac{1}{3}, -4$$

The quadratic formula gives the solutions of the general quadratic equation in standard form. We will present the formula without deriving it. The solutions of the quadratic equation

$$ax^2 + bx + c = 0 \quad \text{where } a \neq 0$$

are

$$x = \frac{-b + \sqrt{b^2 - 4ac}}{2a} \quad \text{and} \quad x = \frac{-b - \sqrt{b^2 - 4ac}}{2a}$$

In shortened form, the solutions are

$$x = \frac{-b \pm \sqrt{b^2 - 4ac}}{2a}$$

To use the quadratic formula, we first write the equation in standard form, then identify the coefficients a, b, and c and substitute those numbers into the formula.

EXAMPLE D-14

To solve the quadratic equation

$$x^2 = 3 - 2x$$

by the quadratic formula we write the equation in standard form.

$$x^2 + 2x - 3 = 0$$

Thus $a = 1$, $b = 2$, and $c = -3$. Substituting these values into the formula yields

$$x = \frac{-2 \pm \sqrt{4 - 4(1)(-3)}}{2(1)}$$

$$= \frac{-2 \pm \sqrt{16}}{2}$$

$$= \frac{-2 \pm 4}{2} = 1, -3$$

The equation in Ex. D-14 could also be solved by factoring. However, the equation in the next example would most certainly require the quadratic formula.

EXAMPLE D-15

To solve the quadratic equation

$$5x^2 = 3x + 1$$

by the quadratic formula we first put the equation in standard form.

$$5x^2 - 3x - 1 = 0$$

Thus $a = 5$, $b = -3$, $c = -1$. Hence the solutions are

$$x = \frac{-(-3) \pm \sqrt{9 - 4(5)(-1)}}{2(5)}$$

$$= \frac{3 \pm \sqrt{29}}{10}$$

EXERCISE D-5

Solve by factoring.
- (1) $x^2 - 1 = 0$
- (2) $x^2 - 4 = 0$
- (3) $x^2 - 9 = 0$
- (4) $x^2 - 16 = 0$
- (5) $x^2 + 5x + 6 = 0$
- (6) $x^2 + x - 2 = 0$
- (7) $x^2 + x - 6 = 0$
- (8) $2x^2 - 3x - 2 = 0$
- (9) $2x^2 + x - 3 = 0$
- (10) $2x^2 + 3x - 5 = 0$
- (11) $3x^2 + 5x - 2 = 0$
- (12) $3x^2 + x - 2 = 0$

Solve by the quadratic formula.
- (1) $x^2 + x - 1 = 0$
- (2) $x^2 + x - 3 = 0$
- (3) $x^2 + 4x + 2 = 0$
- (4) $x^2 - 5x - 1 = 0$
- (5) $x^2 + 6x + 1 = 0$
- (6) $2x^2 + 7x + 1 = 0$
- (7) $2x^2 - 2x - 3 = 0$
- (8) $2x^2 - 5x + 1 = 0$
- (9) $3x^2 + x - 1 = 0$
- (10) $3x^2 + 5x + 2 = 0$
- (11) $4x^2 + x - 2 = 0$
- (12) $4x^2 - 3x - 1 = 0$

D-6 CANCELLING ALGEBRAIC EXPRESSIONS

An algebraic expression can often be simplified by cancelling a factor which is common to the numerator and the denominator.

EXAMPLE D-16

The algebraic expression

$$\frac{(x+1)(x-1)}{x-1}$$

has the common factor $x - 1$. We can cancel it and write

$$\frac{(x+1)(x+1)}{x-1} = \frac{(x+1)\cancel{(x-1)}}{\cancel{x-1}} = x + 1 \quad \text{for } x \neq 1$$

Note that the two expressions are not equal for $x = 1$.

EXAMPLE D-17

The numerator and denominator of the expression

$$\frac{x^2 - 1}{x^2 + 3x + 2}$$

SEC. D-6   Cancelling Algebraic Expressions

can be factored and we can write

$$\frac{x^2 - 1}{x^2 + 3x + 2} = \frac{(x+1)(x-1)}{(x+1)(x+2)}$$

$$= \frac{\cancel{(x+1)}(x-1)}{\cancel{(x+1)}(x+2)}$$

$$= \frac{x-1}{x+2} \quad \text{for } x \neq -1$$

For $x = -1$ the last expression equals -2, whereas the original expression is not defined for $x = -1$.

EXAMPLE D-18

To simplify the expression

$$\frac{x^3 - x^2 - 6x}{x^3 - 2x^2 - 3x}$$

we factor the numerator and denominator to get

$$\frac{x^3 - x^2 - 6x}{x^3 - 2x^2 - 3x} = \frac{x(x-3)(x+2)}{x(x-3)(x+1)}$$

$$= \frac{\cancel{x}\cancel{(x-3)}(x+2)}{\cancel{x}\cancel{(x-3)}(x+1)}$$

$$= \frac{x+2}{x+1} \quad \text{for } x \neq 0, 3$$

EXERCISE D-6

Simplify the expressions.

(1) $\dfrac{x^2 - 4}{x + 2}$

(2) $\dfrac{x^2 - 9}{x - 3}$

(3) $\dfrac{x^3 - 16x}{x - 4}$

(4) $\dfrac{x^2 + 4x + 3}{x^2 - 1}$

(5) $\dfrac{x^2 + x - 2}{x^2 + 5x + 6}$

(9) $\dfrac{x^2 + 7x + 12}{x^2 + 2x - 3}$

(7) $\dfrac{x^3 - x^2}{x^4 - x^3}$

(8) $\dfrac{x^3 + x^2 - 6x}{x^3 - 2x^2}$

(9) $\dfrac{2x^2 - x + 1}{6x^2 + x - 1}$

(10) $\dfrac{x^4 + x^3 - 2x^2}{x^4 - 2x^3}$

D-7 INEQUALITIES

In this section we discuss an alternate method of solving inequalities. One technique for solving inequalities was presented in Sec. 1-1. The reader may find the method outlined in this section easier.

EXAMPLE D-19

To solve the inequality

$$(x-1)(x-3) > 0$$

we solve the equation arrived at by replacing the inequality sign by an equals sign. Hence we solve

$$(x-1)(x-3) = 0$$

so that $x = 1, 3$. We plot $x = 1$ and $x = 3$ on the real line and thereby subdivide the real line into three regions: the region to the left of $x = 1$, i.e., $(-\infty, 1)$; the region between 1 and 3, i.e., $(1,3)$; and the region to the right of 3, i.e., $(3, \infty)$. (See Fig. D-1.) The inequality will be satisfied either by all of the numbers in a given region or by none of them. This is true for each region. Thus to see whether $(-\infty, 1)$ is part of the solution set, it is sufficient to select any one number in the set and see whether it is a solution. We select $0 \in (-\infty, 1)$ and substitute $x = 0$ into the inequality and get

$$(0-1)(0-3) = (-1)(-3) = 3 > 0$$

so that $x = 0$ is a solution and hence $(-\infty, 1)$ is part of the solution set. Next we select $2 \in (1, 3)$ and then substitute $x = 2$ into the inequality and get

$$(2-1)(2-3) = (1)(-1) = -1 \not> 0$$

and so $x = 2$ is not a solution and hence no element in $(1, 3)$ is in the solution set. Finally we select $4 \in (3, \infty)$ and consider

$$(4-1)(4-3) = (3)(1) = 3 > 0$$

and hence $x = 4$ is a solution and therefore $(3, \infty)$ is part of the solution set. Thus the solution set is $(-\infty, 1) \cup (3, \infty)$.

The solution set for $(x - 1)(x - 3) > 0$

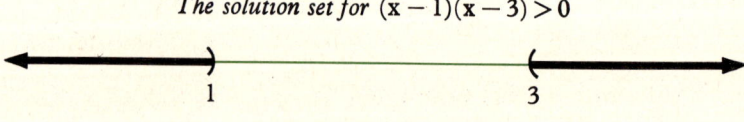

FIGURE D-1

EXAMPLE D-20

To solve the inequality

$$(x-4)(x+2) \leq 0$$

we solve

$$(x-4)(x+2) = 0$$
$$x - 4 = 0$$
$$x + 2 = 0$$
$$x = -2, 4$$

We consider the three regions $(-\infty, -2)$, $[-2, 4]$, and $(4, \infty)$. (See Fig. D-2). Note that the intervals are closed, i.e., the endpoints are included, because the original inequality involves an equals sign and hence the endpoints are solutions. We select $-3 \in (-\infty, -2)$, $0 \in [-2, 4]$, and $5 \in (4, \infty)$. Recall that any number (except an endpoint) contained in the interval will suffice. We check to see if they are solutions

$$(-3-4)(-3+2) = (-7)(-1) = 7 \not\leq 0$$
$$(0-4)(0+2) = (-4)(2) = -8 \leq 0$$
$$(5-4)(5+2) = (1)(7) = 7 \not\leq 0$$

Since -3 and 5 are not solutions, neither $(-\infty, -2)$ nor $(4, \infty)$ are part of solution set. Since 0 is a solution, the interval $[-2, 4]$ is the solution set.

The solution set for $(x - 4)(x + 2) \leq 0$

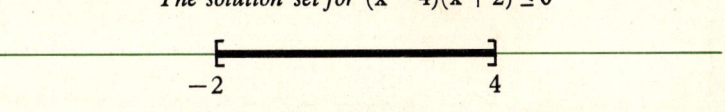

FIGURE D-2

EXAMPLE D-21

To solve the inequality

$$2x^3 > 8x - x^2$$

we first rearrange the terms to get

$$2x^3 + x^2 - 8x > 0$$

and then factor.

$$x(2x-5)(x+3) > 0$$

We next solve the corresponding equality.

$$x(2x-5)(x+3) = 0$$
$$x = 0$$
$$2x - 5 = 0$$
$$x + 3 = 0$$
$$x = -3, 0, \tfrac{5}{2}$$

Hence the real line is subdivided into the four regions; $(-\infty, -3)$, $(-3, 0)$, $(0, \tfrac{5}{2})$ and $(\tfrac{5}{2}, \infty)$. (See Fig. D-3.) We select one element from each, say $-4 \in (-\infty, -3)$, $-1 \in (-3, 0)$, $1 \in (0, \tfrac{5}{2})$, $3 \in (\tfrac{5}{2}, \infty)$. We then substitute them into the inequality and get

$$(-4)(2(-4)-5)(-4+3) = (4)(-13)(-1) = -52 \not> 0$$
$$(-1)(2(-1)-5)(-1+3) = (-1)(-7)(2) = 14 > 0$$
$$(1)(2(1)-5)(1+3) = (-3)(4) = -12 \not> 0$$
$$(3)(2(3)-5)(3+3) = (3)(1)(6) = 18 > 0$$

Hence the solution set is $(-3, 0) \cup (\tfrac{5}{2}, \infty)$.

The solution set for $2x^3 > 8x - x^2$

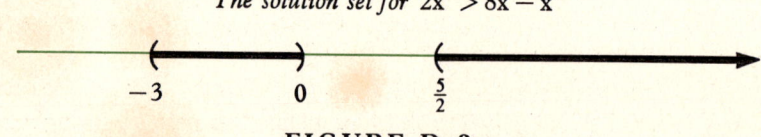

FIGURE D-3

EXERCISE D-7

Solve the inequalities.
(1) $(x-1)(x-4) > 0$
(2) $(x+2)(x-5) \geq 0$
(3) $(x+3)(x+1) < 0$
(4) $(x+5)(x+1) \leq 0$
(5) $x^2 + 5x + 6 < 0$
(6) $x^2 + 6x + 5 > 0$
(7) $x^2 - 3x + 2 \leq 0$
(8) $x^2 - 4 \geq 0$
(9) $9 - x^2 \leq 0$
(10) $4 + 3x - x^2 \geq 0$
(11) $5 - 4x - x^2 < 0$
(12) $(2x-3)(3x+1) > 0$
(13) $(5x-1)(2x+3) < 0$
(14) $(x-1)(x-2)(x-3) < 0$
(15) $x^3 + 5x^2 + 6x \leq 0$
(16) $3x^2 - 2x < x^3$

Chapter Tests

CHAPTER 1

1. Find $A \cup B$ and $A \cap B$ for $A = [0,3)$ and $B = [2,4]$.

In problems 2, 3, and 4 solve the inequalities.

2. $(x-2)(x+4) > 0$
3. $(2x+1)(4-x) \geq 0$
4. $|2x+1| > 2$
5. Is y a function of x?
   (a) $x^2 + y^2 = 4$
   (b) $x^2 + y = 3$
6. Find the equation of the line passing through the points $(-1,0)$ and $(2,6)$.
7. Find the slope and the y-intercept of the line passing through $(1,1)$ and parallel to $y = 3x+1$.
8. Find the vertex and the line of symmetry of the parabola $y = x^2 + 4x + 5$.

In problems 9 and 10 find the inverse of $y = f(x)$.

9. $f(x) = 3x + 7$
10. $f(x) = x/(x+3)$

CHAPTER 2

In problems 1 through 6 calculate the limits if they exist.

1. $\lim\limits_{x \to 1} \dfrac{x^2 - 1}{x + 1}$
2. $\lim\limits_{x \to 2} \dfrac{x - 2}{x^2 - 4}$
3. $\lim\limits_{x \to -1} \dfrac{x^2 - 1}{x + 1}$
4. $\lim\limits_{x \to 3} \dfrac{x^2 - 3x}{x^2 - 9}$

In problems 5 and 6 let $f(x) = \begin{cases} x + 2 \text{ for } x \leq 1 \\ 4x^2 - 1 \text{ for } x > 1 \end{cases}$.

5. $\lim\limits_{x \to 1} f(x)$
6. $\lim\limits_{x \to 2} f(x)$

In problems 7 and 8 determine if the function is continuous at $x = 1$.

7. $f(x) = \dfrac{x^2 - 1}{x + 1}$

8. $f(x) = \dfrac{x^2 - 1}{x - 1}$

In problems 9 and 10 let $\lim\limits_{x \to 3} f(x) = 3$ and $\lim\limits_{x \to 3} g(x) = -4$.

9. Find $\lim\limits_{x \to 3} (f(x) + g(x))$ and $\lim\limits_{x \to 3} \dfrac{f(x)}{g(x)}$.

10. Find $2 \cdot \lim\limits_{x \to 3} (5f(x) + 2g(x))$.

CHAPTER 3

1. Use the definition of the derivative to find $f'(3)$ where $f(x) = 3x + 1$.
2. Find the equation of the tangent line to the curve $f(x) = 2x^2 + x^3$ at $x = 2$.
3. At what point do the functions $f(x) = 2|x| + 1$ and $g(x) = |2x + 1|$ fail to have a derivative?
4. An object is thrown upward with position function $s(t) = -16t^2 + 32t + 10$. Find the velocity $v(t)$ and acceleration $a(t)$ functions. How far up will the object travel?

In problems 5 through 10 find $f'(x)$.

5. $f(x) = 2x^5 - \tfrac{2}{5}x^4 + x^2 - 10 + x^{-1/2}$
6. $f(x) = x^2 + \sqrt{2x + 1}$
7. $f(x) = (x + x^{-3})^{1/3}$
8. $f(x) = \dfrac{(x^2 - 5x^{-1})^3}{(x + 2x^{-2})^{1/2}}$
9. $f(x) = e^{x^2} \ln(x^4 + 1)$
10. $f(x) = \ln(3x + 2e^{3x})$

CHAPTER 4

In problems 1 and 2 use the First Derivative Test to find the extrema.

1. $y = 2x^3 - 9x^2 - 18x + 1$

2. $y = x^3 - 3\ln x$
3. Use the Second Derivative Test to find the extrema of $y = x^3 - 6x + 9x + 2$.

In problems 4 and 5 find y using implicit differentiation.

4. $xy + y^2 + e^{xy} = 0$
5. $xy^3 + \ln(x^2 + y^2) = 3$
6. Water is being pumped into a cylindrical tank at a rate of 16 ft^3/min. How fast is the water level rising when the height of the water is 5 ft?
7. Verify that $f(x) = x^3 - 3x$ satisfies the hypothesis of the Mean Value Theorem in the open inverval (0,2) and find the number which satisfies the conclusion.
8. Evaluate $\lim\limits_{x \to 1} \dfrac{x^2 \ln x}{x - e^{x-1}}$.

In problems 9 and 10 find the antiderivative of $f(x)$.

9. $f(x) = 9x^3 - x^{-3/5}$
10. $f(x) = 7x^{1/2} + 2x^{-1} + 3e^{4x}$

CHAPTER 5

1. Use the definition of the definite integral to compute $\displaystyle\int_0^1 2x\, dx$.

In problems 2 through 8 evaluate the integrals.

2. $\displaystyle\int_1^2 (x + 3x^2 + x^{-2})\, dx$

3. $\displaystyle\int_0^1 x(1 + x^2)^{1/2}\, dx$

4. $\displaystyle\int \dfrac{(x+1)\, dx}{(x^2 + 2x)^3}$

5. $\displaystyle\int \dfrac{dx}{x^2 + 2x}$

6. $\displaystyle\int \dfrac{\ln x\, dx}{x^2}$

7. $\displaystyle\int \dfrac{x^3\, dx}{x^2 + 2x}$

8. $\displaystyle\int x^3 (x^2 + 1)^{-1/2}\, dx$

In problems 9 and 10 use the table of integrals to evaluate the integrals.

9. $\displaystyle\int \frac{dx}{1+4x^2}$

10. $\displaystyle\int \frac{dx}{x\sqrt{1+4x^2}}$

CHAPTER 6

In problems 1 and 2 use the Trapezoidal Rule to approximate the integral with $n = 4$.

1. $\displaystyle\int_0^1 x^2\, dx$

2. $\displaystyle\int_0^4 x^3\, dx$

3. Use Simpson's Rule to approximate $\displaystyle\int_0^4 x^2\, dx$ with $n = 4$.

4. Evaluate $\displaystyle\lim_{x\to\infty} (x^{-3}+x^{-4})$.

In problems 5, 6, and 7 evaluate the improper integrals.

5. $\displaystyle\int_1^\infty (x^{-3}+x^{-4})\, dx$

6. $\displaystyle\int_0^\infty xe^{-x^2}\, dx$

7. $\displaystyle\int_0^1 x^{-3}\, dx$

In problems 8 and 9 find the volume of the solid generated by rotating the area bounded by the given curves about the x-axis.

8. $y = 3x$, $y = 0$, $x = 2$
9. $y = x^2$, $y = 0$, $x = 1$
10. Find a number C such that the function $f(x) = C(x+1)^{-2}$ is a probability density function in the interval $[3,4]$.

CHAPTER 7

In problems 1 and 2 let $f(x) = \displaystyle\int_1^x (1+t^2)\, dt$.

1. Find $f(1)$, $f(2)$, $f(3)$, and $f(-1)$.
2. Find another formula for $f(x)$.
3. Use the tables to find $\ln 5.1$ and $\ln 310$.

4. Use the tables to evaluate $\int_2^5 t^{-1}\, dt$.

5. Use the tables to find $e^{6.1}$ and e^{310}.

6. Use logarithms to evaluate $\dfrac{(2.12)(6.01)}{10.3}$

7. Suppose $\ln a = 6$, $\ln b = 5$, and $\ln c = 4$. Calculate $\ln ab^2c^3$.

In problems 8 and 9 find y'.
8. $y = \ln(e^x + x^2 \ln x)$
9. $y = e^{x \ln(x^2+1)} + (x^2+1)^x$

10. Evaluate $\int (\ln e^{x^2} + e^{2 \ln x})\, dx$.

CHAPTER 8

In problems 1 and 2 show that the function y satisfies the differential equation.
1. $y = (1-x^3)^{-1}$, $y' = 3x^2y^2$
2. $y = e^x - 2xe^x$, $y'' - 2y' + y = 0$

In problems 3, 4, and 5 find the general and the particular solution when $y(0) = 0$.
3. $y' = 3x^2 + 4x$
4. $y' + 2e^x = 3(x+1)^2$
5. $3y^2 y' = 2x$

In problems 6 through 10 find the general solution.
6. $xy' = 2x^2y + y + x^{-1}y$
7. $y' + 3x^2y = e^{-x^3}$
8. $2yy' = 3x^2(y^2+1)$
9. $x^2 y' = y + 1$
10. $xy' + y = x^2$

CHAPTER 9

1. Use the definition of continuity to show that $f(x,y) = 4$ is continuous at $(0,0)$.
2. Explain why $\lim_{(x,y) \to (0,0)} (x+y)^{-1}$ does not exist.

In problems 3 and 4 find f_x and f_y.
3. $f(x,y) = x^2 + y^2 + xy^2$
4. $f(x,y) = xe^{xy^2} + (x+y)^3$

In problems 5 and 6 find the extrema.

5. $f(x,y) = x^2 + y^2 + 4x + 6y$
6. $f(x,y) = y^3 + 6x^2 - 3y + 6x$
7. Use Lagrange multipliers to find the extrema of the function $f(x,y,z) = (x^2+y^2+z^2)^{1/2}$ subject to the constraint $g(x,y,z) = x+3y-z+2 = 0$.

In problems 8 and 9 evaluate the integrals.

8. $\int_0^1 \int_x^{x^2} xy \, dy \, dx$

9. $\iint_A xy \, dA$ where A is the region bounded by $y = x^2 - 2$ and $y = x$.

10. Use the method of least squares to find the line of regression of the points $(0,1), (1,1), (2,3), (3,2), (4,3)$.

CHAPTER 10

1. Convert the numbers $\pi/6, 3\pi/2, -7\pi/6$ from radians to degrees.
2. Evaluate $\sin 7\pi/6$, $\cos 7\pi/6$, and $\tan 7\pi/6$.

In problems 3 and 4 find y'.

3. $y = \sin(x^2+1) + x\cos^2(x^3+2)$
4. $y = \tan(\ln \cos x) + \sin^{-1} 2x$
5. Evaluate $\sin^{-1}(-1/2) + 3\tan^{-1}(\cos 2\pi)$

In problems 6 through 10 evaluate the integrals.

6. $\int x \sin x^2 \cos^2 x^2 \, dx$

7. $\int \sec^2 x \tan^3 x \, dx$

8. $\int (4x^2 - 9)^{-1/2} \, dx$

9. $\int \frac{dx}{9 + x^2}$

10. $\int (9 - 4x^2)^{-1/2} \, dx$

Solutions to Selected Exercises

CHAPTER 1

Section 1-1

1. **(a)** Yes **(b)** No **(c)** No **(d)** Yes **(e)** Yes

3. **(a)** $\{3, -3\}$ **(c)** $\{1, 2, 3, 4\}$
   (b) $\{1, -1\}$ **(d)** $\{2, 3, 5, 7, 11, 13, 17, 19\}$

5. $A = B = C = D$

7. $A \cup B = \{1, 2, 3, 4, a, b\}$
   $B \cup C = \{3, 4, a, b, c, d\}$
   $A \cap B = \{3, 4\}$
   $C \cap A = \emptyset$

9. $A \subset B, B \not\subset A, C \not\subset D, D \subset C$

10. **(a)** $\{-5, -4, -3, -2, -1, 0, 1, 2, 3, 4, 5\}$
    (b) $\{-2, -1, 0, 1, 2, 3\}$

11. **(a)** A is the empty set **(b)** B is J, the set of integers

13. $2/0$ is meaningless, $0/0$ is meaningless since $x \cdot 0 = 0$ is true for all numbers x. $(x-1)/(x-1)$ is 1 when $x = 2$ and is meaningless when $x = 1$.

15. **(a)** $a \in A$ is correct **(c)** $\{a\} \in A$ is not correct
    (b) $a \subset A$ is not correct **(d)** $\{a\} \subset A$ is correct

17. A, B, and D are finite

Solutions to Selected Exercises

18. (a) $[0, 3)$ (e) $(0, 1]$
    (c) $(1, 4]$ (g) $(1, 3)$

19. $B \subset D$

21. (a) p and q have no common factor, $4/6$ is not in lowest terms since 2 is a common factor
    (b) Odd

23. (a) $(5, \infty) \cup (-\infty, 0)$ (g) $(-2, 4)$
    (c) $(0, 2)$ (i) $(1, \frac{5}{3})$
    (e) $[-\frac{5}{3}, 6]$

Section 1-2

1. (a) All real numbers   (e) All real numbers except 2
   (b) $[-1, \infty)$ (f) $(-\infty, 1] \cup [2, \infty)$
   (c) $(-\infty, 3)$ (g) All real numbers except $\frac{1}{2}$ and 3
   (d) All real numbers except 0

3.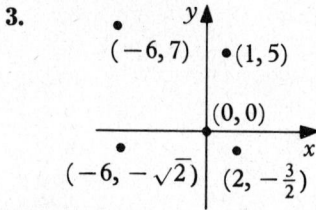

4. (a) 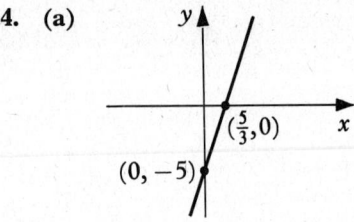   (c)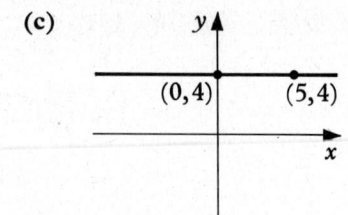

   (e)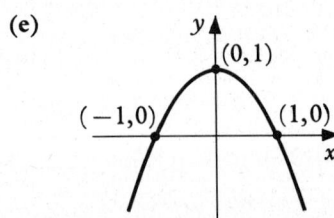

396

5. The first and fourth are functions.

7. (a) $\{(1,3)(1,4)(2,3),(2,4)\}$

Section 1-3

1. (a) 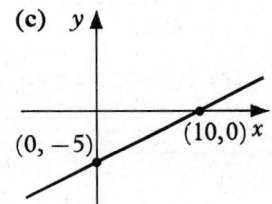 (c)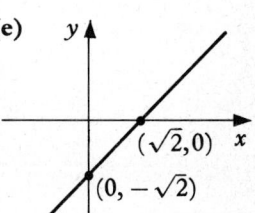

   (b) 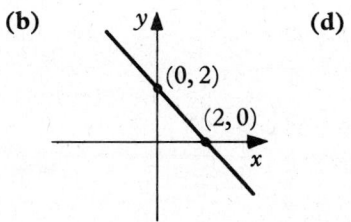 (d)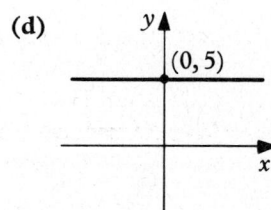

   (e) [graph with $(\sqrt{2},0)$ and $(0,-\sqrt{2})$]

2. (a) Slope is $\frac{2}{3}$, y intercept is $-\frac{1}{3}$
   (c) Slope is zero, y intercept is $\frac{5}{3}$

3. (a) $y = 1$ (c) $x = 1$ (d) $y = 3x + 2$
   (b) $y - 6 = \frac{5}{3}(x-2)$ (e) $x = 6$ (f) $y = 2x$

5. (a) $0°C$ (c) $\frac{340}{9}°C$ (e) $-\frac{70}{3}°C$
   (b) $\frac{160}{9}°C$ (d) $100°C$ (f) $\frac{130}{3}°C$

7. $\dfrac{f(2.1) - f(2)}{0.1}$

8. (a) $(\frac{4}{11}, -\frac{1}{11})$ (c) $(\frac{2}{7}, \frac{18}{7})$

Section 1-6

1. $1, 3, 9, 27, 81, \sqrt{3}, \sqrt[3]{3}, 1/3, 1/9, 1/27$
5. $0, 1, 2, -1, -2$

Section 1-7

1. $g(f(x)) = g(5x+3) = \dfrac{(5x+3) - 3}{5} = x = f(g(x))$

Solutions to Selected Exercises

3. (a) $(x-1)/2$
   (b) $(x-7)/3$
   (c) $\frac{1}{2}x^{1/3}$
   (d) $\left(\dfrac{x+5}{2}\right)^{1/3}$
   (e) $(2-x)^{1/3}$
   (f) $(\frac{1}{3}x)^{1/5}$
   (g) $\left(\dfrac{x+7}{5}\right)^{1/5}$
   (h) $x^3 - 3$
   (i) $(x^5+5)^3$
   (j) $(x^{1/3}-7)^{1/2}$
   (k) $(x^{1/5}-5)^3$
   (l) $\left(\dfrac{x^5+2}{7}\right)^{1/3}$
   (m) $\dfrac{3x+1}{5+x}$
   (n) $\dfrac{1+7x}{5x-2}$

5. (a) $[0, \infty)$
   (b) $[0, \infty)$
   (c) $[-1, \infty)$
   (d) $[0, \infty)$
   (e) $[0, \infty)$
   (f) $[0, 1]$

CHAPTER 2

Section 2-1

1. (a) -1
   (b) -59
   (c) 1
   (d) 2
   (e) 0
   (f) 1
   (g) 0
   (h) Does not exist

3. (a) 0
   (b) Does not exist
   (c) Does not exist

5. 42

7. 200

8. (a) $\lim_{x \to 2} f(x)$ does not exist
   (c) $f(2)$ is not defined
   (e) $\lim_{x \to 2} f(x)$ does not exist

9. (a) $f(x)$ is continuous
   (b) $f(x)$ is not continuous at 3
   (c) $f(x)$ is not continuous at zero
   (d) $f(x)$ is not continuous at 1, 2, or 3
   (e) $f(x)$ is continuous

11. a, b, c, e, f, g, h, j

13. (a) $f(x) + g(x) = 2x + 1$
    (b) $f(x) - g(x) = 3x - 1 - x^2$
    (c) $f(x)g(x) = 3x^4 - x^3 + 3x^2 - x$
    (d) $f(x)/g(x) = 1/(x^3+x)$
    (f) $f(x)g(x) = xe^x + e^{2x}$
    (h) $f(x)g(x) = x^2 - (\ln x)^2$

15. (a) 4   (c) 25   (e) $\frac{31}{4}$
    (b) -6 (d) 0   (f) $\frac{2}{7}$

17. (a) 0   (g) $\ln 3$ (m) 2
    (b) 8   (h) $4\ln 2$ (n) 4
    (c) 264   (i) 0   (o) -6
    (d) e^2 (j) $-4/3e$ (p) -3
    (e) e^5 (k) 6   (q) -1
    (f) $\ln 2$ (l) 8   (r) 2

CHAPTER 3

Section 3-1

2. (a) $f'(1) = 3, f'(x) = 3$ (e) $f'(2) = 12, f'(x) = 6x$
    (c) $f'(2) = 4, f'(x) = 2x$ (g) $f'(1) = -1, f'(x) = -1/x^2$

3. (a) $x = 0$ (b) $x = 0$ (c) $x = 0$ (d) $x = 0$ (e) $x = 1$

5. 3,3,3

7. 3,3,3

Section 3-2

1. (a) 0   (i) $2x + \frac{1}{2}x^{-3/2}$
    (b) 5   (j) $2x - 1 - \frac{5}{8}x^{-3/8}$
    (c) $16x - 4$ (k) $1 + e^x$
    (d) $18x^2 - 10x$ (l) $5e^x$
    (e) $36x^3 - \frac{9}{2}x^2 + 10x - 1$ (m) $2x + 1/x$
    (f) $10x^4 + 2\frac{24}{7}x^3 - 2x$ (n) $6x + 5x^{-1}$
    (g) $-2x^{-3} - 12x^{-5}$ (o) $e^x - 1/x$
    (h) $\frac{3}{2}x^{1/2} + \frac{3}{4}x^{-3/4} - \frac{5}{8}x^{-9/8}$ (p) $3e^x - 8x^{-1}$

3. (a) $y = 0$ (i) $y = \frac{5}{2}(x-1)$
    (c) $y - 5 = 12(x-1)$ (k) $y - (1+e) = (1+e)(x-1)$
    (e) $y - \frac{29}{2} = \frac{81}{2}(x-1)$ (m) $y - 1 = 3(x-1)$
    (g) $y - 4 = -14(x-1)$ (o) $y - e = (e-1)(x-1)$

5. $\frac{2}{3}(9)^{-1/3} + 1$

7. $P(x) = x^3 - 7.4x^2 - 7x - 10$
    $P'(x) = 3x^2 - 14.8x - 7$

Solutions to Selected Exercises

Section 3-3

1. (a) $3(x+1)^2$
   (b) $6(2x+1)^2$
   (c) $15(5x+3)^2$
   (d) $20(5x+3)^3$
   (e) $4x(x^2+1)$
   (f) $2x \ln x + (x^2+1)x^{-1}$
   (g) $e^x(x^2+2x+1)$
   (h) $3x^2(e^x + \ln x) + (x^3+1)(e^x + x^{-1})$
   (i) $2(x^3 + 2x - 1)(3x^2 + 2)$
   (j) $(6x-5)e^x + (2x^3 - 5x + 1)e^x$
   (k) $2(x^4 - 3x^{1/2})(4x^3 - \frac{3}{2}x^{-1/2})$
   (l) $-(x+1)^{-2}$
   (m) $\dfrac{e^x(2x-3)}{(2x-1)^2}$
   (n) $(x+1)^{-2}$
   (o) $\dfrac{12x^2(x+1)}{(3x+2)^2}$
   (p) $\dfrac{x^{-1}(x^2-1) - 2x \ln x}{(x^2-1)^2}$
   (q) $\dfrac{(e^x + 1/x)(2x^2 + x) - (4x+1)(e^x + \ln x)}{(2x^2 + x)^2}$
   (r) $\dfrac{-2x^2 + 2x - 1}{(x^2 + x)^2}$

3. $C'(x) = 1 + 4x(x^2+1)$
   $R'(x) = 3(x^{1/2} + x)^2(\frac{1}{2}x^{-1/2} + 1)$

Section 3-4

1. (a) $6(3x+1)$
   (b) $9(3x+1)^2$
   (c) $30(3x+1)^9$
   (d) $-3(3x+1)^{-2}$
   (e) $-30(3x+1)^{-11}$
   (f) $16x(2x^2+1)^3$
   (g) $(-16x+1)(8x^2-x)^{-2}$
   (h) $(-x^{-1/2} - 2)(x^{1/2} + x)^{-3}$
   (i) $(2x+1)^{-1/2}$
   (j) $(x+1)(x^2+2x+1)^{-1/2}$
   (k) $(3x-5)^{-2/3}$
   (l) $\frac{4}{5}x(x^2+1)^{-3/5}$
   (m) $-\frac{1}{6}x^{-1/2}(x^{1/2}+2)^{-4/3}$
   (n) $(2x+1)^{-1/2} + \frac{5}{2}(5x+1)^{-1/2}$
   (o) $\frac{3}{2}(3x-1)^{-1/2} + \frac{5}{2}(5x+1)^{-1/2}$
   (p) $2 + \frac{3}{2}(3x+1)^{-1/2}$
   (q) $4xe^{2x^2}$
   (r) $3x^2 e^{x^3}$
   (s) $(2x+1)e^{x^2+x}$
   (t) $\dfrac{3x+2}{x^2+x}$
   (u) $\dfrac{1+e^x}{x+e^x}$

400

2. (a) $-4(3x-1)^{-2}$
   (c) $(3x^2-1)(x^2+x)^{1/2} + \frac{1}{2}(x^3-x+1)(2x+1)(x^2+x)^{-1/2}$
   (e) $-2(6x+x^{1/2})^{-3}(6+\frac{1}{2}x^{-1/2})(x^{1/3}+2)^{-3}$
   $\qquad + (-3)(x^{1/3}+2)^{-4}(\frac{1}{3})(x^{-2/3})(6x+x^{1/2})^{-2}$
   (g) $3(8x+1)(6x+5)^{1/2} + 8(3x-1)(6x+5)^{1/2}$
   $\qquad + 3(6x+5)^{-1/2}(3x-1)(8x+1)$
   (i) $\dfrac{1+2xe^{x^2}}{x+e^{x^2}}$

3. (a) $y-4=9(x-1)$ (b) $2y-1=-(x-1)$

Section 3-5

1. (a) $12x^2+6$ (g) $-\frac{1}{4}x^{-3/2} - 18x^{-4}$
   (b) $20x^3 - 12x$ (h) $-\frac{2}{9}x^{-4/3} - 60x^{-5}$
   (c) $6 + x^{-3/2}$ (i) $8(2x+1)^{-3}$
   (d) $10 - x^{-3/2}$ (j) $6(x^2+1)^{-5}(7x^2-1)$
   (e) $6x^2 + \frac{4}{25}x^{-9/5}$ (k) $x^{-3}(x^{-1}+x)^{-1/2} - \frac{1}{4}(x^{-1}+x)^{-3/2}(-x^{-2}+1)^2$
   (f) $2x^{-3} + 24x^{-5}$ (l) $6(x-1)^{-3}$

3. (a) 24 (d) $120x^{-6}$
   (b) $360x^2$ (e) $-\frac{45}{16}x^{-7/2}$
   (c) 0 (f) $24(x+1)^{-5}$

5. $f''(x) = g''(x)h(x) + 2g'(x)h'(x) + g(x)h''(x)$

CHAPTER 4

Section 4-1

1. (a) Minimum at $(0,-4)$
   (b) Minimum at $(-2,-6)$
   (c) Minimum at $(-\frac{1}{2},-\frac{1}{4})$
   (d) Minimum at $(\sqrt{3}/3, -2\sqrt{3}/9)$, maximum at $(-\sqrt{3}/3, 2\sqrt{3}/9)$
   (e) Maximum at $(-3,0)$, minimum at $(-\frac{1}{3}, -\frac{256}{27})$
   (f) No extrema
   (g) Minimum at $(-1,0)$ and $(2,0)$, maximum at $(\frac{1}{2}, \frac{81}{16})$
   (h) Minimum at $(0,1)$
   (i) No extrema
   (j) Minimum at $(2, \frac{8}{3})$, maximum at $(-3, \frac{47}{2})$
   (k) Minimum at $(1,0)$
   (l) Minimum at $(\frac{3}{2}, -\frac{5}{4})$

Solutions to Selected Exercises

      (m) Minimum at (1,2)
      (n) Maximum at (1,20.1)
      (o) Maximum at (4,201.6)
      (p) Minimum if $a > 0$, maximum if $a < 0$ at $\left(-\dfrac{b}{2a}, \dfrac{4ac - b^2}{4a}\right)$

2. (a) No extrema   (c) Minimum at (1,2)

3. 200 and 100

5. Total revenue is $50,000

9. 1000 ft wide and 2000 ft long

11. 15 trees per acre

13. 7500

15. $2\pi^{-1/3}$

17. Length and width and height are 4 in.

21. $x = 9.9$

23. $(\sqrt{65} + 7)/2$, which is approximately 7.5

Section 4-2

1. (a) Minimum at (1,0)
   (b) No extrema, point of inflection at $(0, -4)$
   (c) Minimum at $(2\sqrt{3}/3, -16\sqrt{3}/9)$, maximum at $(-2\sqrt{3}/3, 16\sqrt{3}/9)$, point of inflection at (0,0)
   (d) No extrema, point of inflection at (1,0)
   (e) Minimum at $(\frac{10}{3}, -\frac{500}{27})$, maximum at (0,0), point of inflection at $(\frac{5}{3}, -\frac{250}{27})$
   (f) Minimum at $x = (5+\sqrt{7})/3$, maximum at $x = (5-\sqrt{7})/3$, point of inflection at $x = \frac{5}{3}$
   (g) Minimum at (3,0) and (5,0), maximum at (4,1), points of inflection at $x = 4 \pm \sqrt{3}/3$
   (h) Minimum at $x = 0$ and $x = (9+\sqrt{17})/8$, maximum at $x = (9-\sqrt{17})/8$ points of inflection at $x = (9 \pm \sqrt{33})/18$
   (i) Minimum at $(-\frac{1}{2}, -\frac{43}{8})$, point of inflection at $(0, -5)$
   (j) Minimum at $(-1, -4)$
   (k) Minimum at $(\frac{5}{3}, -\frac{4}{27})$, maximum at (1,0), point of inflection at $(\frac{4}{3}, -\frac{2}{27})$
   (l) Minimum at (2,0), maximum at (0,4), point of inflection at (1,2)
   (m) Minimum at $(\frac{5}{3}, -\frac{256}{27})$, maximum at $(-1,0)$, point of inflection at $(\frac{1}{3}, -\frac{128}{27})$
   (n) Minimum at (3,0), maximum at $(\frac{5}{3}, \frac{32}{27})$, point of inflection at $(\frac{7}{3}, \frac{16}{27})$

3. 90 and 30
5. 3 hr or 4 P.M.
7. $5 \times 10 \times 60$
9. 650
11. 5

Section 4-3

1. (a) $\dfrac{-1}{1+2y}$

   (b) $-4/3$

   (c) $\dfrac{-1-2x}{3y^2+1}$

   (d) $\dfrac{-3}{10y}$

   (e) $\dfrac{-8}{y^{-1/2}+4y}$

   (f) $\dfrac{-1-y}{1+x}$

   (g) $\dfrac{-1-y^2}{2xy+2y}$

   (h) $\dfrac{10xy^2-2x}{1-10x^2y}$

   (i) $\dfrac{2y^{1/2}-x^{-1/2}}{2-xy^{-1/2}}$

   (j) $\dfrac{1+y^{-2}}{y^{-2}+2xy^{-3}}$

   (k) $\dfrac{-y-2xy^2+3x^2y^{1/2}}{x+2x^2y+\frac{1}{2}x^3y^{-1/2}}$

   (l) $-1-2(x+y)^{1/2}$

   (m) $\dfrac{3x-2y}{x-2(x-y)^{1/2}}$

   (n) $\dfrac{-1-y(x-y)^{1/2}-\frac{1}{2}xy(x-y)^{-1/2}}{x(x-y)^{1/2}-\frac{1}{2}xy(x-y)^{-1/2}}$

3. $f(x)=\sqrt{1-x^2}$ and $g(x)=-\sqrt{1-x^2}$ are two choices.

5. (a) $y-1=2(x-1)$
   (b) $y-1=2(x-3)$
   (c) $y-2=-\frac{1}{2}(x-1)$
   (d) $y-4=-\frac{33}{7}(x-1)$
   (e) $y-4=2(x-1)$
   (f) $y-1=3(x-1)$
   (g) $y-1=-5(x-3)$
   (h) $y-3=-\frac{31}{11}(x-1)$

Section 4-4

1. 40π ft^2/sec

3. $9/20\pi$ cm/min

5. -0.03

7. $\frac{1}{12}$ ft/min

9. $\frac{1}{3}\sqrt[3]{\dfrac{5}{4\pi}}$ cm/sec

11. $1/80\pi$ in/sec

Solutions to Selected Exercises

13. $\frac{5}{6}$ ft/sec

15. $\frac{3}{125}$

Section 4-5

1. (a) Any $x_0 \in (-3, 8)$ (c) 2   (e) $-\frac{1}{3}$ (g) $\sqrt{2}$
   (b) 0   (d) $\frac{1}{3}$ (f) $\sqrt{2}$

5. Any linear function, say, $y = x$

7. Let $f(x) = x + 1$, $g(x) = x + 2$

Section 4-6

1. (a) 6   (e) $\frac{4}{3}$
   (b) $\frac{1}{2}$ (f) $\frac{1}{2}$
   (c) 0   (g) $\frac{1}{4}$
   (d) Does not exist   (h) $\frac{1}{2}$

Section 4-7

1. (a) $\dfrac{5x^2}{2} + 7x + C$ (g) $2x^{1/2} - \frac{14}{3}x^{3/2} + C$
   (b) $\frac{8}{3}x^3 - 3x^2 + 2x + C$ (h) $\frac{1}{3}x^3 + \frac{12}{7}x^{-7/2} + C$
   (c) $\frac{9}{4}x^4 - \frac{2}{3}x^{3/2} + C$ (i) $\frac{1}{6}e^{6x} + C$
   (d) $\frac{1}{5}x^5 - \frac{2}{3}x^3 + x + C$ (j) $\frac{1}{8}\ln x + 8\ln x + C$
   (e) $\frac{1}{6}(x^2 + 1)^3 + C$ (k) $\frac{1}{2}e^{2x} + \ln x + C$
   (f) $\frac{5}{8}x^{8/5} - 4x^{3/2} + C$ (l) $-\frac{1}{8}e^{-8x} + \pi x + C$

4. $C(x) = \dfrac{x^3}{3} + x^2 + \ln x + 10$

5. $R(x) = \dfrac{x^4}{4} - \dfrac{x^2}{2} + 10x$

7. $f(t) = \dfrac{kt^{m+1}}{m+1}$

9. $c(t) = \dfrac{k}{v}\left(\dfrac{t^3}{3} + \dfrac{3t^2}{2}\right) + c_o$

11. $l(t) = \dfrac{1}{k}\left(\dfrac{t^3}{3} - \dfrac{t^2}{2}\right)$

CHAPTER 5

Section 5-1

2. (a) $\frac{23}{4}$ (c) 34   (e) 78
3. (a) 6   (c) $\frac{21}{2}$ (e) $\frac{4}{3}$
   (b) 12   (d) $\frac{8}{3}$ (f) $\frac{5}{3}$

Section 5-2

1. (a) 4   (f) $\frac{14}{3}$ (k) $\frac{11}{12}$
   (b) 6   (g) $-\frac{14}{3}$ (l) $e^2 - 1$
   (c) $\frac{15}{2}$ (h) $\frac{255}{4}$ (m) $2(e^2 - 1)$
   (d) 1   (i) $\frac{117}{4}$ (n) $e - \frac{1}{2}$
   (e) 0   (j) $\frac{127}{7}$

3. $\frac{208}{3}$
5. $\frac{26}{3}$
7. $\frac{62}{3}$
9. 4
11. $\frac{1}{6}$
13. $\frac{1}{24}$
17. $0.52(10^{4/3} - 9^{4/3}) + 1.2$
19. $\pi k R^4 + \frac{CR^2}{2}$
23. $-\frac{33}{2}$

Section 5-3

1. (a) $x^4/4 + x^2/2 + C$ (c) $\frac{1}{3}(x^4 + x)^3 + C$
   (b) $\frac{1}{9}(x^3 + 1)^3 + C$ (d) $\frac{1}{18}(2x^3 + 3x^2)^3 + C$
2. (a) $\frac{2}{9}(1 + x^3)^{3/2} + C$ (i) $\frac{1}{3}e^{x^3 + 3x}$
   (c) $\frac{3}{10}(x^2 + 1)^{5/3} + C$ (k) $(\ln x)^2$
   (e) $\frac{8}{7}(4t^3 + t)^{7/8} + C$ (m) $\frac{3}{4}\ln(x^4 + 1)$
   (g) $\frac{3}{28}(x^{4/3} + 1)^7 + C$
3. (a) $\sqrt{3}$ (c) $2\sqrt{2} - \sqrt{3}$ (e) $\frac{1}{3}(e^2 - 1)$
   (b) $\frac{1}{4}(2^{4/3} - 1)$ (d) $\ln 5$ (f) $(\ln 2)^4/4$
5. 14

Section 5-4

1. (a) $\frac{x}{2}e^{2x} - \frac{1}{4}e^{2x} + C$

Solutions to Selected Exercises

(b) $\dfrac{x^2}{4}(2\ln x - 1) + C$

(c) $\dfrac{2x}{3}(x+1)^{3/2} - \dfrac{4}{15}(x+1)^{5/2} + C$

(d) $\dfrac{2x^2}{3}(x+1)^{3/2} - \dfrac{8x}{15}(x+1)^{5/2} + \dfrac{16}{105}(x+1)^{7/2} + C$

(e) $x^3 e^x - 3x^2 e^x + 6xe^x - 6e^x + C$

(f) $\tfrac{1}{9}x^3(3\ln x - 1) + C$

(g) $\tfrac{2}{3}x^{3/2}\ln x - \tfrac{4}{9}x^{3/2} + C$

(h) $\dfrac{4x^{3/2}}{9}(\tfrac{3}{2}\ln x - 1) + C$

(i) $-2x(2-x)^{1/2} - \tfrac{4}{3}(2-x)^{3/2} + C$

(j) $-e^x(e^x + 1)^{-1} + \ln(e^x + 1) + C$

2. (a) 1
(c) $1 - 3e^{-2}$
(e) $3 \cdot 10^{3/2} - \tfrac{2}{15} \cdot 10^{5/2} + \tfrac{2}{15}$

Section 5-5

1. (a) $-\tfrac{1}{3}\ln|2x+1| + \tfrac{1}{3}\ln|x-1| + C$
(b) $-\tfrac{1}{6}\ln|2x+1| + \tfrac{2}{3}\ln|x-1| + C$
(c) $\ln|x+1| - (x+1)^{-1} + C$
(d) $\tfrac{1}{12}\ln|3x-4| + \tfrac{1}{4}\ln|x| + C$
(e) $-\ln|x-1| + 2\ln|x-2| + C$
(f) $-\tfrac{1}{5}\ln|x| + \tfrac{1}{5}\ln|x-5| + C$
(g) $\tfrac{1}{2}x^2 + x - \ln|x| + 2\ln|x-1| + C$
(h) $\tfrac{1}{3}x^3 + \tfrac{3}{2}x^2 + 7x + 16\ln|x-2| - \ln|x-1| + C$
(i) $\ln|x| - \ln|x-1| - (x-1)^{-1} + C$
(j) $\tfrac{3}{4}\ln|x| - \tfrac{3}{4}\ln|x+2| + \tfrac{1}{2}(x+2)^{-1} + C$
(k) $\ln|x-1| - \ln|x-2| - 2(x-2)^{-1} + C$
(l) $\ln|x-1| - (x-1)^{-1} - \tfrac{1}{2}(x-1)^{-2} + C$

Section 5-6

1. (a) $\tfrac{1}{3}\ln\left|x + \tfrac{1}{3}\sqrt{1+9x^2}\right|$
(b) $\tfrac{1}{3}\ln\left|x + \tfrac{1}{3}\sqrt{9x^2-1}\right|$
(c) $-\ln\left|\dfrac{1+\sqrt{1-9x^2}}{3x}\right|$
(d) $-\ln\left|\dfrac{1+\sqrt{1+9x^2}}{3x}\right|$
(e) $-\tfrac{1}{3}\ln\left|\dfrac{3+\sqrt{9+16x^2}}{4x}\right|$

(f) $\dfrac{1}{24} \ln \left| \dfrac{3 + 4x}{3 - 4x} \right|$

(g) $-\dfrac{1}{12} \ln \left| \dfrac{2x + 3}{2x - 3} \right|$

(h) $\dfrac{x}{2} - \dfrac{3}{4} \ln |2x+3|$

(i) $\dfrac{3}{4(2x+3)} + \dfrac{1}{4} \ln |2x+3|$

(j) $-\dfrac{1}{(5x-1)} + \ln \left| \dfrac{x}{5x-1} \right|$

(k) $-\ln \left| \dfrac{x}{5x-1} \right|$

(l) $\frac{1}{2} x \sqrt{16x^2 + 3} + \frac{3}{8} \ln |x + \frac{1}{4} \sqrt{16x^2 + 3}|$

3. (a) $-\dfrac{1}{2} \ln \left| \dfrac{1 + \sqrt{17}}{2 + 2\sqrt{5}} \right|$

(b) $\frac{1}{2} - \frac{1}{4} \ln 3$

(c) $\ln \frac{9}{8}$

(d) $\frac{14}{3} + \ln 3$

(e) $\frac{1}{12} \ln 7$

(f) $\frac{243}{2} \ln 3 - \frac{728}{25}$

CHAPTER 6

Section 6-1

1. (a) 377/1296
   (b) 4941/16000
   (c) 67/25
   (d) 1782/200

3. (a) 41/90
   (b) 1031/1680
   (c) 50012/45045
   (d) 6939/5236

Section 6-2

1. (a) 1/3
   (b) 1/3
   (c) 125/3
   (d) 125/3

Solutions to Selected Exercises

3. (a) 11/10
   (b) 1483/600
   (c) .748
   (d) .393

Section 6-3

1. (a) 0
   (b) 0
   (c) 0
   (d) Does not exist
   (e) 2
   (f) $\frac{5}{2}$
   (g) -1
   (h) 0
   (i) Does not exist
   (j) Does not exist
   (k) 1
   (l) 0
   (m) 0
   (n) 0
   (o) $\frac{1}{2}$
   (p) $-\frac{7}{2}$

2. (a) 3
   (c) $\frac{1}{2}$
   (e) $\frac{1}{2}$

3. (a) Does not exist
   (b) 0
   (c) Does not exist

4. (a) $\frac{16}{3}$

Section 6-4

1. (a) $32\pi/3$
   (b) $\pi/5$
   (c) $4\pi/5$
   (d) 8π
   (e) 8π
   (f) $\pi/2$
   (g) $16\pi/15$
   (h) $\pi/7$
   (i) $2\pi/15$
   (j) $\frac{1}{2}(e^2 - 1)$

3. 3,750 ft/lb

5. 300 ft/lb

Section 6-5

1. (a) $1, 2, 3, 4, 5, 6; \frac{1}{6}$
   (c) $1, 2, 3, 4, 5, 6, 7, 8, 9; \frac{1}{13}:10; \frac{3}{13}$
   (e) $0; \frac{1}{8}:1; \frac{3}{8}:2; \frac{3}{8}:3; \frac{1}{8}$
   (g) $0; \frac{3}{10}:1; \frac{3}{15}:2; \frac{1}{10}$

4. $c = \dfrac{1}{\int_a^b f(x)\,dx}$

5. (a) $\frac{35}{2}$
   (c) 2
   (e) $\frac{3}{34}$

7. 300

9. 0.105, 0.865, 0.815

Section 6-6

1. (a) $\frac{35}{4}, \frac{5}{4}$ (b) 35, 10 (c) $\frac{40}{3}, 3(9-7^{2/3}) - \frac{40}{3}$
3. (a) $x_0 = 2, CS = \frac{44}{3}, PS = \frac{64}{3}$

CHAPTER 7

Section 7-1

1. $\ln 4 \cong 89/64$
   $\ln \frac{1}{2} \cong -51/64$
   $\ln 5 \cong 103/64$
3. $f(1) = 0, f(2) = 3, f(3) = 8, f(4) = 15$
5. $f(1) = 0, f(2) = 73/12, f(3) = 86/3, f(4) = 339/4, f(x) = \frac{1}{12}(4x^3 + 3x^4 - 7)$
7. (a) 1.38629 (d) 2.20166
   (b) 1.60944 (e) 2.30259
   (c) 2.08443 (f) 3.21888

Section 7-2

1. (a) $\dfrac{2}{x}$ (e) $2x + \ln(x^3 + 1) + \dfrac{3x^3}{x^3 + 1}$
   (b) $\dfrac{2x}{x^2 + 1}$ (f) $\dfrac{2\ln x}{x}$
   (c) $\dfrac{6x + 1}{3x^2 + x}$ (g) $\ln x$
   (d) $\dfrac{3x^2 + 5}{x^3 + 5x}$ (h) $\dfrac{1}{x \ln x}$

3. (a) $2\ln|x|$ (d) $\frac{1}{2}\ln|1 + x^2|$
   (b) $\ln|x + 1|$ (e) $\frac{1}{3}\ln|1 + x^3|$
   (c) $\frac{1}{2}\ln|2x + 5|$ (f) $\ln \ln x$

5. (a) 2.78740 (e) -6.72545
   (b) 2.82832 (f) -0.57047
   (c) 6.16052 (g) -4.69730
   (d) 12.76571 (h) -9.70139

Solutions to Selected Exercises

7. (a) 1.3   (e) 1.2
   (b) 2.5   (f) 2.6
   (c) 3.1   (g) 0.42
   (d) -0.5

Section 7-3

1. 7.29, 19.7, 1.6, 0.37, 0.14, 8.48

3. (a) e^2 (c) e^{-1} (e) 3
   (b) e^3 (d) 2

Section 7-4

1. (a) 1.150   (f) 20.086
   (b) 1.271   (g) 44.701
   (c) 2.226   (h) $(7.389)^3$
   (d) 3.004   (i) $(20.086)^2$
   (e) 12.182   (j) $(20.086)^4$

3. (a) $e^{\sqrt{2}\ln 2} \cong e^{0.99} \cong 2.7$ (d) $e^{3.14} \cong 22.3$
   (b) $e^{e \ln 3} \cong e^{2.97} \cong 20.0$ (e) $e^{e \ln \pi} \cong e^{8.0} \cong 405$
   (c) $e^{\sqrt{2}\ln \pi} \cong e^{1.6} \cong 4.9$

Section 7-5

1. (a) $2e^x + 1$
   (b) $15e^x$
   (c) $e^x + xe^x$
   (d) $2xe^{x^3} + 3x^4 e^{x^3}$
   (e) $e^{x^2} + 2x^2 e^{x^2} - 2e^{-2x}$
   (f) $\frac{1}{2}(1+3e^{3x})(x+e^{3x})^{-1/2}$
   (g) $\dfrac{1+e^x}{x+e^x}$
   (h) $\dfrac{e^x + xe^x}{xe^x} = \dfrac{1+x}{x}$
   (i) $e^x \ln x + \dfrac{1}{x}e^x$
   (j) $\dfrac{e^{x^2}(1+5e^{5x})}{x+e^{5x}} + 2xe^{x^2}\ln(x+e^{5x})$

3. (a) $\frac{1}{3}(e-1)$ (d) $\frac{1}{3}(e^8-1)$
   (b) $\frac{1}{6}(e^6-1)+\frac{1}{2}$ (e) $\frac{1}{2}(e^2-e)$
   (c) $\frac{1}{2}(e^4-e)$ (f) e

Section 7-6

1. (a) $\dfrac{1}{(x+5)\ln 5}$
   (b) $\dfrac{3x^2+1}{(x^3+x)\ln 7}$
   (c) $\dfrac{5(2x+3)}{(x^2+3x)\ln 10}+\dfrac{4}{(4x+1)\ln 8}$
   (d) $5^{x-1}\ln 5$
   (e) $3x^2 8^{x^3}\ln 8$
   (f) $(6x+1)\,10^{3x^2+x}\ln 10$
   (g) $5^x 8^{x-1}(\ln 5+\ln 8)$
   (h) $5^x \log_8 x \ln 5+\dfrac{5^x}{x\ln 8}+8^x \log_5 x \ln 8+\dfrac{8^x}{x\ln 5}$
   (i) $\dfrac{2x\ln 5}{\ln 10}+5^{x^2}+2x^2 5^{x^2}\ln 5$

3. (a) 0.70 (c) 1.56
   (b) 1.31 (d) 2.52

CHAPTER 8

Section 8-1

1. (a) One (c) One (e) One
   (b) Two (d) One

3. (a) $y=t^3-t+C$, $y_1=t^3-t$, $y_2=t^3-t+2$ (for example)
   (b) $y=2t^2+e^t+C$
       $y_1=2t^2+e^t+1$
       $y_2=2t^2+e^t+5$
   (c) $y=\frac{2}{9}(t^3-1)^{3/2}+C$, $y_1=\frac{2}{9}(t^3-1)^{3/2}$, $y_2=\frac{2}{9}(t^3-1)^{3/2}-5$
   (d) $y=\dfrac{t^5}{20}+C_1 t+C_2$
       $y_1=\dfrac{t^5}{20}+t$
       $y_2=\dfrac{t^5}{20}$

Solutions to Selected Exercises

(e) $y = \frac{4}{15}t^{5/2} + \frac{t^3}{6} + C_1 t + C_2$

$y_1 = \frac{4}{15}t^{5/2} + \frac{t^3}{6} + t$

$y_2 = \frac{4}{15}t^{5/2} + \frac{t^3}{6} + 3t - 8$

5. $E(t) = \frac{3b}{5}t^{5/3} + Ct + E_0$

Section 8-2

1. (a) $y = Ce^{2x}$
   (b) $y = Ce^{x^2/2}$
   (c) $y = Ce^{x^3/3}$
   (d) $y^2 = 4x + C$
   (e) $y^2 = x^2 + C$
   (f) $y = (\frac{3}{2}x^2 + C)^{1/3}$
   (g) $\frac{1}{3}y^3 + y = \frac{1}{3}x^3 - 2x + C$
   (h) $2y^2 - y^4 = 2x^2 - x^4 + C$
   (i) $y = \ln(e^x + C)$
   (j) $y = -\ln(C - \frac{1}{2}x^2)$
   (k) $y^2 = -\ln(\frac{2}{3}x^3 + C)$

2. (a) $y = \frac{2}{3 - x^2}$
   (c) $y = \frac{3}{2}\sqrt{2 + 2x^2}$

3. $x(t) = \frac{P}{1 + Ce^{Pat}}$

5. $x(t) = \frac{a}{1 + Ce^{akt}}$

Section 8-3

1. (a) $y = 1 + Ce^{-x}$
   (b) $y = Ce^x - x - 1$
   (c) $y = -\frac{1}{2}e^{-x} + Ce^x$
   (d) $y = Ce^{-x^2/2} + 2$
   (e) $y = 3 + Ce^{-x^3/3}$
   (f) $y = 1 - Ce^{x^3/3 + x}$
   (g) $y = \frac{x^4}{3} - 5x^2 + Cx$
   (h) $y = Ce^x - 5x - 5$
   (i) $y = -2 + Ce^{x^3/3}$
   (j) $y = \frac{\ln x}{2} + \frac{C}{\ln x}$

2. (a) $y = 2 - e^{1-x}$
   (c) $y = \frac{x^3}{2} - x \ln x + \frac{x}{2}$

CHAPTER 9

Section 9-1

5. (a) True (f) True (k) True
    (b) False (g) False (l) True
    (c) True (h) False (m) True
    (d) True (i) True
    (e) True (j) True

Section 9-2

1. (a) 0   (g) 0
    (b) 1   (h) 2
    (c) -11 (i) 10
    (d) $\frac{3\,3}{2}$ (j) $-\frac{5\,5}{4}$
    (e) 13   (k) 37
    (f) All ordered pairs of real numbers   (l) All ordered pairs of real numbers

3. A surface $z = f(x,y)$ is a function if and only if no line perpendicular to the xy-plane intersects the surface in more than one point.

5. (a) xy-plane
    (b) xy-plane except x-axis and y-axis
    (c) xy-plane except the points on the lines $x = 1$ and $y = 2$
    (d) xy-plane
    (e) The points within the circle with center at (0,0) and radius 2

7.

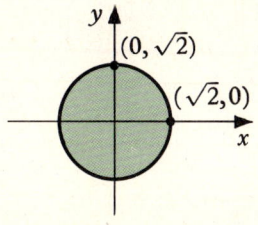
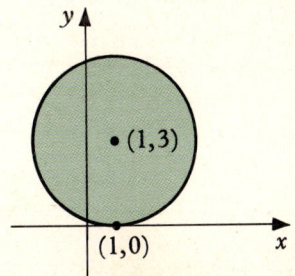

Solutions to Selected Exercises

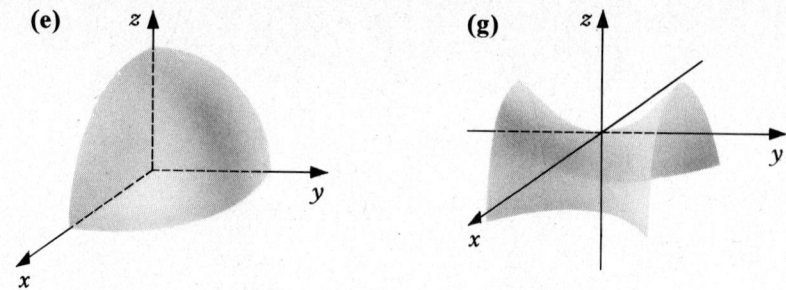

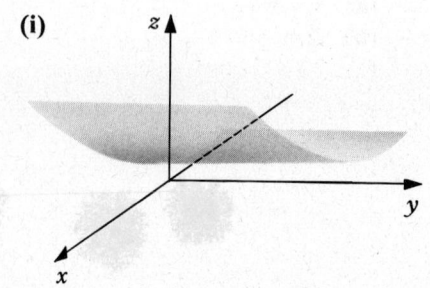

Section 9-3

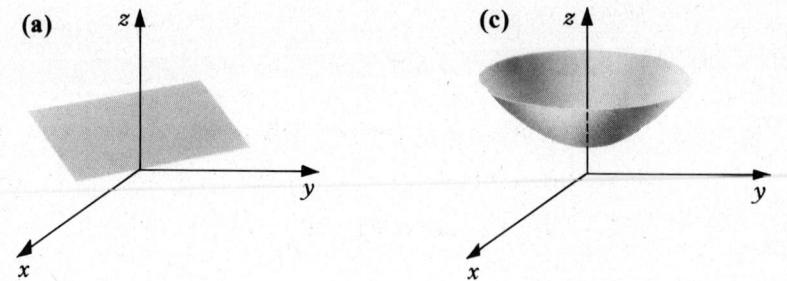

Section 9-4

1. (a) $f_x = 2, f_y = 3$
   (b) $f_x = 5 + 6y, f_y = 6x - 2$
   (c) $f_x = 2x + y, f_y = x + 2y$
   (d) $f_x = 3x^2 + 4xy + y^2, f_y = 2x^2 + 2xy + 9y^2$

(e) $f_x = 2x(y^2+1)$, $f_y = (x^2+1)2y$
(f) $f_x = x(y^2+1)(x^2y^2+x^2+y^2+1)^{-1/2}$
$f_y = y(x^2+1)(x^2y^2+x^2+y^2+1)^{-1/2}$
(g) $f_x = e^x \ln y$, $f_y = \dfrac{e^x}{y}$
(h) $f_x = y^2 e^{xy} + 2x \ln y$
$f_y = e^{xy} + xye^{xy} + x^2 y^{-1}$
(i) $f_x = \dfrac{4xe^{x^2}+y}{2e^{x^2}+xy}$, $f_y = \dfrac{x}{2e^{x^2}+xy}$

3. $g_s(1,-1) = 2 + \dfrac{1}{e}$, $g_r(0,1) = 1$

5. (a) $6x + 4y$
   (b) $2 \ln xy + 5$
   (c) $(2x^{-3} - 2x^{-2}y + x^{-1}y^2) e^{xy}$

8. $\dfrac{\partial f}{\partial x} = \lim\limits_{x \to a} \dfrac{f(x,b,c) - f(a,b,c)}{x-a}$

9. $f_y = 2y + xz$, $f_z = xy$

10. (a) $f_x = 2xy + yz$
    $f_y = x^2 + xz + z^2$
    $f_z = xy + 2yz$
    (c) $f_x = ye^{yz}$
    $f_y = xe^{yz} + xyze^{yz}$
    $f_z = xy^2 e^{yz}$

Section 9-5

2. $f(a,0) = 2a^2 > f(0,0) = 0$ but $f(0,a) = -a^2 < f(0,0)$

3. (a) $(-1,0)$ (b) $(0,2), (-2,2)$ (c) $(-\tfrac{1}{2},0)$

5. $f(0,0) = 0$ is a minimum

6. $(0,0)$

7. $x = y = z = 10$

Section 9-6

1. (a) $(\tfrac{1}{2}, \tfrac{1}{2})$ (c) $(-1,0), (-\tfrac{5}{2}, \tfrac{9}{4})$ (e) $(0,0,0)$
   (b) $(-1, \tfrac{1}{2})$ (d) $(0,0,0)$

3. $\dfrac{20}{7}$

Solutions to Selected Exercises

5. 11

7. $x = \dfrac{30}{\sqrt{29}}, y = \dfrac{20}{\sqrt{29}}, z = \dfrac{20}{\sqrt{29}}$

Section 9-7

1. (a) 1   (e) $-\frac{1}{12}$ (h) $\frac{1}{80}$
   (b) $-\frac{5}{24}$ (f) $\frac{4}{35}$ (i) $\frac{1}{2}(e-1)$
   (c) $\frac{4}{15}$ (g) $\frac{844}{15}$ (j) $\frac{28}{15}$
   (d) $-\frac{1}{3}$

2. (a) $\frac{1}{6}$ (c) $\frac{74}{105}$

3. $\frac{1}{12}$

5. $\frac{500}{3}$

7. $\frac{1}{6}$

9. 350

Section 9-8

1. $y = 0.9x - 0.8$

3. $37y = -3x + 45$

5. $5y = -8x + 114$

CHAPTER 10

Section 10-1

1. (a) $\pi/9$ (e) $5\pi/3$
   (b) $2\pi/9$ (f) $20\pi/9$
   (c) $3\pi/4$ (g) $31\pi/36$
   (d) $7\pi/6$ (h) $40\pi/9$

3. (a) 540°   (e) 72°
   (b) 1°   (f) 135°
   (c) 10°   (g) 210°
   (d) 36°   (h) 105°

5. 2.5π ft and 12.5 ft^2

Section 10-2

1.

x	$\sin x$	$\cos x$	$\tan x$	$\cot x$	$\sec x$	$\csc x$
0	0	1	0	u	1	u
$\pi/2$	1	0	u	0	u	1
π	0	-1	0	u	-1	u
$3\pi/2$	-1	0	u	0	u	-1
2π	0	1	0	u	1	u
$-\pi/2$	-1	0	u	0	u	-1
$-\pi$	0	-1	0	u	-1	u
$-3\pi/2$	1	0	u	0	u	1
$5\pi/2$	1	0	u	0	u	1

3.

		sin	cos	tan
(a)	$\pi/4$	$\sqrt{2}/2$	$\sqrt{2}/2$	1
(b)	$-\pi/4$	$-\sqrt{2}/2$	$\sqrt{2}/2$	-1
(c)	$3\pi/4$	$\sqrt{2}/2$	$-\sqrt{2}/2$	-1
(d)	$-3\pi/4$	$-\sqrt{2}/2$	$-\sqrt{2}/2$	1
(e)	$\pi/3$	$\sqrt{3}/2$	$1/2$	$\sqrt{3}$
(f)	$-\pi/3$	$-\sqrt{3}/2$	$1/2$	$-\sqrt{3}$
(g)	$2\pi/3$	$\sqrt{3}/2$	$-1/2$	$-\sqrt{3}$
(h)	$-2\pi/3$	$-\sqrt{3}/2$	$-1/2$	$\sqrt{3}$
(i)	$\pi/6$	$1/2$	$\sqrt{3}/2$	$\sqrt{3}/3$
(j)	$-\pi/6$	$-1/2$	$\sqrt{3}/2$	$-\sqrt{3}/3$
(k)	$5\pi/6$	$1/2$	$-\sqrt{3}/2$	$-\sqrt{3}/3$
(l)	$-5\pi/6$	$-1/2$	$-\sqrt{3}/2$	$\sqrt{3}/3$

Section 10-3

1. (a) π (d) $2\pi/5$
   (b) 4π (e) 1
   (c) π

Section 10-4

1. To prove (8) use (5) and (6), i.e.,
$$\sin((\pi/2)+x) = \sin((\pi/2)-(-x)) = \cos(-x) = \cos x.$$
To prove (9) use (4) and (7), i.e.,
$$\cos((\pi/2)+x) = \cos((\pi/2)-(-x)) = \sin(-x) = -\sin x.$$
To prove (10) use (8) and (9), i.e.,
$$\sin(\pi+x) = \sin((\pi/2) + ((\pi/2)+x)) = \cos((\pi/2)+x) = -\sin x.$$

To prove (11) use (9) and (8), i.e.,
$$\cos(\pi+x) = \cos((\pi/2) + ((\pi/2)+x)) = -\sin((\pi/2)+x) = -\cos x.$$
To prove (12) use (10) and (4), i.e.,
$$\sin(\pi-x) = \sin(\pi+(-x)) = -\sin(-x) = -(-\sin x) = \sin x.$$
To prove (13) use (11) and (5), i.e.,
$$\cos(\pi-x) = \cos(\pi + (-x)) = -\cos(-x) = -\cos x.$$

3.

		sin	cos
(a)	$3\pi/4$	$\sqrt{2}/2$	$-\sqrt{2}/2$
(b)	$5\pi/4$	$-\sqrt{2}/2$	$-\sqrt{2}/2$
(c)	$7\pi/4$	$-\sqrt{2}/2$	$\sqrt{2}/2$
(d)	$5\pi/6$	$1/2$	$-\sqrt{3}/2$
(e)	$7\pi/6$	$-1/2$	$-\sqrt{3}/2$
(f)	$11\pi/6$	$-1/2$	$\sqrt{3}/2$
(g)	$2\pi/3$	$\sqrt{3}/2$	$-1/2$
(h)	$4\pi/3$	$-\sqrt{3}/2$	$-1/2$
(i)	$-\pi/4$	$-\sqrt{2}/2$	$\sqrt{2}/2$
(j)	$-3\pi/4$	$-\sqrt{2}/2$	$-\sqrt{2}/2$
(k)	$-5\pi/6$	$-1/2$	$-\sqrt{3}/2$
(l)	$-4\pi/3$	$\sqrt{3}/2$	$-1/2$

5. $\cos x = -\frac{4}{5}$, $\tan x = 1$, $\sec = -\frac{5}{3}$, $\csc x = -\frac{5}{4}$, $\cot x = 1$

Section 10-5

1. (a) $2\cos 2x$
   (b) $5\cos 5x$
   (c) $3x^2 \cos x^3$
   (d) $8x \cos 4x^2$
   (e) $2e^x \cos 2e^x$
   (f) $(1+e^x)\cos(x+e^x)$
   (g) $(2x+(1/x))\cos(x^2+\ln x)$
   (h) $-(2x+1)\sin(x^2+x)$
   (i) $-4(2x+1)\sin(2x+1)^2$
   (j) $-2e^{2x}\sin e^{2x}$
   (k) $-2xe^{x^2}\sin e^{x^2}$
   (l) $3\sin^2 x \cos x$
   (m) $-4\cos^3 x \sin x$
   (n) $-x\cos^{-1/2} x^2 \sin x^2$
   (o) 0

(p) $2x \sec^2 x^2$
(q) $2 \tan x \sec^2 x$
(r) $-\cos(\cos x) \sin x$
(s) $\sec^2(\sin x) \cos x$
(t) $-\csc x \cot x$

3. (a) $-2 \cot x \csc^2 x$
   (b) $2x \sec x^2 \tan x^2$
   (c) $-(2x+1) \csc(x^2+x) \cot(x^2+x)$
   (d) $-\csc x$
   (e) $e^{\sec x} \sec x \tan x$
   (f) $2 \sec^2 x \tan x$
   (g) $-\csc^2(\sec x) \sec x \tan x$
   (h) $e^x \cot x^2 - 2x e^x \csc^2 x^2$
   (i) $-2(1+2x) \csc^2(x+x^2) \cot(x+x^2)$
   (j) $-\cos(\csc x) \csc x \cot x$

Section 10-6

1. (a) $-\frac{2}{3} \cos 3x + \frac{5}{2} \sin 2x + C$
   (b) $\frac{1}{2} \ln|\sec 4x| + \frac{3}{2} \ln|\sec 2x + \tan 2x| + C$
   (c) $-\frac{1}{4} \cos(4x+5) + C$
   (d) $\frac{1}{5} \sec 5x + C$

3. (a) $-\frac{1}{7} \cos^7 x + C$
   (b) $\frac{1}{3} \sec^3 x + C$
   (c) $\frac{1}{3} \tan^3 x + C$
   (d) $-\frac{1}{5} \cot^5 x + C$
   (e) $\frac{2}{5} \sqrt{3} \sin^{5/2} x + C$
   (f) $\frac{1}{4} \cos^{-2} 2t + C$

5. $4/3$

Section 10-7

1. (a) $\frac{1}{2} x \sqrt{9+x^2} + \frac{9}{2} \ln|x + \sqrt{9+x^2}| + C$
   (b) $\frac{1}{2} x \sqrt{16+x^2} + 8 \ln|x + \sqrt{16+x^2}| + C$
   (c) $\frac{3}{2} x \sqrt{\frac{1}{9}+x^2} + \frac{1}{6} \ln|x + \sqrt{\frac{1}{9}+x^2}| + C$
   (d) $\frac{1}{2} x \sqrt{x^2-9} + \frac{9}{2} \ln|x + \sqrt{x^2-9}| + C$
   (e) $x \sqrt{x^2-\frac{9}{4}} + \frac{9}{4} \ln|x + \sqrt{x^2-\frac{9}{4}}| + C$
   (f) $-\frac{1}{3}(16-x^2)^{3/2} + C$
   (g) $-\frac{1}{12}(9-4x^2)^{3/2} + C$
   (h) $\frac{1}{3} \ln|x + \frac{1}{3}\sqrt{9x^2-16}| + C$
   (i) $\frac{1}{2} \ln|x + \frac{1}{2}\sqrt{1+4x^2}| + C$

Section 10-8

1. (a) $\pi/3$
   (b) $-\pi/3$
   (c) 0
   (d) π
   (e) $\pi/2$
   (f) $\pi/2$
   (g) $2/\sqrt{3}$
   (h) 0.1

3. (a) $\dfrac{6}{\sqrt{1-9x^2}}$
   (b) $\dfrac{2x}{\sqrt{1-x^4}}$
   (c) $\dfrac{2x+5}{\sqrt{1-x^4-10x^3-25x^2}}$
   (d) $\dfrac{24x^2}{1+4x^6}$
   (e) $\dfrac{-1}{1+x^2}$
   (f) $\sin^{-1}2x + \dfrac{2x}{\sqrt{1-4x^2}}$
   (g) $\dfrac{\sec^2 x}{\sqrt{1-\tan^2 x}}$
   (h) $\dfrac{1}{(\sin^{-1} x)\sqrt{1-x^2}}$

Selected Bibliography

ANDREWORTHA, H. G., and L. C. BIRCH. *The Distribution and Abundance of Animals.* Chicago: University of Chicago Press, 1954.

ATKINSON, R. C., ed. *Studies in Mathematical Psychology.* Stanford: Stanford University Press, 1964.

ATKINSON, R. C., G. H. BOWER, and E. J. CROTHERS. *Mathematical Learning.* New York: Wiley, 1965.

BAILEY, N. T. *The Mathematical Approach to Biology and Medicine.* New York: Wiley, 1967.

BASS, F. M., ed. *Mathematical Models and Methods in Marketing.* Homewood, Ill.: R. D. Irwin, 1961.

BAUMOL, W. J. *Economic Theory and Operations Analysis.* Englewood Cliffs, N.J.: Prentice-Hall, Inc., 1961.

BENAVIE, A. *Mathematical Techniques for Economic Analysis.* Englewood Cliffs, N.J.: Prentice-Hall, Inc., 1972.

BREMS, H. *Quantitative Economic Theory: A Synthetic Approach.* New York: Wiley, 1968.

BUSH, R. R., and W. K. ESTES, eds. *Studies in Mathematical Learning Theory.* Stanford, Calif.: Stanford University Press, 1959.

CARRIER, G., and C. PEARSON, *Ordinary Differential Equations* Waltham, Mass: Blaisdell, 1968.

CHARLESWORTH, J. C., ed. *Mathematics and the Social Sciences.* Philadelphia: Temple University Press, 1963.

CHIANG, A. C. *Fundamental Methods of Mathematical Economics.* New York: McGraw-Hill, 1967.

COHEN, J. E. *A Model of Simple Competition.* Cambridge, Mass: Harvard University Press, 1966.

COLEMAN, J. S. *Introduction to Mathematical Sociology* New York: Free Press, 1964.

COOMBS, C. H., R. M. DAWES, and A. TVERSKY. *Mathematical Psychology: An Elementary Introduction.* Englewood Cliffs, N.J.: Prentice-Hall, 1970.

HOWELL, J. E., and D. TEICHROEW. *Mathematical Analysis for Business Decisions.* Homewood, Ill.: R. D. Irwin, 1963.

HOWLAND, J. L., and C. A. GROBE. *A Mathematical Approach to Biology.* Lexington, Mass.: D. C. Heath, 1972.

KEMENY, J., and J. L. SNELL. *Mathematical Models in the Social Sciences.* New York: Blaisdell, 1962.

KEYFITZ, N. *Introduction to the Mathematics of Population.* Reading, Mass.: Addison-Wesley, 1968.

KLINE, M. *Mathematics: A Cultural Approach.* Reading, Mass.: Addison-Wesley, 1962.

LOTKA, A. *Elements of Mathematical Biology.* New York: Dover Publications, 1956.

LUCE, R. D., R. R. BUSH, and E. GALANTER, eds. *Handbook of Mathematical Psychology.* New York: Wiley, 1965.

———. *Readings in Mathematical Psychology*, Vol. I. New York: Wiley, 1963.

MARGALEF, R. *Perspectives in Ecological Theory.* Chicago: University of Chicago Press, 1968.

MASSARIK, F., and P. RATOOSH, eds. *Mathematical Explorations in Behavioral Science*, Homewood, Ill.: R. D. Irwin, 1965.

MCADAMS, A. K. *Mathematical Analysis for Management Decisions.* New York: Macmillan, 1970.

MCGINNIS, R. *Mathematical Foundations of Social Analysis.* Indianapolis: Bobbs-Merrill, 1965.

PAYNE, D., and R. MCMORRIS, eds. *Educational and Psychological Measurement.* Waltham, Mass.: Blaisdell, 1967.

PIELOU, E. C. *An Introduction to Mathematical Ecology.* New York: Wiley, 1969.

RASHEVSKY, N. *Mathematical Biology of Social Behavior.* Chicago: University of Chicago Press, 1959.

RESTLE, F., and J. G. GREENE. *Introduction to Mathematical Psychology.* Menlo Park, Calif.: Addison-Wesley, 1970.

SHELFORD, V. E. *The Ecology of North America.* Urbana, Ill.: University of Illinois Press, 1963.

SLOBODKIN, L. B. *Growth and Regulation in Animal Populations.* New York: Holt, Rinehart & Winston, 1961.

STEDRY, A. *Budget Control and Cost Behavior.* Englewood Cliffs, N.J.: Prentice-Hall, 1960.

THOMPSON, H. E. *Applications of Calculus in Business and Economics.* Menlo Park, Calif.: W. A. Benjamin, 1973.

THRALL, R. M., J. A. MORTIMER, K. R. REBMAN, and R. F. BAUM, eds. *Some Mathematical Models in Biology.* Ann Arbor, Mich.: University of Michigan, 1967.

TINBERGER, J., and H. C. BOS. *Mathematical Models of Economic Growth.* New York: McGraw-Hill, 1962.

Index

A

Absolute value, 9
Absolute value function, 30
Angle, 311
Antiderivative, 133
Arccosecant, 350
Arcsine, 346
Arctangent, 349
Autocatalytic equation, 247
Average value of a function, 155

B

Base:
    of exponential function, 38
    of a logarithmic function, 38
Binomial expansion, 77
Biological association, 244
Boundary value, 238

C

Cardiology, 144
Cartesian coordinate system, 16
Centigrade, 23
Chain Rule, 82
Common logarithm, 40, 234
Complete the square, 383
Composition of functions, 44, 83
Concavity, 105
Constant function, 20
Constraint, 286
Consumer's surplus, 212
Continuous, 53
Contraction of a muscle, 249
Coordinate plane, 254
Cosecant, 317
Cosine, 316
Cost:
    variable, 24
    fixed, 24
Cotangent, 317
Critical points, 96
Critical values, 96
Curve fittings, 33

D

Decay process, 36
Decibel, 41
Decreasing function, 92
Definite integral, 140
Degree, 311
Demand function, 33
Derivative, 63
Descartes, René, 16
Difference quotient, 65
Differentiable function, 66
Differential equation, 136
Domain of a function, 13
Double integral, 297

E

e, 62, 226
Elastic limit of a spring, 199
Element of a set, 1
Elliptic paraboloid, 263
Empty set, 3
Entropy, 51
Epidemic model, 58
Equilibrium price, 213
Escape velocity, 200
Exponential function, 229
Extension of a muscle, 249
Extremum, 94

F

Factoring, 379
Fahrenheit scale, 23
Fick's law, 239
First derivative test, 97
Function:
    of more than one variable, 259
    of one variable, 12
Fundamental Theorem of Calculus, 152

G

Galileo, 27, 73
General solution to a differential equation, 238
Glucose, 250
Graph of a function, 16
Greatest integer function, 31
Group behavior model, 158
Growth equation, 137, 242

H

Half-life, 137
Hemoglobin, 139
Hooke's law, 198
Hullian model, 251
Hydrogen ion concentration (pH), 41

I

Image, 13
Implicit function, 117
Improper integral, 192
Increasing function, 92
Inequality, 5, 389
Infinite limit, 189
Inflection point, 106
Infusion, 139
Initial condition, 238
Integers, 3
Integrating factor, 248
Integration:
    by partial fractions, 168
    by parts, 164
Intersection:
    of lines, 22
    of sets, 7
Intervals, 6
Inventory control, 114
Inverse, 144
Irrational numbers, 3

K

Kelvin, 52
Kinetic energy, 201

L

Lagrange multipliers, 286
Laminar flow, 157

423

Index

Learning curve, 109
Least squares, 306
Left-hand limit, 62
L'Hôpital's Rule, 129
Limit, 49, 269
Line of regression, 307
Linear function, 19
Logarithmic function, 37, 218
Lotka, 359

M

Marginal:
   analysis, 69
   cost, 69
   demand, 71
   physical productivity, 210
   revenue, 72
   supply, 71
Marginal analysis model, 69
Mathematical induction, 385
Maximum:
   of a function of more than one variable, 279
   of a function of one variable, 94
Mean value of a function, 155
Mean Value Theorem, 127
Melting point, 51
Minimum:
   of a function of more than one variable, 279
   of a function of one variable, 94

N

Natural logarithm, 40
Newton's Second Law of Motion, 200
Newton's Universal Gravitation Law, 201

O

Objective function, 292
Order of a differential equation, 237
Origin:
   of a coordinate system, 15
   of a real line, 4
Oscillatory, 354

P

Parabola, 25
Partial derivative, 272
Partial fractions, 168
Particular solution of a differential equation, 238
Partition, 148
Periodic, 323
Point-slope form of a straight line, 21, 65
Poiseuille's Law, 243
Polynomial function, 30
Power function, 30
Power rule, 72
Probability density function, 204
Producer's surplus, 212
Proportion, 43

Q

Quality control model, 85
Quadrants, 18
Quadratic function, 24
Quadratic formula, 386
Quotient rule, 80

R

Radian, 312
Range of a function, 13
Rate of change, 63
Ratio, 43
Rational function, 31
Rational numbers, 3
Real line, 4
Related rates, 121
Rent, 211
Reynold's number, 110
Right-hand limit, 311

S

Saddle point, 279
Scientific notation, 223
Secant, 317
Second derivative test, 107
Second partial derivative test, 283
Section of a surface, 263
Sensitivity to a dose of a drug, 74
Set, 1
Shadow price, 294
Simple harmonic motion, 355
Simpson's Rule, 187
Sine, 316
Sinusoidal motion, 354
Slope-intercept form of a straight line, 21
Slope of a straight line, 20
Snell's Empirical Law of Refraction, 101
Sound, 356
Specific gravity, 110
Standard position, 311
Step function, 33
Straight line, 19
Subjective brightness, 75
Subset, 3
Substitution of variable, 161
Supply function, 35
Symmetry, 25

T

Tangent: 317
   to a curve, 53
   to a surface, 301
Threshold, 101
Trace of a surface, 263
Trapezoidal Rule, 181
Trigonometric substitution, 343

U

Union of sets, 7

V

Van't Hoff's Law, 239
Vertex of a parabola, 25
Viscosity, 110
Volterra, 359
Volume, 194

W

Work, 198

Y

y-intercept, 20

BIOLOGY

Hydrogen ion concentration, 41
Entropy, 51
Spread of a disease model, 68, 108, 114, 247
Sensitivity to a dose of a drug, 74, 115, 138
Subjective and objective brightness, 75
Central nervous system, 100
Excitation of a nerve fiber, 105
Reynold's number, 110
Infusion, 139
Hemoglobin concentration in blood, 139
Operation of a muscle, 139
Cardiology, 144
Arteriole blood flow, 157
Regeneration of organisms, 160, 164
Diffusion across a membrane, 239
Van't Hoff's equation, 239
Reproduction of a bacteria culture, 243
Poiseuille's law, 243
Blood flow through the aorta, 243
Autocatalytic reaction, 247
Contraction and extension of a muscle, 249
Addition of one solution to another, 250

SOCIOLOGY

Community decisions based on consumer's surplus, 215
Diffusion of information, 245

ECOLOGY

Pollution of a stream, 82
Predator-prey model, 102, 124, 358
Archaeological analysis, 137, 139
Land productivity, 206
Principle of biological association, 244
Pollutants, 250
Ecological growth, 247
Lotka and Volterra's model, 358

INDEX TO MATHEMATICAL MODELS